Power Systems in Emergencies

Power Systems in Emergencies

From Contingency Planning to Crisis Management

U. G. Knight

Honorary Research Fellow,
Imperial College of Science, Technology and Medicine,
London, UK

JOHN WILEY & SONS, LTD
Chichester • New York • Weinheim • Brisbane • Singapore • Toronto

National 01243 779777
International (+44) 1243 779777
e-mail (for orders and customer service enquiries): cs-books@wiley.co.uk

Visit our Home Page on http://www.wiley.co.uk or http://www.wiley.com

Other Wiley Editorial Offices

John Wiley & Sons, Inc., 605 Third Avenue,
New York, NY 10158-0012, USA

Wiley-VCH Verlag GmbH, Pappelallee 3,
D-69469 Weinheim, Germany

Jacaranda Wiley Ltd, 33 Park Road, Milton,
Queensland 4064, Australia

John Wiley & Sons (Canada) Ltd, 22 Worcester Road,
Rexdale, Ontario M9W 1L1, Canada

John Wiley & Sons (Asia) Pte Ltd, Clementi Loop #02-01,
Jin Xing Distripark, Singapore 129809

Library of Congress Cataloguing-in-Publication Data

Knight, U. G. (Upton George)
Power systems in emergencies: from contingency planning to crisis management / U. G. Knight.
p. cm.
Includes bibliographical references and index.
ISBN 0-471-49016-4
1. Electric power systems – Reliability. 2. Emergency management. I. Title.
TK1010.K55 2000
621.31 – dc21 00-043413

British Library Cataloguing in Publication Data
A catalogue record for this book is available from the British Library

ISBN 0 471 49016 4

Typeset in $10\frac{1}{2}$/13pt Sabon by Techset Composition Ltd
Printed and bound in Great Britain by Bookcraft (Bath) Ltd
This book is printed on acid-free paper responsibly manufactured from sustainable forestry, in which at least two trees are planted for each one used for paper production.

Contents

Preface

An author will consider four questions when planning a book – subject matter, why write it, when should it be available, and will it find a market. Many books have been written on the planning and operation of power systems, but these concentrate on the system in normal and near normal conditions, or deal with very specific abnormalities such as short-term instabilities as one-off events. Very few books provide a comprehensive account of power systems in severely disturbed conditions. This book aims to fill that gap, from planning to meet such contingencies, to managing and documenting the crises in operation and supply which can result.

Most of my career has been spent in system planning, operation and control, and within these broad areas I have had particular interest in optimization of the power system and its control during emergencies. Privatization may alter the emphasis, but does not eliminate the need for such work, in particular the latter – man may propose but God will dispose. It seems to be acknowledged now that extreme weather conditions, one of the commonest causes of power system disturbances, are becoming more frequent, emphasizing the value of an account of emergency control at this time. Reviewing the literature, one could say that this subject has been somewhat neglected in recent years, with the interest in reorganization, privatization and restructuring of the supply industry in many parts of the world.

I had thought that writing this book would not entail much work, having published numerous papers over the years, and with the advantage of contacts with many people working in this and related areas. This expectation was sadly adrift. I hope the end result of this work in the shape of this book will appeal to all who have a direct or peripheral interest in the subject – power system operators, planners and managers, the financial community, the control community, manufacturers and, not least, government – as an account of the vital contribution by the power supply industry worldwide to keeping the lights on and the power flowing.

My background and experience in this subject has come from years of working on system planning, then operations, in the UK supply industry, followed by a long association with the energy section of the Department of Electrical and

Electronic Engineering at Imperial College of Science, Technology and Medicine. During these years I have been able to participate in numerous international activities, such as conferences and committees and working groups of CIGRE (the International Conference of Large Electrical Systems). Not least, this has contributed to the international view on the subject of emergency control presented in this book.

My thanks are due to past and present colleagues in these organisations and in particular to Miss K. Hancox of I. C. whose contribution in moving from manuscript to publisher-ready text was invaluable.

U. G. Knight

1
Introduction and Contents

1.1 REVIEW OF CONTENTS

It is surprising that as of the later 1990s, no comprehensive account of the subject of emergency control of power systems has appeared in book form. This is in spite of its importance in the planning and operation of power systems, and hence to the integrity of supply to consumers. The purpose of this book is to provide such a review. It is hoped that it will be valuable not only to engineers with direct responsibilities for emergency control in power system design and operation, but also to others associated with the power industry, for instance system planners and operators, consultants, plant engineers, station staff, R&D staff, manufacturers, even commercial and financial interests.

Redundancy will be built into the system structure, and sometimes the individual components, of power systems, but there cannot be any guarantee that problems will remain within the levels of contingency allowed in the design margins, that is, that the redundancies will be sufficient to maintain supplies. If worse problems should occur, any degradation in the quality of the electricity supply, whether of continuity, voltage, frequency or waveform, should be of such short duration or small magnitude as to be acceptable to all classes of consumer, and to the supply industry itself. The body of theory, practice and experience assembled in the pursuit of this objective has been termed 'emergency control'. In short, emergency control is the assembly of measures provided to ensure continuing stable operation, and then recovery, towards meeting the normal demand should abnormal system conditions develop. The questions and requirements posed by this statement will be taken up in the text, and will include:

- What is system failure and what forms can it take? What is meant by 'a disturbance'?
- The different forms of disturbance, e.g. sudden as a result of environmental conditions, or foreseen as a result of shortage of resources; how do these develop?
- What severity of disturbance should be covered by normal protection and control, leaving more severe disturbances to be handled by emergency control facilities?

- The measures that can be taken in planning and operation to minimize the impact of disturbance.
- The restoration of normal conditions following a disturbance.
- The training of staff to handle disturbed conditions.
- A review of some of the major disturbances that have occurred worldwide; environmental factors in disturbances.
- The costs and benefits of emergency control.
- Emergency control in the future.

Although the basic measures will be common to most systems, the detailed design and application will be tailored to the characteristics of the individual systems. These characteristics will change as systems get larger and become more interconnected, new types of primary plant are introduced, the characteristics of the primary plant change, and protection and control systems evolve. The emergency control measures should keep pace with the net effect of all these changes. Checking that this is happening requires experienced engineers with a critical 'what-if...?', even sceptical approach, who will regularly review the contingencies studied, the system conditions assumed, and the adequacy of the models used, this to be done for the present, near and longer term futures. Experience from other countries and utilities will be valuable.

The chapter by chapter contents of the book are as follows:

Chapter 2 – Disturbances in Power Systems and their Effects

This chapter reviews the disturbances which may confront a power system and the potential impact of these on its viable operation. Disturbances are classified as '*sudden*', that is there is no warning of their onset, or '*predictable*'/'*foreseen*', and possible causes for each are outlined. The possible forms of system failure are then reviewed – plant loading and other operating parameters outside limits, instabilities, system separations – and an outline of analytical techniques applicable to their evaluation described. The chapter concludes with views on trends in the development of analytical techniques.

Chapter 3 – Some General Aspects of Emergency Control

Definitions, concepts and standard terminology used in the literature on emergency control are introduced in this chapter. The impact of disturbances is pursued further, including the preferred corrective actions, the possible consequences if the actions are insufficient, and the ways in which disturbances can then typically develop to, in extreme cases, a complete loss of supply. Following a brief comment on the effect of system structure on the form of emergency control, design criteria for emergency control facilities are proposed.

Chapter 4 – The Power System and its Operational and Control Infrastructure

This chapter provides a system and operational background for the remainder of the book, covering the following main issues:

- structure, function and alternatives for main transmission, including direct current transmission and FACTS devices;
- standards of security and quality of supply in planning and operation;
- timescales and tasks in system operation and control;
- SCADA facilities – functions, structure, performance criteria, data and human – computer interface;
- energy management systems;
- communications, telemetry and telecommand; and
- distributed generation.

Chapter 5 – Measures in the Planning and Operational Timescales to Minimize the Impact of Sudden Disturbances and of Foreseen Disturbances

One of the core topics of emergency control will be reviewed in this chapter, namely what measures should be taken in the management, planning and operation of power systems to minimize the effects of disturbances on their viable operation. The objectives of the measures should be to reduce both the frequency of such disturbances and their harmful effects if they do occur. The chapter opens with an assessment of factors affecting the onset, severity and propagation of a disturbance. Measures to minimize the risk are then discussed, for both the planning and operational timescales.

The measures surveyed include:

- generation margins, demand adjustment, under frequency relays and load shedding, operational memoranda and procedures, on-line security assessment, special protection schemes and co-ordinated defence plans used in several countries;

- general measures such as rapid fault clearance; and
- handling predicted disturbances, including natural phenomena, incipient plant breakdown, industrial action outside and within the supply industry.

Chapter 6 – The Natural Environment, Some Disturbances Reviewed

Nature imposes an environment on power systems which man can influence in the long term – usually it seems for the worse – but hardly at all in the short term (some of the 'killer smogs' that occurred in the UK in the early 1960s and continue in other parts of the world are perhaps exceptions to this generalization). Also, global warming seems to be happening at an increasing rate, with effects perceived in decades rather than centuries as in the past, and in part blamed on human activities. Extreme weather and other environmental conditions will determine many of the plant and system criteria. Hence, it is important to have an appreciation of the weather conditions which may occur. The first part of this chapter reviews these, mainly qualitatively, whilst the second part gives brief descriptions of disturbances from around the world. Datewise, the disturbances range from the mid-1970s to the end of 1999. Some have involved virtually the total loss of supply to whole countries, and others have been 'near misses'. Where available, the lessons learnt from the disturbances have been included.

Chapter 7 – Restoration

The objective of power system restoration is to bring the system to a point at which as much demand as possible, within the capacity of the generation and transmission remaining after the disturbance, is supplied at acceptable frequency, voltage and security levels. In practice, restoration will be a combination of operator decisions and automatic control actions. Following an appreciation of the factors which define the range and severity of a disturbance, the general issues which must be settled before a restoration strategy can be built up are listed – priorities in restoration, etc. Actions for restoration from a localised failure are described, followed by an extended review of the 'black start' situation. Aids to the restoration process are listed. The chapter concludes with descriptions of some of the problems which hinder restoration.

Chapter 8 – Training and Simulation in Emergency Control

The human component in decision making is more important in real time system operation, particularly during disturbed conditions, than in most areas of system engineering, and it is appropriate to discuss the training of system operators in this book. The general approach to training adopted in the supply industry is outlined followed by descriptions of need, content and forms of training for system operators.

Training usually requires access to an operational or mock up control engineer's desk (increasingly called workstation), and the ways in which this can be provided are outlined – for instance, use of a standby desk (sometimes even a standby control room) and supporting computer systems when not required for operation, or a stand-alone simulator and workstation. The chapter includes short descriptions of training simulators installed by the National Grid Company (England and Wales), Electricité de France (France), Svenska Kraftnet (Sweden) and EPRI (USA), and concludes with statistics on the use of dispatch training simulators.

Chapter 9 – Plant Characteristics and Control Facilities for Emergency Control and Benefit to be Obtained

I had originally intended to provide a simple comparison of the cost of emergency control facilities against an estimate of the benefits to be achieved by their installation. Several factors noted in the chapter precluded this, and instead the first part of this chapter reviews total facilities and characteristics for emergency control which should be considered by a utility, with emphasis on functions and relationships to normal control, whilst the second part discusses benefits in qualitative and quantitative terms.

The characteristics of plant, system and demand of particular importance in emergency control are considered first, followed by views on the system control facilities specifically provided to handle emergencies. Qualitative and quantitative benefits of emergency control are discussed, and the chapter ends with a brief comment on the question '*Is emergency control worthwhile?*'.

Chapter 10 – Systems and Emergency Control in the Future

This is a wide ranging chapter which attempts a forecast of the role of emergency control in the future. This is considered from two aspects – organizational changes and facilities – noting that restructuring and unbundling have occurred in numerous countries. The regulatory aspects are illustrated by reference to several utilities, the current and future regulatory background being described. Short descriptions of some of the relevant international organizations are included – the European Union, UCPTE and CIGRE. The regulatory framework of several countries is also summarized.

The second part of this chapter describes some of the expected trends in organizations, systems, manpower, supply standards and plant. The major part is on control plant developments covering static var compensators, series compensators, unified power flow controllers, new type storage systems, FACTS devices in general, etc.

In view of its growing importance worldwide, the possible impacts of privatization and restructuring within the industry are discussed in this chapter.

Appendix 1 – Some Major Interconnected Systems Around the World: Existing and Possible Development

Perhaps more than at any other time, emergencies demonstrate the values of interconnection in providing mutual support between utilities. Hence, this appendix outlines some of the intranational and international interconnections that have been formed, some almost piecemeal and others through development policies.

Appendix 2 – Glossary of Useful Terms

Power system engineers have assembled their own concise vocabulary to describe conditions and events within a power system, and this appendix provides a comprehensive glossary of terms found in this book, and the literature in general.

Appendix 3 – Modelling

The power system analysis techniques used in planning and operation for normal conditions are applicable in emergency control, although there may be increased emphasis on obtaining rapid solutions. This has meant that in the past, approximations have been used to achieve these. The continuing improvement in the performance of computers has now decreased the importance of these. As many descriptions of models and analytical formulations have been published, particular examples have been summarized. Mention is also made of the 'slick' handling of data and the incorporation of results into operational decisions.

1.2 GENERAL APPROACH OF THE BOOK

The approach is practical, describing the criteria and means adopted by utilities to prevent and control emergency conditions. Mathematical details are kept to a minimum and are mainly concentrated in Appendix 3.

2
Disturbances in Power Systems and their Effects

The term 'disturbance' will be used frequently in this book. The connotation is usually of some quite serious incident, such as the tripping of a circuit due to a fault, occurring unexpectedly. In this book, however, it will be taken to mean any event, unexpected or foreseen, which requires corrective action to be taken. The possible disturbances which planners and operators have to consider are discussed below.

The more severe disturbances may well impact on the ability of the system to continue supplying all its consumers at satisfactory frequency and voltages. The relationships between the potential disturbances and the potential forms of system failure are described.

2.1 SUDDEN DISTURBANCE

Sudden disturbances on power systems may result from factors external to the system itself, such as weather or environment, or internal factors such as insulation failure on some item of plant.

2.1.1 Weather

As a general comment, the design criteria for the plant and system will take some account of the ambient weather conditions, for instance:

- selection of insulation levels taking account of isokeraunic levels and pollution;
- adoption of thermal ratings with reference to conductor cross sections which take account of ambient temperatures, wind and solar radiation conditions.

Nevertheless, weather remains one of the main causes of equipment failures; a worldwide survey over several years in the 1980s indicated that about 20 percent of failures were attributable to weather conditions, higher than any other identifiable cause.

Some of the ways in which weather can cause problems in supply include the following:

- conductor failure on overhead line – snow and ice loading. The broken ends may touch the ground, causing an earth fault, as well as an open circuited phase;
- joint failure on overhead line – snow and ice loading; as with conductor failure;
- conductor clashing on overhead line – wind or loss of snow/ice load;
- tower failure on overhead line – tower collapse due to snow/ice loads on conductors, conditions worsened by high winds. Sometimes a run of several towers can be affected;
- insulator flashover on overhead line or outdoor substation – current leakage across dirty insulator surfaces due to moisture, condensation, freezing fog, lightning. The dirt may be caused by industrial pollution or, in coastal regions, wind blown salt deposits;
- conductor heating on overhead line – a combination of weather conditions such as low wind speed, high ambient temperatures, perhaps high solar radiation leading to higher than expected conductor core temperatures at specified current flows;
- conductor sag on overhead line – may be caused by conductor heating or by mechanical loading from snow/ice/freezing fog;
- conductor overheating in cables – apart from the obvious cause of excessive current flows (low voltages could contribute to these), higher than expected temperature could be caused by increased soil thermal resistivity following very dry weather.

2.1.2 Environment

Some of the more frequent factors which may cause disturbances, together with the parts of the system most likely to be affected are:

- flashover to vegetation from overhead lines;
- falling trees or windblown materials (including kites!) contacting overhead lines causing flashovers, usually to earth;

- ground subsidence affecting overhead line towers and resulting in contact between conductors and earth;
- ground excavation or subsidence damaging cables;
- smoke and fire products, usually from grass or forest fires, blowing across overhead lines and resulting in flashovers; the heat and smoke can produce conducting paths between the line conductors anywhere along the line span.

The impact of these environmental effects on the power system will usually be limited. Other, hopefully less frequent but with potentially much wider consequences, will include earthquakes, flooding and tornadoes. Solar magnetic disturbances can induce electric potentials in the earth causing quasi-d.c. earth currents to flow. Transformer cores may be damaged, protective gear operations caused and communications disrupted. Effects reported from a major disturbance in 1989 [2.1] were fading of microwave and carrier communications and loss of telemetry. Serious interference can be caused between utilities and others using the same frequencies when using mobile radio VHF; signals can be propagated over abnormally long distances (over 3000 km) due to ionospheric scatter; high voltages can be induced in wire-based communication systems; the signal-to-noise ratio in power line carrier communications is likely to decrease; terminal equipment in fibre-optic systems may be susceptible to problems, although the cable itself will not be affected. Geomagnetic storms peak at intervals of some 10 years [2.1–2.5].

2.1.3 Balance between Demand and Generation

Sudden changes of demand or generation resulting in imbalance between the two can result from numerous causes: loss of transfers in to/out of external systems or lower voltage networks; transmission circuit trippings isolating parts of the system with embedded generation or demand; etc. As a general comment, this form of disturbance can be one of the most dangerous to system viability, and also one of the most frequent, in that many disturbances may result in some imbalance during their development.

2.1.4 Plant Failure

Generation failures differ from other plant failures in several respects:

(1) The generation margin is shared system-wide.

(2) As a result, a generation loss will be felt to a greater or lesser extent across the whole system.

(3) The loss of output from a generator may be partial, say as a consequence of the failure of some auxiliary plant.

(4) There is often more warning of an incipient reduction in output than of decrease in throughput capacity such as experienced with lines, and transformers (or of var output, as with reactive compensation devices).

(5) Generation failures are one of the commonest forms of plant failure.

2.1.5 Human Error

Human error on the part of utility personnel can occur at all stages of the production of electricity – planning/design, plant manufacture, plant installation, maintenance/testing, operation. Members of the public may be involved inadvertently (e.g. kite flying) or more deliberately (e.g. tower climbing, illegal entry into substations).

Of all the possible forms of human error, the ones that immediately come to mind are switching errors, in decision or execution, and mistakes involving protection. Whereas the results of the first will often be immediately apparent, those of the second may not be apparent for months or years. Mistakes here can range from the specification of an inappropriate type of protection, through installation, calculation of setting, setting the relay on site and testing. Testing may not identify that an inappropriate type of protection or incorrect setting is being used. Because of this, it is quite possible that the first time such decisions can be validated will be on the occurrence of a fault. Another consequence is that a fault on its own can result in a trip; it does not necessarily need to be accompanied by a protective gear maloperation. Insufficient liaison between neighbours can also reduce the security of supply. Reliability targets such as 'one failure in 10,000 days' may, in the author's view, discount the possibility of 'dormant' faults which in simple terms can turn a double contingency event into a single contingency event. Beware correlations when calculating probabilities!

2.2 PREDICTABLE DISTURBANCES

Stochastic events are predictable in mass but not in detail. Some disturbances, however, are predictable with warning times of hours, days, or even months. These are predictable in the sense that although the event is involuntary, the lead times will often give time for preventive action to be taken.

Examples of predictable disturbances are described below.

2.2.1 Shortage of Plant Capacity

Generating plant on load will sometimes exhibit symptoms such as increasing shaft vibrations, which indicate that it must be shut down or its output decreased quickly. The system situation will often be correctable by adjusting the plant commitment, rescheduling transfers with neighbours, changing the operating regime of the pumped storage plant, etc. Incipient problems on transmission plant may be evident from excessive corona or noise. Cable problems can be detected by power factor measurements, although these will require the circuit to be off load. Observations of oil and winding temperatures, and winding resistances, will provide a check on the health of transformers and reactors. The integrity of the insulation on plant in general can be monitored by power factor measurements.

2.2.2 Shortage of Fuel

Shortage of fuel will develop from various environmental or man-made causes, for instance:

- coal fired station
  - extreme weather preventing the lifting of coal from coal stocks or interrupting coal deliveries to stations;
  - rationing of coal deliveries as a result of action by miners or transport staff;
  - rationing of coal deliveries as a result of currency problems if the coal is not indigenous.
- oil fired stations
  - extreme cold weather interrupting oil flows;
  - rationing of oil deliveries as a result of currency problems if the oil is not indigenous;
  - rationing of oil deliveries as a result of industrial action by oil company or transport staff.
- gas fired stations
  - rationing of gas deliveries as result of currency problems if the gas is not indigenous;
  - rationing of gas deliveries as a result of industrial action by staff of gas company;
  - potentially, breakdown of gas distribution network.
- hydro stations
  - poor hydraulicity conditions leading to low reservoir levels or river flows.

2.2.3 Shortage of 'Ancillary' Supplies

Stations will usually require various ancillary supplies, without which the operation of the station cannot be guaranteed. The type of station will dictate those additional supplies, but the following can be mentioned:

- hydrogen – for cooling rotating machines;
- pure (ion free) water – for condensate make up and for use in cooling circuits. Various chemicals will be required to treat the incoming mains, sea or river water;
- lighting up oil – for firing boilers on start up;
- lubricating oil – as required;
- insulating oils – for use in transformers, cables, and other plant;
- nitrogen – where inert atmospheres are required;
- diesel oil – for diesel engines; diesel driven generators may be installed in thermal stations as the second level of auxiliary supply within the station site, i.e. if there is a complete shut down of the station and no external supplies are available, the restoration chain may comprise (a) batteries power up a diesel which powers up a gas turbine, which starts up the main plant; (b) diesel driven generators may also be the main source of on-site auxiliary power in transmission substations.

2.2.4 Shortage of Operating Staff

In keeping with industry in general, the perceived trend is to reduce staffing levels, for example through remote control and automation, and to diversify staff skills. This can contribute both towards and against plant availability, in that station operation is less likely to be affected by the absence of critical groups of staff, but it is more likely to be affected by the absence of specific members of staff.

It is judged that routine transmission district work is less dependent on staff availability than station work. Not least, it is likely to be less time critical. In adverse weather conditions when the need for repair teams may well outstrip the availability of field staff and equipment, the less hard hit utilities have loaned resources to their suffering neighbours ('there but for the grace of God go I!').

2.2.5 Shortage of Control Staff

Control centres are often continuously staffed – 24 hours a day, every day of the year – although the level will vary with the expected activity. Much of the repairs, maintenance and new construction activity on the system will be done during daylight on working days, requiring extra staff in the control room for switching duties, backed up by network studies to ensure that security of supply is adequate. Sometimes these will be engineers rotated from other duties (e.g. operational planning), also broadening the experience of those involved. The core staff will, however, be on shift, with four of five staff needed to cover each position on the rota. By using overtime and deferring days off, the risk of a critical shortage of control room staff is likely to be small.

2.3 FORMS OF SYSTEM FAILURE

The potential causes of system failure will be manifold, some stemming directly from plant failures and others from system effects, either as a result of operating conditions or as a consequence of plant failures. The conditions which may lead to a lesser or greater system failure will include:

- overloads
- voltages outside limits
- frequency outside limits
- instability (transient, dynamic, voltage)
- disconnection of substation or generating station
- system splitting.

Some of the incidents which could lead to these conditions are:

- fault on primary equipment (E)
- protective gear maloperation (E/S)
- communications (e.g. intertrip) maloperation (E/S)
- protective gear settings exceeded (S)
- equipment ratings exceeded (S)
- voltages outside limits (S)
- steady state limits exceeded (S)
- transient stability limits exceeded (S)
- dynamic oscillations (S)
- voltage decay/collapse (S).

(E) indicates incidents which will initially at least primarily affect items of plant (but which may spread to the system), whilst (S) indicates incidents which occur because of system conditions, and are potentially more serious. In general, conditions conducive to system failure should not occur during normal or 'credible' conditions on the system.

2.3.1 Thermal Overloads

'Overload rating' is the excess rating that can be carried by equipment above the continuous thermal rating for defined short time periods. Overload ratings of overhead lines will depend upon the pre-fault flows carried in the immediate past, and the duration of these flows. In the UK, values have been quoted for 3, 5, 10 and 20 minute periods with different pre-loads, for example Table 2.1 (see *Modern Power Station Practice Volume L* (MPSP-L) [2.6]). As a general comment, the author has noted that, over the years, utilities tend to assign higher ratings as knowledge of plant performance increases.

2.3.2 Switchgear Ratings, Excessive System Fault Levels

Fault levels on a power system are closely related to the power density and the network configuration [2.7], but not (surprisingly) to the geographical size of the system. The existence of a 'terminal fault level' (that is, the fault level that would

Table 2.1 Example of the Variation of Overhead Line Thermal Ratings over the Year (Note: the bracketed figures are limits imposed by other than line ratings (in this example by protection)

		Winter	Spring/Autumn	Summer
Pre-fault continuous flow (amps)		1670	1550	1340
Post-fault continuous flow (amps)		1960	1830	1580
Short term overloads (amps):				
for pre-fault flows 85 percent of continuous				
Duration (mins)	20	2050	1900	1640
	10	2210	2040	1750
	5	2520	2320	1970
	3	(2730)	2640	2230
for pre-fault flows 60 percent of continuous				
Duration (mins)	20	2200	2030	1740
	10	2590	2380	2020
	5	(2730)	(2730)	2520
	3	(2730)	(2730)	(2730)

exist on a network of infinite extent) can be demonstrated analytically for regular networks, such as those in Figures 2.1 (a and b). An example is shown in Figure 2.1(c). Terminal fault levels will be reached on quite small networks (Figure 2.1(d)), are comparatively insensitive to the magnitudes of fault infeeds at each node (Figure 2.1(e)), but are strongly dependent on the power density, the network and its voltage. These considerations lead to the concept of 'break even' fault levels. If a sudden change of network voltage to a higher voltage is imagined with the same total generation but sited to suit the new voltage, the new network will, in spite of the greater nodal spacing, have a higher fault level. Thus, a fault level of 10,000 MVA at 220 kV would be equivalent to 23,000 MVA at 400 kV, and 25,000 MVA at 400 kV to 60,000 MVA at 765 kV (Figure 2.1(f)). Unless remedial action is taken, as the density (σ) increases, so will the fault level (Figure 2.1). An effective way to contain increasing fault levels is to switch the network – the 'overlay' system shown in Figure 2.2 is very effective. In effect, this uses the two circuits of double circuit lines and separate busbars at double-busbar switching stations to form two networks, which are virtually mirror images of each other, coupled via bus coupler breakers at selected substations. If, however, the system is disturbed, for example by involuntary tripping of circuit XY in Figure 2.2(b), it may be necessary to close some of the open circuit breakers (say C and D), resulting sometimes in fault levels in excess of the nominal switchgear breaking capacity at some substations (with due precautions taken).

In another example, the network may be sectioned to prevent overloading of the circuits, with reswitching necessary if faults or significant changes in load or generation should occur. Thus, it may be infrequently necessary to operate switchgear at fault levels slightly in excess of the assigned rating. In such cases, conditions will be defined under which the switchgear may be operated so as to ensure that effective arrangements and procedures are in place to safeguard personnel, for instance by barring personnel from the vicinity of the switchgear.

2.3.3 Voltage Outside Limits

Voltage tolerances (maximum deviations from nominal voltage) are quite small, as are the allowable times for which the voltages may be outside the limits. The criteria used may be quite complex, depending, for instance, on the system nominal voltage, the planned configuration and whether the configuration is normal or depleted, etc. There will also be criteria regulating voltages at points of supply to lower voltage networks, and for station auxiliary supplies. IEC regulations, for instance, have included recommended tolerances of +6 percent, −10 percent at the low voltage level, with absolute tolerances of +10 percent and −20 percent.

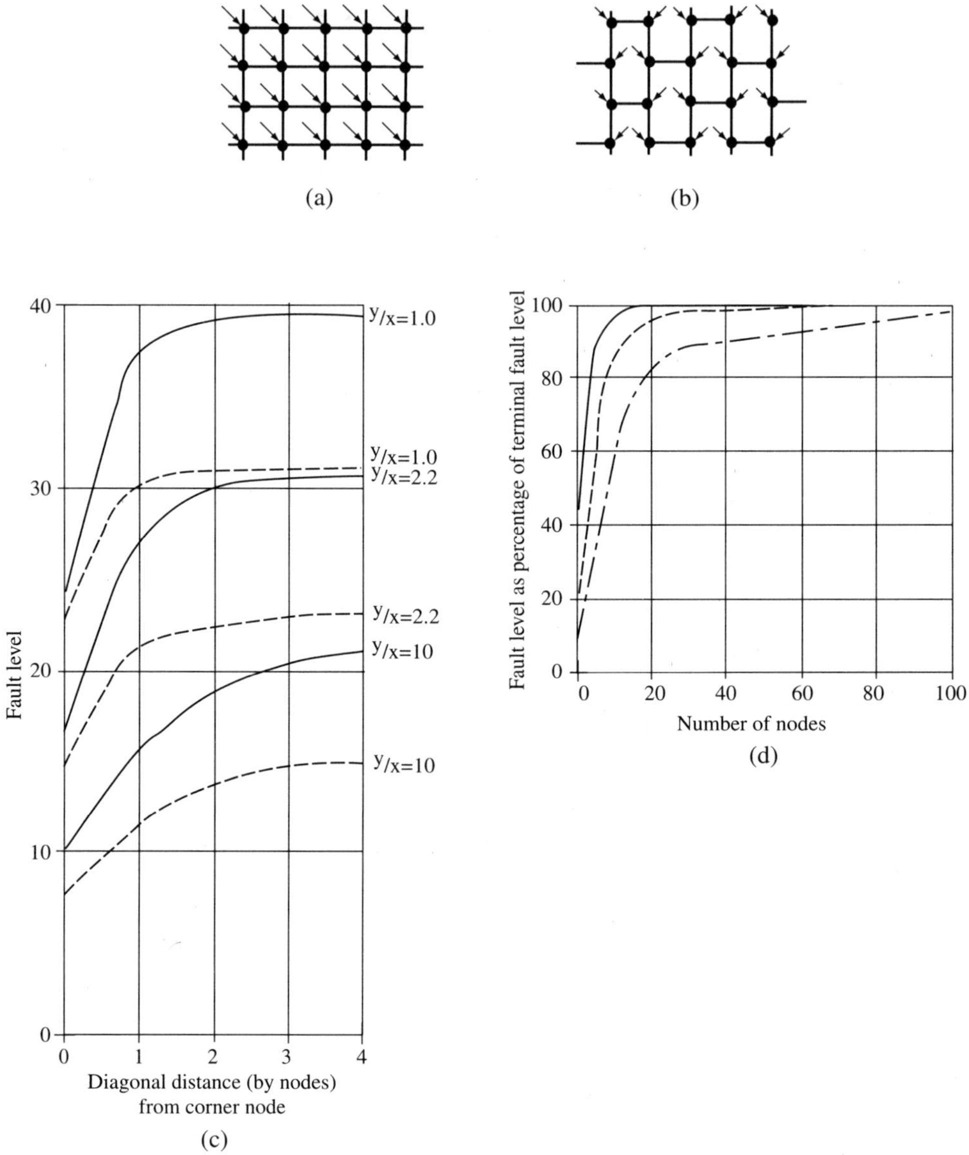

Figure 2.1 Power density configuration and fault level. Reproduced by permission of IEE from [2.7]

Limits are also set on the permissible frequency of voltage variations in terms of the magnitude of the variation. Indicative figures are given in Table 2.2.

Low voltage is the more usual problem, potential causes being high network loadings in relation to network capacities, caused, for instance, by the tripping of

(e)

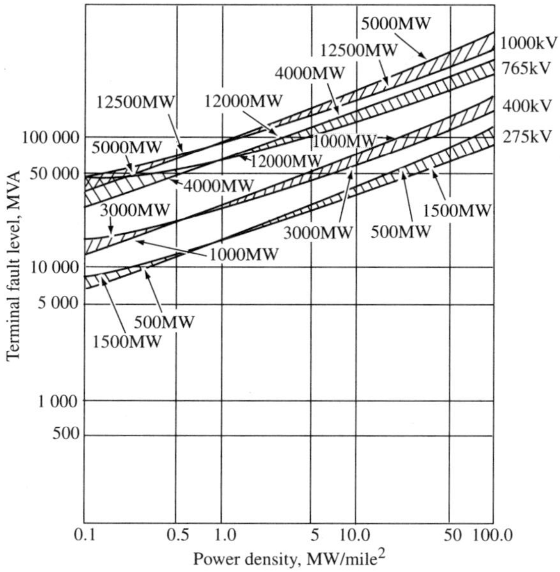

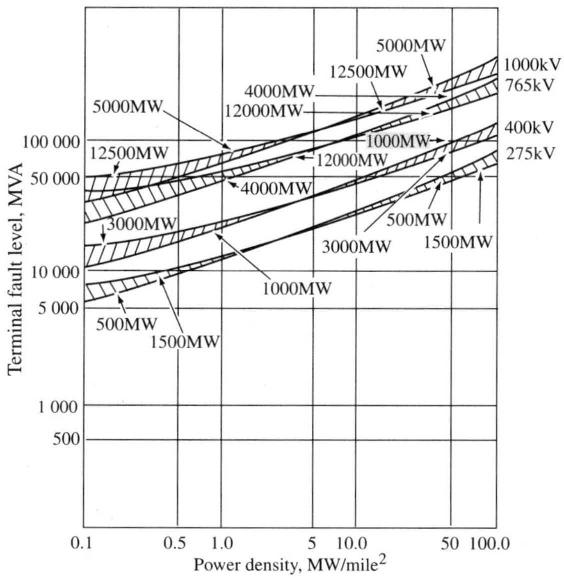

(f)

Voltage V_1	Voltage V_2	Assumed fault level at V_1	Approximate break-even fault level at V_2
kV	kV	MVA	MVA
132	220	3000	6000
132	275	3500	10 000
220	440	10 000	23 000
275	400	15 000	25 000
400	765	25 000	60 000
400	765	35 000	85 000
400	1 000	35 000	125 000
765	1 000	100 000	145 000

Figure 2.1 (*continued*)

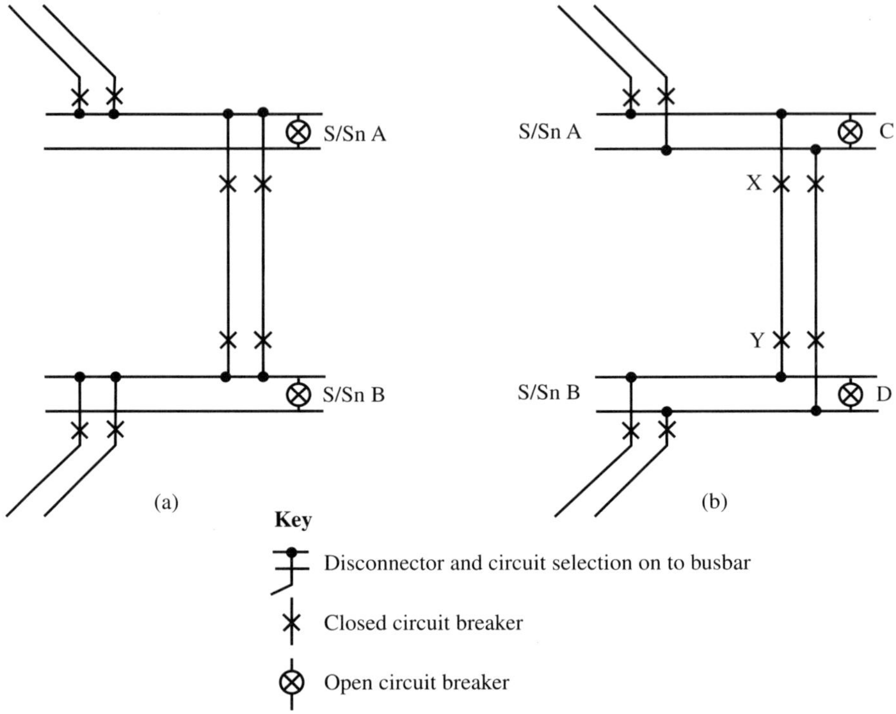

Figure 2.2 Network switching for control of fault levels. (a) Solid, (b) overlay

Table 2.2 Examples of voltage tolerances

System configuration	Tolerances in planning	Comment
Normal	95–97.5 percent	Depending on nominal voltage
Outage	90–95 percent	Depending on nominal voltage
Outage	90 percent	Peripheral parts of system
Normal	105–102.5 percent	Depending on nominal voltage
System configuration	Tolerances in operation	Comments
All	88 percent	
All	109–105 percent	Depending on nominal voltage. The 105 percent limit may be increased to 110 percent for 15 minutes maximum.

a circuit or shortage of reactive compensation. Low voltages will increase the current loading of equipment, and hence losses, and reduce the thermal capability. The operator will work to minimize the duration of such conditions. Conversely, the allowable duration of high voltages will be set mainly by manufacturers, taking account, for instance, of possible overfluxing of transformer cores.

2.3.4 Frequency Outside Limits

Except during short periods of rapid change, when there may be significant imbalances between demand and generation, the frequency averaged over short time periods across a system will be constant, although there will be phase angles between nodal voltages. System frequency is the most important single variable indicating the viability of the operating state of a power system. Acceptable departures from nominal frequency have been small, for instance ±75 mHz in UCPTE, and ±100 mHz in Great Britain and in Nordel [2.8]. The North American standards required frequency deviations to be corrected within 30 seconds.

The standards achieved in practice will depend primarily upon the frequency control methods used. System size will also have some effect. With some searching, statistics can be found on standard deviation of frequency, frequency and duration of system frequency outside various levels, traces of system frequency during disturbances, etc. Occasionally, protracted periods of low frequency operation will occur as a result of shortage of plant or fuel resources.

The resonance frequencies of turbine generator shafts may occur at frequencies only slightly below these, and to avoid metal fatigue, operation at these frequencies may only be acceptable for minutes over the whole life of the unit.

The duration of disturbed frequency will depend, in practice, very much upon the cause and the general system conditions [2.6], for instance:

System condition: duration of disturbed frequency.

- Loss of generation or import from neighbour: 5 seconds to 1 minute, depending on corrective actions available, minutes to hours if no running spare plant available (depending on the demand profile).
- Reduced generation: minutes to hours if the reduction is caused by shortage of generating plant (depending on the demand profile); up to days if resulting from fuel shortage.

2.3.5 Steady State, Transient and Dynamic Stability

During normal operation, the angles between each pair of generator rotors on a power system will change continuously by small amounts as demand, generator outputs and power flows change. If the configuration of the network is changed, there may be larger changes in angles to new values, which will be reached in some tenths of seconds. Once reached, however, the continuous small change condition will resume. This is not to say that the frequency of the whole system will remain constant; it may rise or fall, with angles between rotors remaining almost constant.

The relative angles can change in three typical ways:

- One or more can approach limiting values at which any small increase will lead to a decrease in incremental power transfer between the associated rotors; this is known as steady state instability. A formal definition of *steady state stability* is 'the ability of all generators to remain in synchronism following a very small increase in power transfer about the operating point'.
- As a result of a sudden large increase in transfer impedances or power flows across the network, the relative angles between two or more rotors increase continuously; this is known as transient instability. A formal definition of *transient stability* is 'the ability of the system to regain synchronism following a large signal disturbance'.
- As a result of interaction between control mechanisms, possible following some change in the system, oscillations of 1 Hz or less can occur in the relative angle between the rotors; if such oscillations are not damped out, the system is dynamically unstable. A formal definition of *dynamic stability* is 'the ability of the system to remain stable following small signal disturbances about the operating point'.

Steady state instability occurs very infrequently; in terms of system effects, it would be characterized by pole-slipping of generators and oscillations of large magnitude in system currents and voltages. These could cause operation of impedance and over-current protection. Station auxiliary power supplies and consumer demands could also be affected. Instability, if it occurred, would probably result from high power transfers, and hence if there were circuit trippings resulting in complete sectioning of the system, there could well be significant imbalances between generation and demand in the separate sections. Prolonged pole-slipping of generators could cause overheating and damage to the rotor.

The system effects of *transient instability* would be similar, since the immediate effect, the pole-slipping of generators, would be the same. Criteria for stability used in the planning and operational planning timescales are given in Haubrich and Nick [2.9] and CIGRE WG37.02 [2.10]. Incidents not covered are usually faults in section or coupler breakers, resulting in the loss of two busbars, delayed fault clearance due to malfunction of protection, and simultaneous circuit trippings.

In general, stability is assessed for the most severe fault possible, i.e. three-phase faults causing circuit trippings at the worst locations. The clearance times used will be the slowest combination of main protection, signalling and circuit-breaker type operating times. Normally, faults close to the generator transformer terminals will be the worst fault position, but on some short feeders with impedance protection, a remote-end fault could be more onerous, because of the additional clearance time required for receipt of the acceleration trip signal. It will be appreciated from this brief comment that considerable experience and judgement are needed to assess the transient stability characteristics of a system.

A number of utilities over the past 20–25 years have experienced spontaneous oscillations in flows between parts of their systems in spite of meeting transient stability criteria. The oscillations are small to start with, often building up over minutes unless corrective action is taken, and have occasionally reached values sufficient to cause protection to operate. The periods of oscillation found are around 0.5–1.5 Hz. Although not confined to systems having very long circuits (it has, for instance, occurred in Great Britain), the oscillations tend to occur between discrete generation/demand groups between which there are appreciable power transfers, or between individual generation complexes and the bulk of the system.

2.3.6 Voltage Instability

The control of voltage and the analysis of the behaviour of systems in respect to voltage have been two of the growth areas in power system engineering in recent years. The subjects are of considerable economic importance, since voltage behaviour may be the factor determining maximum power flows on networks at all voltage levels. In recent years, the subject of voltage collapse, that is an uncontrollable fall in voltage, has come to the fore in both academic and utility circles.

Terms proposed in the IEEE documentation (from the IEEE Working Group on Voltage Stability) and elsewhere [2.11, 2.12], in connection with voltage phenomena are:

- *Voltage stability* is the ability of a system to maintain voltage, so that when local admittance is increased, load power will increase, and so that both power and voltage are controllable.

- *Voltage collapse* is the process by which voltage instability leads to a loss of voltage in a significant part of the system. (Voltage may be lost due to angle instability as well, and sometimes only a careful post-incident analysis can determine the primary cause.)
- *Voltage security* is the ability of a system, not only to operate stably, but also to remain stable (as far as the maintenance of system voltage is concerned) following any reasonable credible contingency or adverse system change.

These are very concise and descriptive definitions. One may query, however, whether they fully cover the oscillatory situations sometimes found in dynamic analysis.

The Mechanism of Voltage Collapse

The concept of voltage collapse and the effect of the load characteristic can be illustrated as follows.

Assume a system is providing voltage and current, V_r, I_r at its terminals, to which are connected a load P_r (Figure 2.3). If there is a small change in the system, then at the terminals, and as seen by the load

$$(\delta P_r)_{system} = V_r \delta I_r - I_r \delta V_r \tag{2.1}$$

i.e. if there is a small increase in load current, the extra power available will be approximately that due to the original voltage, multiplied by the increase in current, less the original current multiplied by the decrease in voltage caused by the increased current flow through the system impedance.

The power demanded by the load will depend upon its characteristics. In particular, the load change $(\delta P_r)_{load}$ for a constant power type load will be zero for any small change, hence the power available from the system following the change must be greater or equal to zero, i.e.

$$(\delta P_r)_{system} = V_r \delta I_r - I_r \delta V_r \geq 0 \tag{2.2}$$

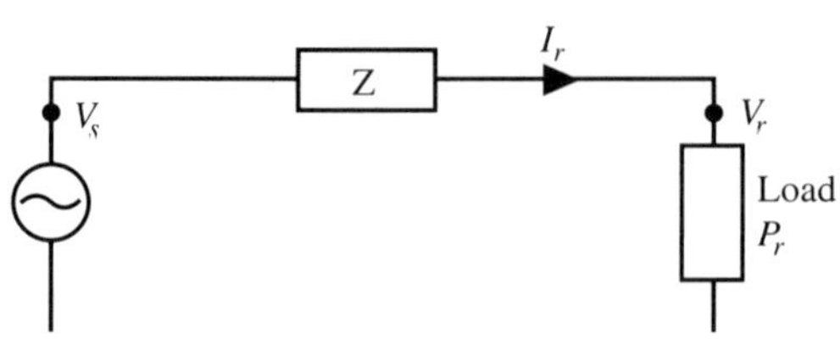

Figure 2.3 Basic circuit

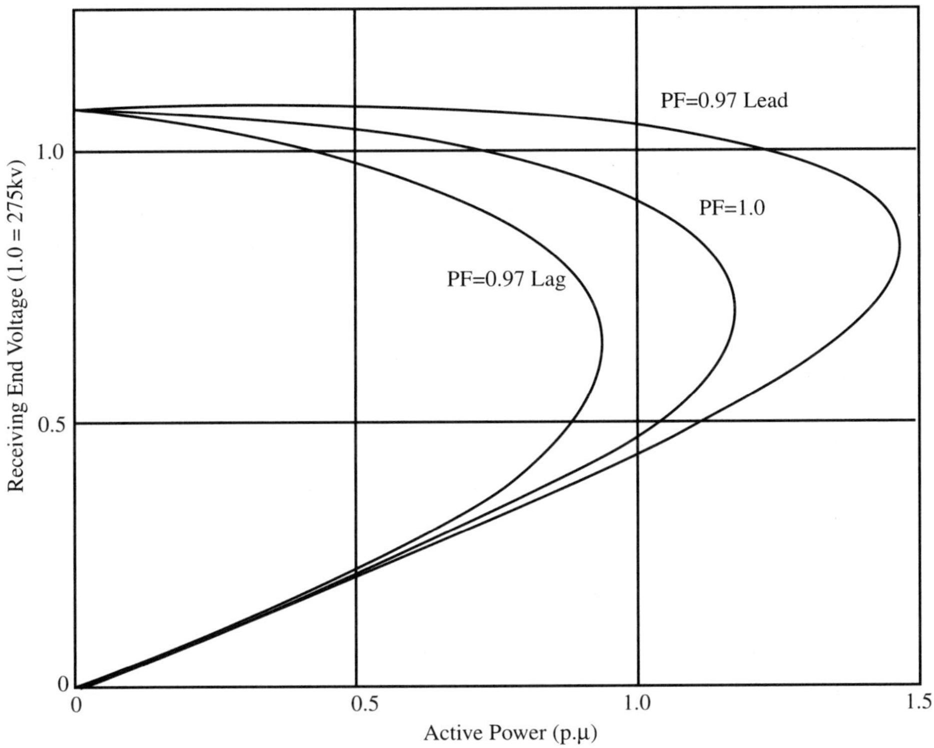

Figure 2.4 Power/voltage characteristic of overhead line (160 km long)

In other words, the extra power available from the increased current must be greater than or equal to the decrease in power resulting from the extra voltage drop in the supply line. There will be no such limit on $(\delta P_r)_{system}$ in the case of a constant impedance load.

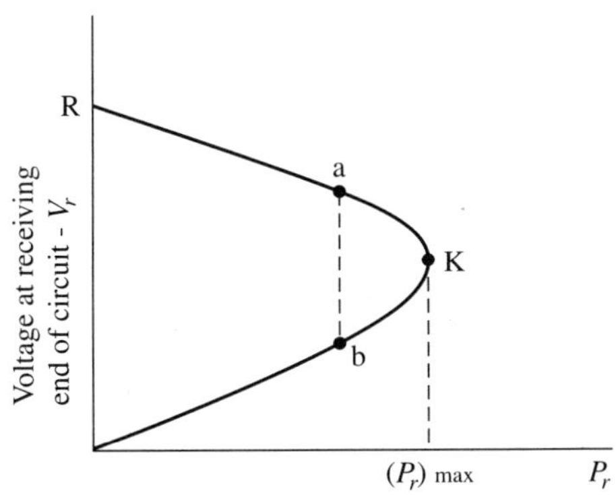

Figure 2.5 Maximum power transfer. In the 'saddle point' bifurcation, the system could in theory jump from stable operating point a to unstable operating point b

For a given connection between sending and receiving ends, there is a family of curves (*P–V* curves) relating the voltage at the receiving end to the receiving end power P_r on each curve, for a given value of received power end factor, as in Figure 2.4. In Figure 2.5, drawn for one particular power factor, operation is possible anywhere on the curve for a constant impedance load. For a constant power load it is only possible between *R* and *K*. The maximum received end power will by $(P_r)_{max}$ for both constant impedance and constant power type loads.

Voltage Oscillations

In spite of its physical simplicity, the analytical solution of the system in Figure 2.3 is complex. The mathematical model contains differential equations representing the dynamics of rotating machines, and algebraic equations representing the network and machine electrical connections. If there are non-linearities present, for instance constant power load, induction motor load, or in the excitation/voltage regulator or tap-changer systems, oscillations in the received end voltage can occur at a certain value of transmitted power. These have been shown conceptually as in Figure 2.6(a). The author is not aware of whether such oscillations have occurred in practice, and it has been suggested [2.13] that they would only be seen at very high values of power transfer. It is also suggested that these die out before the point of voltage collapse (point *K* in Figure 2.6(b)), an effect which can perhaps be explained in terms of consistent values of the eigenvalues at the point of collapse.

The terms *bifurcation* and *chaos* will be found in the literature on voltage collapse. Bifurcation describes the sudden transition from one physical state to another, and follows from the presence of non-linearities in the system. Several

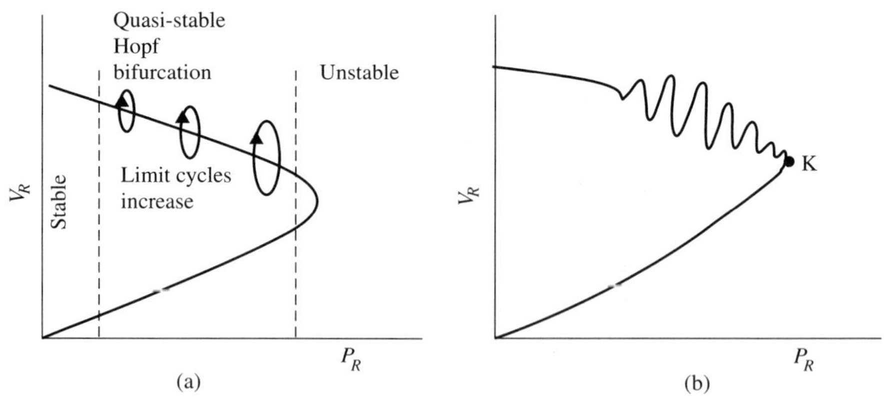

Figure 2.6 (a) Conceptual illustration of Hopf bifurcation (b) Hopf bifurcation

types of bifurcation are possible, depending on the system. Those found in voltage stability/collapse analyses are called the *saddle point bifurcation* and the *Hopf bifurcation*. The saddle point bifurcation describes the transition from a stable operating state, for example, on the upper part of a P–V curve for constant power loads, to an unstable operating point, for example on the lower part of the curve as in Figure 2.5. (It occurs when a real eigenvalue for the operating state of the system becomes positive.)

The Hopf bifurcation describes the transition from a stable, non-oscillatory state to a stable oscillatory state, of constant amplitude at the moment of transition, as in Figure 2.6. (It occurs when the real parts of the conjugate complex eigenvalues pass from positive to negative across the imaginary axis). Whilst there is no doubt that the initial stages of the saddle point bifurcation have occurred in practice (for example, the various system disturbances that have started as sudden rapid falls in voltage and degenerated into voltage collapse), there is little evidence of systems operating on the underside of the P–V curve*. There is also doubt on whether Hopf bifurcations have actually occurred. The possibility of the bifurcation seems to be accepted, but what is doubtful is whether it will occur at parameter values sufficiently close to likely operating values to pose any real risks of occurrence.

The so-called *chaotic state* can develop after the system has passed through a number of bifurcations. Essentially, the system is unpredictable when it has reached this state. Any very minor change can have an unpredictable effect on the future state. A review article in the EPRI journal [2.13] suggests that

> *'if (the possibility of chaos) were confirmed it would at least prompt a fundamental rethinking of the analytical methods used to ensure network stability. At worst, it could mean that power systems harbour an unappreciated potential for voltage oscillations and collapse.' 'These studies do establish the presumption that chaotic behaviour will exist in most power system models. It is not clear however if chaos occurs for parameters in regions sufficiently near (ordinary) operating regimes to affect the stability region (of utility power systems) to a significant extent.'*

Operator's Perception of a Voltage Collapse Situation

Typical features of a voltage collapse situation can be summarized as follows:

* The late Professor J. R. Mortlock once mentioned in a lecture that he had heard of this happening.

Onset of situation

- Often gradual: high transfers (caused variously by high demands, generation loss, circuit loss);
- Indications: increasing reactive power generation; falling voltages;
- System extent: often over a wide area;
- Durations: very variable, from minutes to hours;
- Operator reaction: typically less confidence than in falling frequency or circuit overload problems (there have been cases of operators taking exactly the wrong action).

Collapse

- Timescale: from seconds to minutes;
- Evolution: often complicated by other factors (transient instability, protective gear operations);
- Containment:
  - reduce transfers (increase generation or reduce demands),
  - increase reactive power support,
  - switch in circuits if possible,
  - inhibit sub-transmission and distribution tap change to prevent restoration of distribution voltages,
  - tap change to reduce distribution voltages.

2.4 ANALYSIS TECHNIQUES

The literature on power system analysis is extensive, and it is only intended to discuss user aspects. Some mathematical formulations are summarized in Appendix 3.

2.4.1 Steady State Flows and Voltages

In the following, steady state is used in the sense of pre- and post-contingency conditions when the system is in a steady state (e.g. control actions have been completed) or quasi-steady state (e.g. frequency has been restored, but not tie line flows or economic dispatch, say).

The standard active power (so-called d.c.) and a.c. load flow solutions provide the basic facilities. The d.c. solutions or the power-angle part of the decoupled load flow method will give quite accurate estimates of circuit active power flows, particularly at higher levels of flow. It is widely used in long term and outline development studies, and in optimization studies where its linear form enables it, for instance, to be incorporated in linear programming formulations.

Very large networks can be solved by both d.c and a.c. formulations, although as a personal point of view, the author wonders if there is any real need for the solution of systems of thousands of nodes. For instance, in a reasonably designed and operated system, wide cross-system interaction is unlikely, and if it exists probably depends upon there being two or more simultaneous outages – how to find such conditions anyway? The opposing view that solution of the whole system avoids the need to determine equivalents for parts not modelled in detail becomes more attractive as computer processing power increases. Results from dynamic stability studies may also be more reliable if extensive parts or all of a system are analysed.

Post-contingency demand and generation will have to be calculated first if the contingency disturbs the pre-contingency levels. There may be several alternatives – modelling governor response, tie-line frequency control response, or economic dispatch response. The choice will depend upon whether immediate post-contingency or somewhat later conditions are to be estimated, and also on assumptions concerning transfers of demand at lower voltages between infeeding points from higher voltages. Most stand-alone load flow programs assume constant power and reactive power demands. Studies into extreme system conditions outside the normal limits will need different assumptions, for instance fixed impedance for demands at very low voltages.

Work can be eased and studies made more comprehensive by various routines for data handling and analysis of results, as discussed next.

Some Extensions to the Basic Load Flow Formulation

- *Contingency analysis*: a large number of outages are specified and computed sequentially. The outages may be single, double or multi-circuit. A fault on a tee'd transformer feeder could involve modelling the simultaneous outage of six branches (Figure 2.7(a)), and an apparently simple double circuit line seven double circuit cases (Figure 2.7(b)). Generator and busbar outages may be included. If an outage involves changes in generation or demand, the results of these on the remaining generation and demand have to be specified.
- *Specification of node types in a.c. studies*: in addition to the basic *PV*, *PQ* and slack node types, nodes with specific properties can be provided, e.g. quadrature droop, tap controlled, l.v. generation, etc.
- *Ranging and capability studies*: it is often required to assess the capability of the transmission into part of a system. This can be done by increasing demand

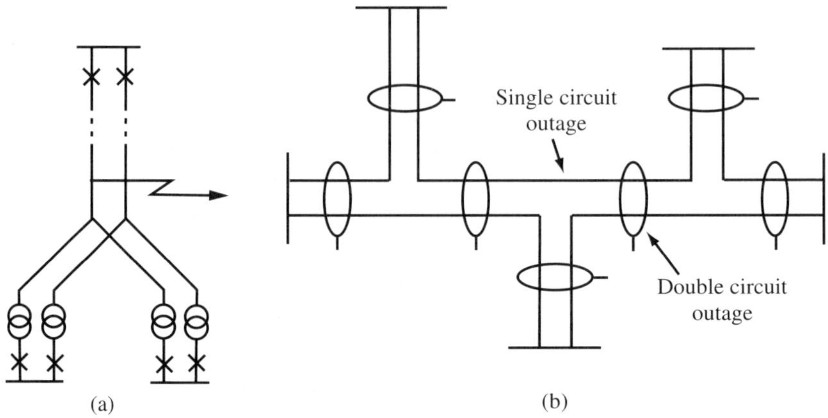

Figure 2.7 (a) Fault on a double circuit tee'd transformer feeder; (b) outage security check on double circuit line – a full check will require five single and seven double circuit outage checks

in one part and decreasing it by the same amount in the remaining part. Simple ratioing procedures can do this.

- *Power and reactive power balances*: balances (demand-generation) of power and reactive power related to transmission and generation capabilities will give a quick appreciation of the situation in specified zones of the system, and can very easily be obtained from load flow solutions.

2.4.2 Fault Levels

Apart from studies to examine reinforcement schemes or switching arrangements, fault level values are most likely to be needed when studying incidents involving suspect operation of protective gear or sequences of circuit breaker trippings. In contrast to the maximum 3-phase and/or 1-phase to earth values needed in planning studies, values may then be required for the generation and transmission conditions at the time of the incidents, for multiple faults, and to include such factors as mutual zero-sequence coupling between circuits sharing towers, and infeeds, including zero sequence, between the relevant protection current and voltage transformers and the point of fault.

2.4.3 Transient Stability

The analysis of transient stability, together with voltage stability, has been an area of extensive study for many years, first in developing and validating the basic models, and then searching for techniques to speed up studies, and thereby enabling more comprehensive examination of worst cases, limiting transfers, and critical clearance times to be made.

The 'step by step' method is the classical method for assessing whether a system will be stable following a particular fault condition, and is still the most widely used. The stages in a study are as follows:

(1) Run a full load flow to obtain the pre-fault operating state of the system.

(2) Determine the positive, negative and zero sequence networks as required; the positive sequence network will be the network used for the load flow with nodal generation and demand transfers replaced by the generator and demand models.

(3) Interconnect the sequence networks to model the type(s) and position(s) of faults.

(4) With the generator internal voltages set at values corresponding to the initial load flow conditions, determine the generation terminal conditions. From these, calculate the fluxes and torques within the generator and the control signals applied to the governor and automatic voltage regulator.

(5) Using a step-by-step integration method, estimate the machine's internal voltage and rotor angle some milliseconds later; the step length may be changed during the study, e.g. 10 ms initially and 100 ms later, when the rates of change of variables have decreased.

(6) Repeat steps (4) and (5), modifying the sequence networks and interconnection at the appropriate times to model changes in network configuration, for example, opening of the faulted circuit(s); the step lengths immediately before and after the switching change(s) are modified to coincide with the changes.

(7) Terminate the iterations either when the differences between rotor angles will clearly converge to steady values (system stable), or when one or more are diverging (system unstable).

Modern transient stability programs are complex. The generator may be represented by some 20 resistance, reactance and time constant values, plus its saturation characteristic; whilst the governor, boiler/turbine and the automatic voltage regulator may require some 20–25 characteristics. Induction motors may be modelled separately, with the remaining demand represented by constant impedances.

The effort in modelling has been complemented by large-scale fault throwing tests and, although it seems that studies give reliable results, alternative methods are being developed. The principal deficiency of the conventional step-by-step transient stability method and programs is that the system is found only to be stable or unstable – there is no quantitative indication of margins. Other problems are the determination of critical cases (demand level and fault location) for analysis, the simplification (reduction) of peripheral systems and possibly the computational load. Considerable efforts have been made to develop fast and approximate (as necessary to meet computation resources) methods to reduce these difficulties. One objective has been to provide to operators a transient stability assessment which can be run as a real-time aid in the control room, and another to facilitate a wide contingency search.

Empirical Methods

In these methods a relationship is sought between some meaningful measure of stability or its impact on system design and a readily calculable variable of the system. One of the best known of these is that due to Hall and Shackshaft [2.14], in which critical clearance time is related to the fault infeed to the faulted generation node from the system after the fault outage (i.e. the fault infeed from local generation is omitted). This quantity is called *post-fault fault infeed*. The paper suggests that

> '*The general shape of the curve does indicate that it is desirable in the basic design of the network to keep the short circuit infeeds above a certain minimum... Of course, this is only a very rough guide to the stability of a power station but it can be used to differentiate between those situations where stability is of no concern and those where stability will need to be frequently reviewed throughout the life of the power station.*'

Pattern Recognition Methods

In these methods a measure of stability (e.g. critical clearance time) is related to a number of network operating variables such as power flows in selected lines. The correlation can be studied by regression analysis between clearance times and operating state variables obtained from a number of stability studies, or other means. Such methods seem essentially to be sophisticated versions of the empirical methods. Their success will depend upon the judgement used in selecting the state variables. The author suggests that care should be taken to avoid fortuitous mathematical correlations, for instance by judging whether any system or physical factors relate the chosen state variables.

Decision Tree Related

In this technique, developed at Liege University in conjunction with Electricité de France [2.15], a decision tree is constructed which progressively defines the operating state of the system and gives a stable/unstable classification. The construction of the decision tree requires many stability studies. It is judged to be more applicable to operational than planning studies.

Equal Area Criterion

The equal area method is an energy-based direct method by which the critical clearance angle and post-fault power limit for a generator connected to an infinite bus can be calculated. The critical clearing time can then be estimated from a swing curve for a sustained fault at this critical angle, using, say, step-by-step time analysis. The method can be extended to two finite machines (these are replaced by an equivalent system with one machine and an infinite bus). It can, with suitable care and experience, be used to assess whether a particular station or group of generation will remain stable with respect to the remainder of the system.

Extended Equal Area Criterion

The extended equal area criterion [2.16] is a further extension of the equal area criterion based on the hypothesis that transient instability in a large system on the occurrence of a fault is identified when the machine angles separate into two groups. One of these, the *critical group*, usually consists of a few generators, with the remaining generators in the other group. The two groups are then replaced by two equivalent machines, and in turn by a single equivalent machine and infinite bus, to which the equal area analysis is applied. A proposal is made to estimate the critical clearing time from the critical clearing angle, which is used to help identify the critical groups of generators. Various refinements are also proposed, including analysis for second swing stability. The method has been tested extensively on systems ranging from six machines (Tunisia and Chinese Regional systems) up to 61 machines (EdF).

Direct Methods

Apart from the equal area criterion, early research into direct methods was aimed at finding a Lyapunov function describing the system dynamics. A general analytical method has not been found, and alternatively, transient energy methods have evolved. It seems that conditions during a fault are in practice such that the necessary mathematical conditions for these methods to be applied are satisfied. The advantage over the earlier Lyapunov function approach is that the energy function can be formed explicitly.

In the transient energy method, the transient energy gain by the system during the fault-on period is evaluated. This remains constant over the post-disturbance period, but with an interchange between kinetic energy and potential energy as the rotor swings. If all the kinetic energy is converted into potential energy the system will be stable, i.e. the transient energy at the end of the fault-on period will be less than a threshold value. Considerable developments have now been made in the model detail. The methods have been tested on a wide range of system sizes and system conditions. Within the limits of the plant models used, answers have compared well with those obtained in time simulation (step-by-step methods). Computation time with the method studied at Imperial College (Potential Energy Boundary Surface – PEBS) has been found to be about one-third of that for the step-by-step method. Furthermore, additional information is available, in particular the margin of stability rather than just the stable/unstable result.

A transient energy function program has been developed under an EPRI contract. This includes a module to compute system energy margins. The program contains some 10 000 lines of Fortran and has been dimensioned for 250 generators. Other workers have studied transient energy methods to assess transient stability loading limits.

2.4.4 Dynamic Stability

A number of power systems have experienced spontaneous oscillations at frequencies around 0.5–1.5 Hz. It is known that oscillations around the steady state operating point in this timescale involve interaction between two natural frequencies. The first results from oscillatory interchange of stored kinetic energy between generator rotors on the system; the second is the natural frequency of the closed loop generator-plus-AVR control system. The oscillations are more likely to occur with high gain settings on automatic voltage regulators, high power transfers between groups of generators and a high resistive component in the demand (it first occurred in Great Britain in 1971 between Scotland and England during a New Year holiday period, when the industrial demand in Scotland was low, and there were also high power flows into England).

As a short-term expedient, the oscillations can usually be damped by reduction of the transfers. AVR gains can also be adjusted and, in the medium term, power system stabilizers fitted. In these, a supplementary signal of generator frequency or power output/speed is added to the terminal voltage signal to control machine oscillation. The best method will depend in part on the electrical strength of the system.

The phenomenon can be analysed by running a transient stability program which includes modelling of the governor and AVR control loops for a system time of, say, 15–20 s. This can be expensive in computer time. The more usual technique is to use the state-space approach in which the generator and network equations are linearised about a given operating point (e.g. summer overnight load level) and the stability of solutions for the system variables (e.g. generator angles) studied by the classical eigenvalue techniques. Both methods have been used in the UK, with similar results obtained.

2.4.5 Medium and Long-term Stability

The methods just described are not satisfactory for frequent applications when knowledge of individual generator or group performance over many seconds is required, as may be the case if the use of low frequency relays for demand disconnection or the sequence of events in a major disturbance is being studied. The step-by-step method gives too much detail, is expensive and may not include all the factors necessary for an extended period of simulation. The state-space method does not give sufficient detail.

Models developed to fill this need can take several forms. Frequently, the inter-machine oscillations are neglected; all generators are assumed to be running at a common speed, calculated by the net accelerating torque on all generator shafts. This means that the faster transient effects can be neglected, and a much longer integration time step (say, 1 s) adopted. Often, the electromechanical generator equations and the algebraic network equations are decoupled, with a new load flow calculated only every few iterations of the generator solution. The generators may be assumed to swing together or, in another approach studied for application in real time simulation in a training simulator, conventional but simplified transient stability equations are used. The integration step length is increased, but a damping factor added to prevent mathematical instability of the solution. Some form of calibration of this model would seem necessary.

2.5 TRENDS IN THE DEVELOPMENT OF ANALYTICAL TECHNIQUES

In a perfect world, analytical techniques would include a model which faithfully reproduces the physical phenomena important in the timescales of the analysis required, algorithms which produce a solution to the model without further simplification and within timescales set by the times available to implement the solutions, and preferably, mechanisms to highlight the important parts of the solution for the planner and operator.

Some compromise may be necessary in practice, particularly between the detail of the model and the range of conditions studied on the one hand, and the solution times on the other. Advances in computer hardware and software technology will contribute towards achieving the ideal models, but there will still be room for improvements in the power systems aspects. For instance, (1) determination of the critical cases to be analysed, taking account of the system loading level, unit commitment and dispatch, the normal network configuration, the fault conditions, the outage cases; and (2) determination of the external system equivalent.

REFERENCES

2.1. Douglas, J., 1989. 'A storm from the sun'. *EPRI Journal*, July/August.
2.2. Kappenman, J. G., 1998. 'Geomagnetic storm forecasting mitigates power system impacts'. *IEEE Power Engineering Review*, November.
2.3. Anon, 1991. 'Solar effects on communications'. *IEEE Power Engineering Review*, September.
2.4. Appell, D., 1999. 'Fire in the sky'. *New Scientist*, February.
2.5. Hay, G., 1999. 'The forecast from space'. *Network*, February.
2.6. *Modern Power Station Practice Volume L*, 1991. British Electricity International.
2.7. Knight, U. G., 1968. 'Study of fault levels on supply networks'. *Proc. IEE*, Vol. 115 (7), July.
2.8. Hagenmeyer, I. E., 1986. 'Operational objectives and criteria'. *Cigre Electra*, No. 108.
2.9. Haubrich, H. and Nick, W., 1993. 'Adequacy of security of power systems at the planning stage'. *Electra*, No. 149.
2.10. CIGRE WG37.02, 1993. 'Review of adequacy standards for generation and transmission planning'. *Electra*, No. 150.
2.11. Mansour, Y. (ed.), 1993. 'Suggested techniques for voltage stability analysis'. *IEEE brochure*, 93TH0620-5 PWR.
2.12. Taylor, C. W., 1991. 'Voltage stability part 1, introduction, definitions, time frames/scenarios and incidents'. Appendix 3, *Survey of the Voltage Collapse Phenomena*. NERC.
2.13. Douglas, J., 1992 "Seeking order in chaos" *EPRI Journal*.
2.14. Hall, J. E. and Shackshaft, G., 1970. 'Developments in the stability characteristics of the power system of England and Wales'. *Cigre*, Paper 32-05.
2.15. Wehenkel, L. and Pavella, M., 1991. 'Decision trees and transient stability of electric power systems'. *Automatica*, Vol. 27 (1).
2.16. Xue, Y. and Pavella, M., 1989. 'Extended equal area criterion: an analytical ultra-fast method for transient stability assessment and preventive control of power systems'. *Journal of Electric Power and Energy Systems*, Vol. 11 (2). (See also *IEEE Trans.*, Vol. 4 (1), 1989).

FURTHER READING

Fouad, A. A. and Vittal, V., 1992. *Power System Transient Stability Analysis Using the Transient Energy Function Method*. Prentice Hall, Englewood Cliffs, NJ.
Wildberger, M., 1994. 'Stability and non-linear dynamics in power systems'. *EPRI Journal*.
Guile, A., Paterson, W., 1977 *Electrical Power Systems Vol 2*, Pergamon.

3
Some General Aspects of Emergency Control

Some of the concepts and definitions used in the context of emergency control of power systems will be described in this chapter. These will be extended to discuss the mechanisms of development of disturbances resulting in system collapse at one extreme to viable operation at the other, providing insights into methods to contain (i.e. to prevent) the spread of a disturbance.

3.1 DEFINITIONS AND CONCEPTS USED IN EMERGENCY CONTROL

3.1.1 Definitions

Control in an emergency can be defined as *the special facilities and procedures provided by a utility to enable it to maintain and restore viable operation following an incident which disturbs the system operating conditions to a point where the available system capacity is no longer sufficient to meet demand in all or parts of the system, or where abnormal splits exist within the network.*

The term *credible contingency* is often used in the area of power systems engineering. It is a contingency or fault which has been specifically foreseen in the planning and operation of the system, and against which specific measures have been taken to ensure that no serious consequences would follow its occurrence. It has sometimes been called a 'defined contingency'. A *non-credible contingency* is one, usually more severe and not specifically defined, for which only general preventive measures are taken. As might be expected, a non-credible contingency is much less likely to occur than a credible contingency. Each utility or interconnected group will adopt its own standards and security criteria reflecting its views and experience on continuity of supply, plant reliability, importance of demand supplied and fault statistics. Typically, the set of credible contingencies will include the loss on fault or otherwise of any circuit or transformer, the largest generator, any busbar, or any reactive source. Three phase faults will usually be assumed for stability assessments, occasionally two phases to earth for systems with weaker networks. Two coincident fault outages are sometimes assumed on well developed systems, perhaps subject to there being bad weather conditions (otherwise, a single outage is assumed). It is important when interpreting such criteria to clarify whether the tripping of a double circuit line will be treated as a double or single contingency.

3.1.2 System States

'System state' is a concise statement of the viability of the system in its current operating mode. Quite often, three are defined – normal, alert and emergency – but the author has suggested that the concept of time-dependent overloads should be introduced [3.1]. This results in four states as follows:

(1) *Normal* – all loadings are within continuous capabilities of the plant, with voltages and frequency within agreed operational limits. System conditions following any credible contingency are acceptable.

(2) *Normal (alert)* – if a credible contingency occurs, action can be taken within the time scales allowed by plant capability to restore the system to a normal state. Very rapid or immediate action is not necessary.

(3) *Alert* – this state requires very rapid or immediate action. If a credible contingency occurs the system will enter the emergency state. Alternatively, the existing conditions are such that action must be taken rapidly to prevent unacceptable overloading, voltage conditions, frequency changes, or plant tripping caused by protective gear operation, or loss of supply, or system split.

(4) *Emergency* – unacceptable loading, voltage or frequency conditions already exist on the system, or demand has been lost, or the system is split. Action must be taken immediately to bring the system to an acceptable state.

The term *restorative state* is also used frequently. This describes the period during which control actions are being implemented to return the system to normal. The term *restorative action* could also be used to denote the return of the system from one of the alert or emergency states to the normal state. In such action, the system is progressed as rapidly as possible from its most abnormal state to a normal state, possibly through a sequence of alert states.

Shortages of primary resources may introduce hazards not adequately covered in the states above. These hazards will usually carry longer term risks, and in the short-term the system may well be able to operate with normal and near normal conditions. Perhaps the term 'extended-alert' would be appropriate in these conditions.

3.1.3 Objectives

The objectives of control during normal operation are to operate the system to meet the accepted security standards at as low a cost (or more generally, use of resources) as possible, and to make provision as necessary to prepare for future operation. The latter will cover such activities as meeting the constraints imposed by generation response limits in the short-term, and releasing plant for maintenance and new construction in the long-term. In contrast, the objectives of emergency control are to implement actions as necessary to prevent a system degenerating into the alert or emergency states, but if this does occur, to minimize disruption and restore normal conditions as quickly as possible, without exposing the plant to non-sustainable overloads or abnormal values of frequency and voltage.

There is clearly an overlap between the two phases of control, but perhaps the main distinguishing features are:

(1) In emergency control in general, the cost of operation is a secondary consideration, not least because the emergency will usually be of short duration; the important issue will be to return to normal operation as quickly as possible.

(2) In normal control, minimum cost of operation and retention of normal operating states will be the main targets; speed of action will be relatively less important.

It may be necessary to qualify the first of these if the emergency is a shortage of resources such as fuel, when the main objective will be to minimize the use of that resource. As an example, expensive oil was burnt to conserve coal during the miners' strikes in Great Britain during the 1970s [3.2].

3.1.4 System States, Contingencies and Types of Control

The system state and control characteristics discussed in Sections 3.1.1–3.1.3 can be interrelated as shown in Figure 3.1(a). It is taken here that a credible contingency, denoted by a single arrow, will degrade a system by one step, i.e. normal to normal (alert), whilst a non-credible contingency will degrade a system by two or more steps, as shown by a double arrow. Restorative actions can be similarly visualized. A similar diagram (Figure 3.1(b)) can be constructed for the three-state definitions of normal, alert and disturbed. Figure 3.1(c) suggests the times available in which to take corrective actions.

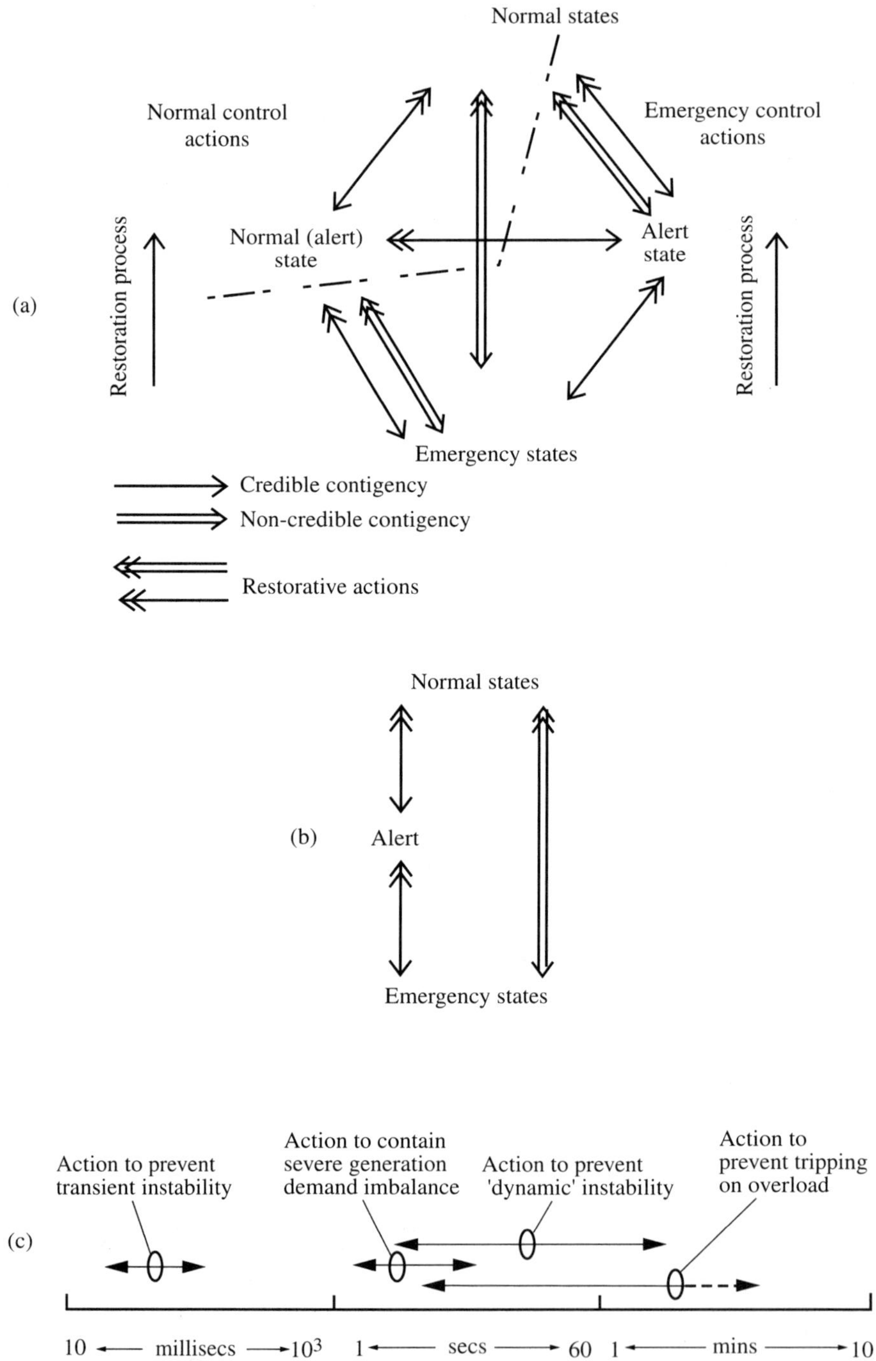

Figure 3.1 System states, contingencies and timescales in emergency control. (a) Four system states, (b) three system states, (c) timescales for actions (Reproduced by permission of Cigre from [3.3])

3.2 SOME STANDARD TERMINOLOGY

Reference will be made in several parts of this book to standards and practices adopted in other countries. Such comparisons are made easier if terms and/or measures are standardized, and two which have some relevance in the emergency control field relate to system structure and to severity of a disturbance. Definitions which were first proposed in a Cigre paper [3.3] are:

For system structure –

U1 – a utility which is part of very much larger interconnection, and occupies a central position in that interconnection

U2 – a utility which is part of a very much larger interconnection, but is on the periphery of that interconnection

U3 – a utility which is not interconnected with neighbours, or is by far the biggest part in any interconnection.

Further classification will indicate broadly whether the utility has a multiply meshed transmission network with thermal rather than stability or voltage limits (sub-classification (a)) or has a lightly meshed or radial transmission network

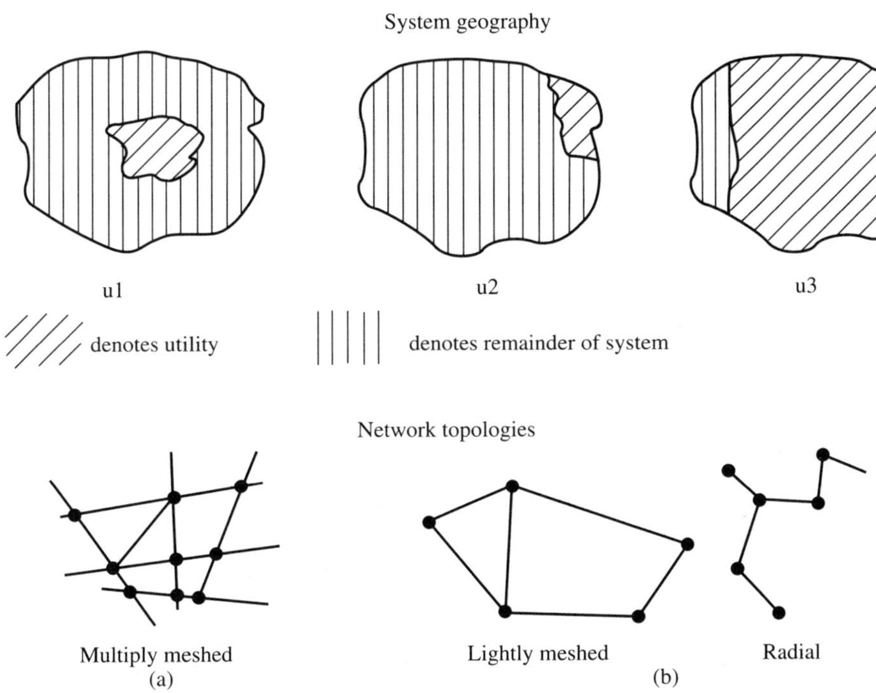

Figure 3.2 Basic geographies and structures

with stability or perhaps voltage rather than thermal limits (sub classification (b)). This gives six possible topologies: U1a, U1b, U2a, U2b, U3a and U3b. These are shown in Figure 3.2.

For severity of a disturbance (see also Table 4.5) –
A useful measure of the severity of system disturbances is in terms of *system minutes* of supply lost [3.4]:

$$\text{Severity of disturbance in terms of system minutes lost} = \frac{\text{energy not supplied due to disturbance (MWh)} \times 60}{\text{maximum system demand met to date (MW)}}$$

3.3 THE EFFECTS OF VARIOUS TYPES OF FAULT OR DISTURBANCE ON SYSTEM PERFORMANCE

Various forms of system failure have been discussed in Chapter 2. It is proposed to extend these by considering what faults or disturbances are most likely to cause such failures, and also how these disturbances could degrade into more severe conditions, including complete system failure. These issues are summarized in Tables 3.1, 3.2 and 3.3.

3.3.1 Sudden Deficit of Generation or Equivalent

This type of disturbance can develop in various ways – most directly through the loss of generation within the system (or alternatively, loss of generation elsewhere in an interconnection), or complete loss of interconnection to neighbours (a partial loss will only result in a redistribution of power flows). The immediate effect will invariably be a drop in frequency by an amount dictated by the spinning spare and the output/frequency stiffness of the system (case A1 in Table 3.1). Other effects may be stressing of the transmission system, resulting in thermal overloads (case A2), or transient or dynamic instability (cases A3 and A4) or excessive voltage drop (case A5). Possible remedial actions and timescales for these are also suggested in the table.

The second stage effects which may develop if the actions are insufficient or are taken too slowly are shown in Table 3.1(b). An unusual effect is covered by case A1. A combination of excessive disconnection of demand and too high gain with poor damping on governors can lead to a succession of under-frequency/demand shed and over-frequency/generation reduction actions, culminating in a complete system collapse. The risk of any major incident, whether it affects generation demand or transmission, can raise one of the important questions in emergency control; that is whether 'system sectioning' protection should be installed, with

Table 3.1 Effect of sudden loss of generation; (b) possible second stage effects if insufficient or incorrect action is taken on sudden loss of generation

(a)

Contingency	Possible results	Containment actions (in order of preference)	Time available to implement action
A. Sudden loss of generation (or import from another part of system)	System frequency fall (1)	Increase generation	1/10's secs to secs
		Reduce demand	1/10's secs to secs
	Transmission overloads (2)	Increase generation Reconfigure network Reduce demand	secs to minutes
	Transient instability (3)	Increase generation Reconfigure network Reduce demand	millisecs
	System oscillations (4)	Increase generation Reconfigure Reduce demand	secs to minutes
	System voltage drop (5)	Increase generation Q and/or P Reconfigure Reduce demand	millisecs/secs to mins (if progressive change)

(b)

Contingency (from Table 3.1(a))	Possible second stage effects
A1	Insufficient demand disconnected: frequency fall not halted – cumulative loss of generation and system collapse Excessive demand disconnected/poor damping of governors – oscillation of frequency/cumulative loss of generation/system collapse
A2	Sequential tripping of overloaded circuits, possibly leading to uncontrolled system split with necessary consequence of generator-demand imbalances (possibly large) in separate sections
A3	System oscillations and tripping of circuits (e.g. on impedance-protection) leading possibly to uncontrolled system split
A4	Build up of oscillations/circuit trippings, up to uncontrolled system split with generation-demand imbalances in separate sections
A5	Cumulative voltage fall as tap changers operate/transmission voltages fall and currents increase, with circuit trippings, generator excitation systems limiting, system voltage collapse and probably system instability

Table 3.2 Effect of a sudden loss of demand; (b) possible second stage effects if insufficient or incorrect action is taken on sudden loss of demand

(a)

Contingency	Possible results	Containment actions (in order of preference)	Time available to implement each action
B. Sudden loss of demand (or export to other part of system)	System frequency rise (1)	Reduce generation	1/10's secs to secs
	System voltage rise (2)	Reduce Q on reactive sources Reduce generation	1/10's secs to secs to minutes (if progressive change)
	Transmission overload (3)	Reduce generation Reconfigure network	secs to mins
	Transient instability (4)	Reduce generation Reconfigure network	millisecs
	System oscillations (5)	Reduce generation	secs to minutes

(b)

Contingency (from Table 3.2a)	Possible second stage effects
B1	Too responsive governors lead to oscillation of frequency with cumulative loss of generation and demand and possibly total loss of system
B2	Voltage rise not halted; if very severe, extensive faults/tripping of circuits possibly resulting in system collapse
B3	Sequential tripping of overloaded circuits leading possibly to uncontrolled system split with necessary consequence of generation-demand imbalances (possibly large) in separate sections
B4	System oscillations and tripping of circuits (e.g. on impedance protection) leading possibly to uncontrolled system split and generation-demand imbalances in separate sections

the primary objective of isolating parts of a system in which operating conditions are badly disturbed from the operationally healthy parts. There are arguments both ways, as discussed further in Chapter 7.

3.3.2 Sudden Deficit of Demand or Equivalent

This type of disturbance can develop either from a direct loss of demand or from export to neighbouring systems. Whatever the cause, the powers involved are likely to be lower and the disturbance is unlikely to be potentially as severe as the

Table 3.3 Effect of sudden loss of transmission; (b) possible second stage effects if insufficient or incorrect action is taken on sudden loss of transmission

(a)

Contingency	Possible results	Containment actions in order of preference	Time available to implement action
C. Sudden loss of transmission (no system split)	Transmission overload (1)	Reconfigure network Adjust generation Adjust generation and demand	seconds to minutes
	Transient instability (2)	Reconfigure network Adjust generation Adjust generation and demand	millisecs
	System oscillations (3)	Reconfigure network Adjust generation Adjust generation and demand	seconds to minutes
	Voltage fall (4)	Reconfigure network Adjust generation power and/or reactive power Adjust generation and demand	seconds to minutes (if progressive change)

(b)

Contingency (from Table 3.3a)	Possible second stage effects
C1	Sequential tripping of overloaded circuits possibly leading to uncontrolled system splits with necessary consequence of generation-demand imbalances (possibly large) in separate sections
C2	System oscillations and tripping of circuits (e.g. on impedance protection) possibly leading to uncontrolled system split with generation demand imbalances in separate sections
C3	Build up of oscillations/circuit trippings possibly leading to uncontrolled system split with generation-demand imbalances in separate sections
C4	Cumulative voltage fall as tap changers operate/transmission voltages fall and currents increase, generation excitation systems limit, system voltage collapse and probably instability

generation loss case; the loss of pumped storage plant when pumping may be an exception to this generalization. The potential immediate and second stage effects are shown in Tables 3.2(a) and 3.2(b).

3.3.3 Sudden Loss of Transmission (Not Resulting in an Immediate System Split)

Overhead lines are more exposed to the vagaries of weather than other plant, and this explains why overhead line faults are more frequent than other equipment faults, and transmission outages are the most frequent source of system disturbances. Other significant factors are that transmission faults may bunch together both in time, because of periods of adverse weather, and in location, because of the local environment or local weather conditions. These effects can be such that security criteria may be modified to counter their effect.

Contingencies with possible results, containment methods and timescales are shown in Table 3.3(a), and possible second order effects in Table 3.3(b).

3.3.4 Sudden Loss of Transmission (Resulting in a System Split)

The conditions in the separate sections formed by the transmission outage will depend upon the pre-outage power flow on the tripped circuits, and will be import/export or float/float. Any immediate actions necessary will be as indicated in Sections 3.3.1 and 3.3.2, supplemented by any circuit switching needed to improve network conditions.

3.4 TYPICAL PATTERN OF THE DEVELOPMENT OF A SUDDEN DISTURBANCE

Although descriptions of actual disturbances indicate many and varied causes, building on the basic developments outlined in the previous section, it is possible to identify a pattern to the way in which many of the large scale disturbances of the past have developed. This is shown in Figure 3.3. (see Reference 3.1).

Some sequence of events (e.g. a simple fault with a compounding factor or a multiple/complex fault with or without a compounding factor – see the upper part of Figure 3.3) leads to a significant imbalance between demand and generation in all or parts of the system. There can be a wide range of compounding factors – inadequate liaison between neighbours, errors in operational planning, errors in control, maloperation of protective gear, telecommunications failure, etc. The system or its separate parts should stabilise at points S1

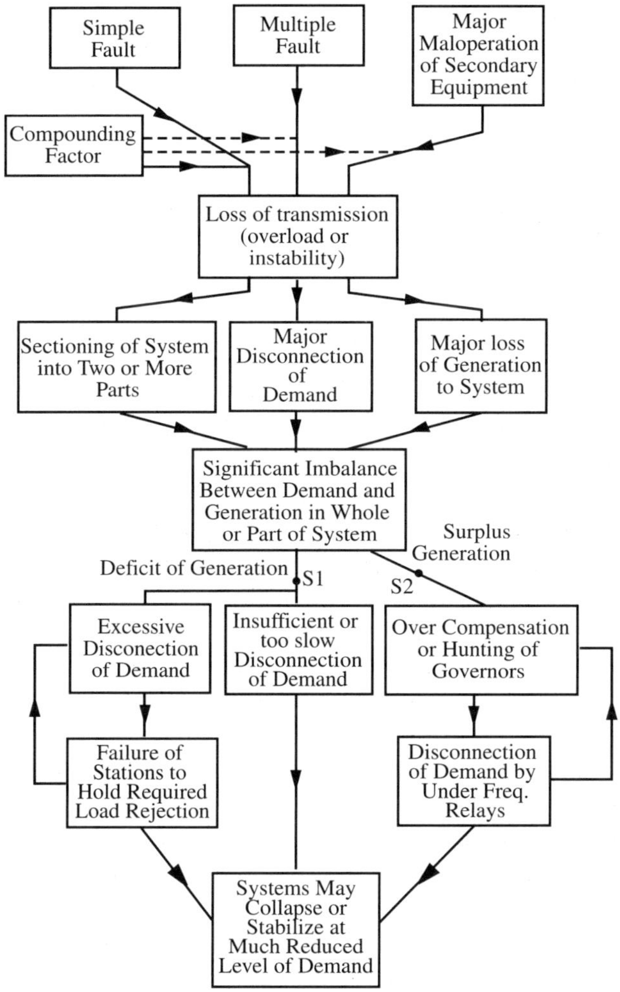

Figure 3.3 Typical mechanisms of large scale system disturbances (Reproduced by permission of Cigre from [3.1])

or S2 through the action of governors and load frequency control, disconnection of demand or less frequently disconnection of generation – all actions intended to eliminate any imbalance between demand and generation, and to restore the frequency to its nominal value. If, however, these control actions are not matched sufficiently closely to the imbalance, further adjustments of generation will be called for, and if these are outside the capability of the plant and its control loops, further disconnection of demand or loss of generation may occur; these possible cases, all of which have occurred in practice, are shown in the lower part of Figure 3.3.

Essentially, the containment of a major disturbance requires that the events shown in Figure 3.3 are halted as early in the sequence as possible. The ideal way to achieve this is to adjust the controlling actions as closely as possible to the magnitude and location of the disturbance. The timescale of the events will frequently be so short that the control actions must be automatic.

3.5 CONCEPTUAL FORMS OF EMERGENCY CONTROL

An essential feature of emergency control is that it is provided to deal with relatively extreme, and hence unlikely, events. These events may be defined as a class, such as the sudden loss of the largest amount of generation possible following a single fault, or as a specific event such as the tripping of defined circuits into a part of the system with limited transmission connections. In the former case, the emergency control is likely to be implemented as *pre-defined logic* (Figure 3.4(a)). In this, an undefined contingency causes system effects which result in frequency, voltage, etc. changes. On detection of these, pre-defined actions are implemented automatically to contain the disturbance.

In the second case, if a pre-defined contingency occurs (see above), pre-defined containment actions are implemented immediately, or perhaps after checks to confirm that action is justified (Figure 3.4(b)). A more advanced concept is shown in Figure 3.4(c). In this, the action taken following an undefined contingency is adapted to the type of contingency and its results. Ideal containment and restorative action are selected based on the observed state of the system. The operator could provide this adaptive feature in an interactive approach, the computer indicating alternative and numerical solutions and the operator making the final choice. Taking the most widely used emergency control technique, under-frequency load shedding, as an example, the simple implementation in which underfrequency relays are installed at fixed points and with fixed settings could be made adaptive by adjusting the location and level of shedding in accordance with power flow and voltage conditions on the transmission network.

It is not a big step to adapt the approach of Figure 3.4(b) to obtain reductions in capital or operating costs, or to compensate for delays in commissioning, as in Figure 3.4(d). If, for instance, because of delays in commissioning a third line, a power station is connected into a system by two double circuit overhead lines (Figure 3.5), with each circuit rated at 40 percent of the station output, and the normal security criterion is to maintain full station output following the loss of any double circuit line, the system operator has three alternatives:

(1) To reduce the station output to 80 percent (1920 MW) of its maximum; the loss of a double circuit line would be accommodated, but with adverse impacts on both the system running cost and system capability.

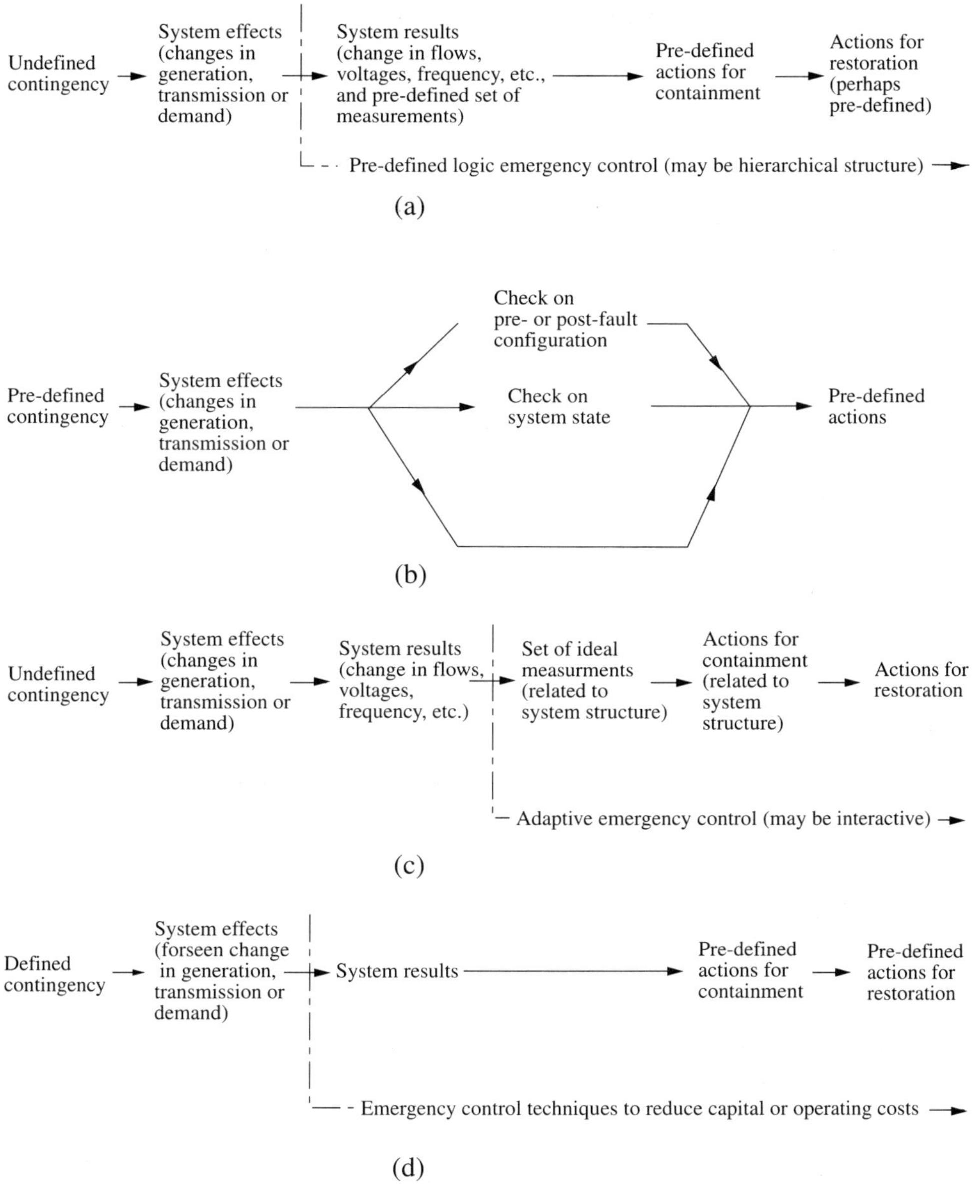

Figure 3.4 Alternative conceptual forms of emergency control. (a) Pre-defined logic, (b) specific contingency, (c) adaptive, (d) emergency control techniques used to reduce capital or operating costs or to compensate for delays in commissioning (Reproduced by permission of Cigre from [3.3])

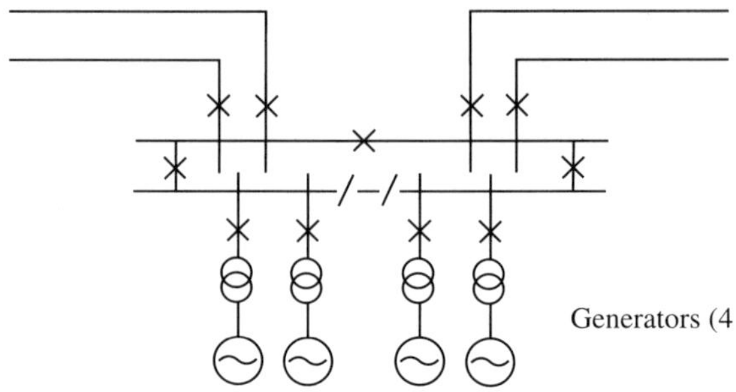

Figure 3.5 Automatic reduction of generation to reduce the capacity of transmission connections

(2) To operate the system to a single rather than double circuit loss criterion; if a double circuit outage then occurred, rapid reduction of the output of the station would be needed (protective gear settings should normally be such that there would be no risk of the remaining circuits tripping on overcurrent protection).

(3) To provide an automatic generation reduction scheme to reduce the station output as necessary on detection of the loss of a circuit(s) into the station. In

Table 3.4 Example of automatic generation reduction at a station with limited transmission connections

System state	Necessary action on circuit loss	Preparation
Four circuits and four generators available		
Generation up to 1920 MW	None	
Generation 1920–2400 MW	Reduce generation to 1920 MW	Arm generation for tripping or reduction of output
Four circuits and three generators available	None	None
Three circuits and four generators available		
Generation up to 960 MW	None	
Generation above 960 MW	Reduce generation to 960 MW	Arm generation for tripping or reduction of output
Four circuits and three generators available		
Generation up to 1800 MW	None	None
Three circuits and three generators available		
Generation up to 960 MW	None	
Generation 960–1800 MW	Reduce generation to 960 MW	Arm generation for tripping or reduction of output

such a case, operating instructions could be provided to the station and control centre to meet a two circuit loss criterion, as shown in Table 3.4.

Generalizing this example, emergency control type facilities can be installed for two broad reasons:

(1) to preserve system integrity following a 'non-credible' contingency:

   (i) to maintain as much as possible of the system in a viable operating state following a disturbance, as a result of which the demand cannot be met by the remaining generation and/or transmission capacity without exposing the plant to non-sustainable overloads or operation at abnormal values of frequency or voltage. (The combination of initial system state and disturbance will together, in general, be more severe than the set of defined contingencies taken for the planning and operation of the system.)

   (ii) following this, to restore a viable operating state, with all demand supplied, as quickly as possible;

(2) as an alternative to system capacity;

   (i) in planning, to reduce installed generation or transmission capacity by transiently equalising demand and system capacity remaining after a defined incident, thereby providing the short time necessary to institute measures to restore a viable operating state with all demand supplied. It seems unlikely that any appreciable reduction in total generation capacity would be achieved, although the plant mix might be affected;

   (ii) in operational planning, to overcome limitations in transmission capacity, in particular caused by delays in consents and construction. The mechanism would be as in (2(i)).

   (iii) in operation, to reduce immediately available operating reserves, particularly of generation. The mechanism would be as in (2(i)).

Although procedures and aids to meet these objectives may be similar, even identical, the design problems of the two applications are different. In the former, any combination of equipment faults, outages or human errors may occur, affecting the whole or part of the system. Neither the resulting system states nor the necessary remedial actions can be determined except in broad terms. Where the distinction is necessary, this is referred to as *true emergency control* in the remainder of this book. In the latter, the abnormal states of the system, and hence the necessary remedial actions, are well defined, as those resulting from one of a number of specific incidents.

3.6 EFFECT OF SYSTEM STRUCTURE ON THE NEED FOR AND IMPLEMENTATION OF EMERGENCY CONTROL

Basic topologies for a transmission system have been outlined in Section 3.2. In practice, the magnitude and distribution of actual and prospective demands, the actual and potential generation sites, the climatic conditions, the terrain and its nature (rural, urban, etc.) will determine the outline interconnections, the structure and the more detailed design of the power system.

Referring back to the six basic topologies of Section 3.2, little correlation is evident between system topology and the necessity for true emergency control. Nevertheless, some generalizations are:

U1(a) – this is potentially a very secure system, but with some risk posed by undefined cross-system flows caused by changes in neighbouring systems. Experience suggests that such risks are not negligible,

U1(b) – there is an obviously greater risk than in U1(a) of such external changes causing problems from cross-system power flows. Such systems are also likely to be less secure than U1(a) against the internal non-credible contingency,

U2(a) – these systems may be susceptible to loss of external interconnectors, and if this risk is to be minimized, there is a need for emergency control of generation and demand which is not dependent on frequency deviations for actuation,

U2(b) – much as U2(a), but with greater risks from internal and non-credible faults,

U3(a) – the security of such a system will depend very much upon its own planning, operational planning and control policies,

U3(b) – this system will be similar to the U3(a) system, but probably less secure against the non-credible contingency.

These general comments assume that the transmission between utilities will include capacity for uncertainties in generation or demand as well as the planned power transfers. This reservation, and the level of uncertainty considered, invalidates any general argument that the larger an interconnection, the more secure it is likely to be.

A quantitative assessment sponsored by a Cigre Working Group [3.5] indicated that interconnection with neighbouring utilities is an important factor in lowering the number of the smaller disturbances, particularly for large systems. Stability limited (type (b)) systems experienced significantly more disturbances at all levels of severity, than thermally limited (type (a)) systems. Systems above 1000 MW in size had similar numbers of disturbances, greater than the number for systems below 1000 MW. No very large disturbances (above 100 system minutes) were reported for these small systems. One might conclude that increasing system size if anything increases the need for emergency control facilities (surely an argument for some form of sectioning very large systems during major disturbances), and that the need is more likely to be evident with stability limited than with thermally limited systems.

It is possible, and sometimes necessary, to consider an interconnected system of several utilities as one utility from the emergency point of view. Conversely, different parts of a single utility can be considered as separate 'utilities' within an 'interconnection', in which case the comments would apply to those parts.

3.6.1 Effect of System Structure on the Form of Emergency Control

System structure and size can significantly alter the form of implementation of emergency control, most obviously in the use of under-frequency disconnection of demand. This will not protect a small part, even a whole utility, within a large interconnection against transmission overloads, excessive voltage changes or instability caused by generation loss; in comparison to the total interconnection capacity, an important local loss of generation is unlikely to influence the system frequency to the extent required to operate under-frequency relays.

An important general question is whether system sectioning should be employed. It can be used to isolate utilities from each other, for instance, on detection of very low frequencies or oscillatory conditions, and/or within large utilities to isolate disturbed from healthy sections of the system.

The form of any transmission constraints are likely to have an important influence. Response times of several seconds to minutes will usually be acceptable for remedial actions to reduce thermal overloads, whilst action to prevent transient instability will need to be taken in milliseconds. Actions to rectify unacceptable voltage conditions should probably be taken in the order of tens of seconds to minutes.

3.7 DESIGN CRITERIA FOR EMERGENCY CONTROL FACILITIES

A number of almost self-evident criteria will establish the basic implementation of an emergency control scheme:

(1) the most appropriate system variable/s should be chosen to initiate emergency action;

(2) the actions taken should be the minimum necessary to contain the disturbance, particularly where adjustment of generation output or disconnection of demand are concerned;

(3) the actions should be implemented at the geographical locations on the system where they are most effective in containing the disturbances, and run least risk of precipitating further problems.

It may be noted that these points are mainly concerned with making the emergency control adaptive to the actual system state.

Other suggested design criteria are:

(4) the emergency control system should have a functional reliability such that the probability of avoiding demand disconnection as a result of its successful operation is several times greater than the probability of demand disconnection as a result of its possible maloperation. An alternative, simpler approach would be to seek a reliability from the emergency control system no worse than that obtained from first line protective systems;

(5) the actions taken and the reasons for these should be indicated to the operator;

(6) alarms should be given when the emergency system is not functioning or its correct operation is doubtful (e.g. suspect data);

(7) it should be possible for the operator to over-ride incoming telemetry known to be incorrect and, where pre-defined logic is used, to select alternatives in line with actual power system conditions;

(8) the control system must be robust, that is it must meet its criteria and objectives whatever the state of the power system;

(9) decision and action times must be less than the time at which further significant degradation of the system would occur.

Utilities generally collect statistics on the performance of their protective systems, which are sometimes extended to include emergency control systems (see Chapter 5).

REFERENCES

3.1. Knight, U. G., 1989. '*The control of power systems during disturbed and emergency conditions.*' Cigre brochure (this brochure collates several papers in the area of emergency control).

3.2. Ledger, F. and Sallis, H., 1995. '*Crisis Management in the Power Industry, An Inside Story*'. Routledge.

3.3. Knight, U. G., 1983. 'The implementation of emergency control'. Paper 207-05, *Cigre IFAC Symposium on Control Applications.*

3.4. Winter, W., 1980. 'Measuring and reporting the overall reliability of bulk electricity systems'. Cigre, Paper 32.13.

3.5. Winter, W. and LeReverend, B., 1986. 'Disturbance performance of bulk electricity systems'. Cigre, Paper 39.05.

4
The Power System and its Operational and Control Infrastructure

It is proposed in this chapter to review briefly the salient features of the design, operation and control of modern day power systems, including recent innovations. This will provide a background for the material on emergency control in the remaining chapters.

4.1 STRUCTURE

The present day structure of most power systems evolved in the middle decades of the 20th century. This period saw the interconnection of separate municipal undertakings to form National or Company utilities. Privatization and restructuring have not of themselves altered the fundamental structures, although voltages and voltage transformations have changed to meet the need for increasing power transfers. For instance, transmission voltages have risen to the current norms of 400–500 kV, with some systems going as high as 1500 kV. Direct current is now a major contender for the transmission of power in terms of a few hundred megawatts over distances of a few hundred kilometres. In an a.c. system, there will typically be one or two intermediate voltage networks between the transmission voltage and the sub-transmission/distribution voltages, as shown in Figure 4.1(a). Many systems will be embraced in the following:

- Transmission voltage level: the voltage used will usually be between, say, 230 and 500 kV, depending on the transmission distance and powers involved. The network will usually be meshed. Large generating stations will be connected to the system at this voltage, as might very large consumers. There will be connections to lower voltage networks.
- Intermediate voltage level-1 (IV1): if there is only one intermediate voltage level, the voltage used will be in the order of 100–150 kV. The network is

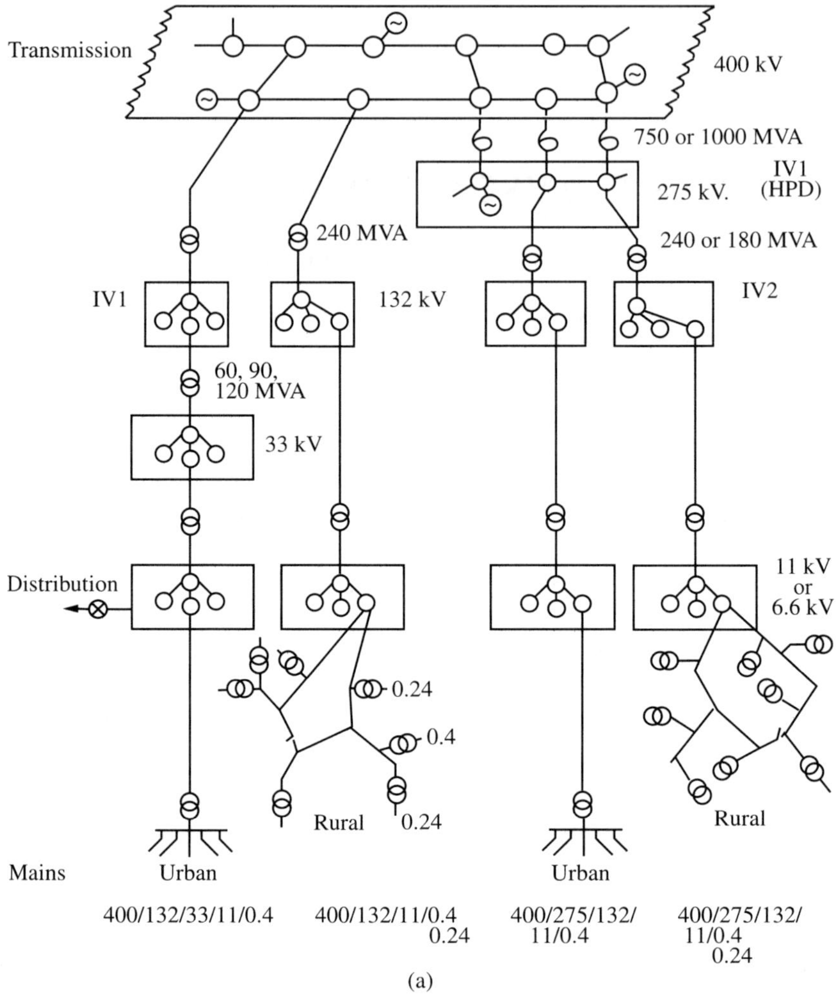

Figure 4.1 (a) Voltage levels in typical a.c. system

likely to be radial, or locally meshed between neighbouring infeed points from the transmission network. Smaller generating stations, including gas turbine and combined cycle stations, and transforming points to lower voltage networks will be connected at this voltage level, which is often called *sub-transmission*.

- If there are two intermediate voltage levels, the higher of these, say, in the range 200–300 kV, may be used for *High Power Distribution* (HPD) in relatively local areas with high densities of demand. Hence, there may be several of these networks, meshed or radial, in and around major cities, each supplied from the transmission network at one or several transforming points.

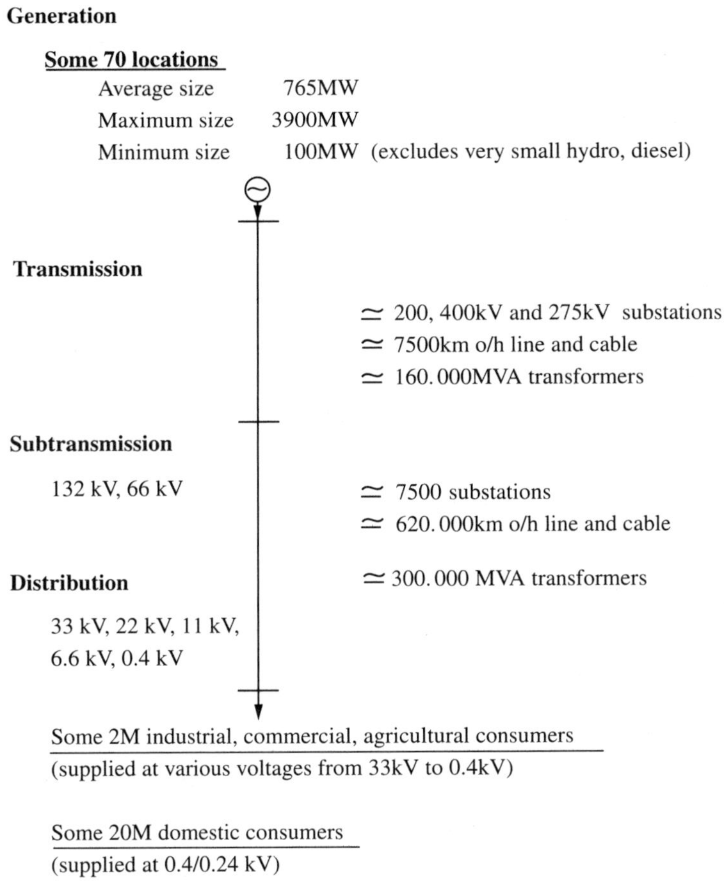

(b)

Figure 4.1 (b) Statistics of electricity supply in the early 1990s

Moderate to large sized generating stations, transforming points to lower voltage networks, and very large consumers may be connected at this voltage. If demands and power transfer(s) in part(s) of a system are comparatively low, this intermediate voltage may be retained as the transmission voltage for some years. This happened in Great Britain where 275 kV was retained in several areas for some years after the core part of the network was uprated to 400 kV.

- The second of the two intermediate voltages (IV2), in the range of, say, 60–150 kV, will be supplied from transforming points on the transmission or intermediate voltage-level-one networks. It is likely to be radial, providing connections to small generating stations and to transforming points to lower voltage networks.

- There may be one, less frequently two, voltage levels between the lowest intermediate voltage level and the mains voltage level. The number of these and the voltages used will again depend upon the magnitude and density of demand and the terrain.

Examples of some of the voltage levels reported in the 1990s are shown in Table 4.1 (it is difficult to label the function of these voltages without a detailed knowledge of the structure of the networks; MV is used in the continental sense of some 20–40 kV).

There is a considerable range in transformer ratings, for example:

- With two intermediate voltages between transmission and medium voltages:
  - transmission to intermediate voltage 1 – mainly in the range 250–1000 MVA
  - intermediate voltage 1 to intermediate voltage 2 – in the range 100–300 MVA
  - intermediate voltage 2 to medium voltage – mostly in range 40–90 MVA, but with smaller sizes down to 10 MVA or less.
- With one intermediate voltage between transmission and medium voltages:
  - transmission to intermediate voltages – mostly in the range 200–400 MVA
  - intermediate voltage to medium voltage – mostly in the range 40–70 MVA.

An idea of the physical scale of modern power systems is apparent from Figure 4.1(b), which shows some of the main statistics of the networks in Great Britain in the early 1990s (the data was obtained from annual reports of the various utilities and the Electricity Council).

Table 4.1 Some standard voltages

Country	Standard voltage levels (kV)
France	400, 225, 90, 63, MV, HV, mains
Great Britain	400, 275, 132, 66, MV, HV, mains
Germany	380, 220, 110, 60, MV, HV, mains
Norway	420, 300, 132, 66, 45, MV, HV, mains
Sweden	400, 220, 130, 70, 40, HV, mains
Finland	400, 220, 110, 45, MV, HV, mains
Netherlands	380, 220, 150, 110, 50, MV, HV, mains
Belgium	380, 225, 150, 70, 36, MV, HV, mains
Italy	380, 220, 150, 132, 60, 50, MV, HV, mains
Russia	1500, 1150, 750, 500, 330, 220, 110, MV, HV, mains

4.1.1 A Theory on the Evolution of Network Voltages

It was suggested by an E de F engineer that transmission systems evolve through a series of intermediate voltage steps to avoid problems of power sharing between a high capacity network and existing lower capacity networks. Thus, 200–300 kV might be introduced as a superimposed network on a 100–132 kV system with the lines upratable to 400 or 500 kV. As the system grows, the majority of the 250–275 kV will be reinsulated to 400–500 kV as the main transmission voltage, perhaps leaving some lines at 250–275 kV for supplies to major cities. In any event, 400 kV and 500 kV are judged to be the most common transmission voltages, with voltages above this restricted to very long distance, high power applications, for instance Itaipu Falls to Sao Paulo in Brazil (a.c. and d.c.), the West Coast interconnection in the USA (700 kV), the Churchill Falls and James Bay transmission to Quebec Province in Canada (700 kV), the Unified Electric Power System (at 1150 kV a.c.) in Russia, and the Inga-Shaba d.c. link in Africa. (See Appendix 1 for information on some major interconnections around the world.)

4.2 THE FUNCTIONS OF INTERCONNECTION

Interconnection within and between power systems will be provided for several purposes:

(1) To provide capacity for planned transfers. Sites for conventional thermal stations in the multi-GW capacity class, meeting the combined requirements of low transport costs for fuel, adequate cooling supplies, low cost ash disposal, environmental acceptability and low connection costs, including low cost access to local demand, will be increasingly difficult to find. Generation costs will far outweigh transmission costs, and as a consequence, low connection costs will be sacrificed for other site benefits. One definition of the planned transfer will be the difference between generation at expected availability and the expected peak demand or the demand at some specified time and date.

(2) To pool generation resources: essentially, it is accepted in deterministic criteria that risks do not increase in proportion to system size. Hence, it is quite usual over a wide range of system sizes to find that a generation margin will be provided to cover the loss of the largest generating unit, or the generation loss which would be incurred by a credible transmission failure. This will apply in both planning and operations.

(3) To improve economic operation. The merit order of generation cost is likely to be a stepwise monotonic function.

(4) To take advantage of demand diversity. Several factors result in diversity in demand:
   - on a geographically small to medium scale; changing working and social habits including those between countries;
   - on a geographically large scale: latitudinal – climatic changes; longitudinal – time changes (15° is equivalent to one hour, although clock-time changes will be banded).

(5) To utilize hydro capability fully. The characteristics of large hydro schemes (location, power and energy capacity, storage and output cycles) are dictated more by site than system considerations. Thus, transmission will be required to connect such schemes to systems large enough to absorb these fluctuations.

(6) To provide support in emergencies. Typical examples will be power imports/exports to compensate for large and unexpected changes in generation and demand; or perhaps to provide power for a blackstart.

(7) To provide supply to peripheral areas. Occasionally, it will be cheaper to supply pockets of demand on the periphery of a system from a neighbouring system rather than from the parent system.

The capacity and security of the transmission links will obviously depend upon their purpose. Links serving *planned transfer* needs will require the same standards as transmission internal to the utilities. The levels of transfer will usually be modest – a few percentage points at most of the partners' own demands/capacities. One explanation for this is that electricity is a strategic commodity, the supply of which countries will wish to retain within their own jurisdiction.

4.2.1 Exchanges Between Neighbours

The types of significant exchanges found between interconnected partners are as follows [4.1]:

1. Firm energy – which may, however, include interruptible power (functions 1 and 5 above), usually on long-term agreements.
2. Marginal exchange of spinning reserves (functions 2 and 3 above), usually on short-term agreements.

3. 'Economy energy exchanges' with no guarantee of capacity (functions 3, 4 and 5 above), usually on short-term agreements.
4. Back-up exchanges for emergency power (function 6 above).
5. 'Compensation exchanges' made in kind to bring the exchange over an agreed time period to the target figure.

4.3 THE ALTERNATIVES FOR MAIN TRANSMISSION [4.2, 4.3]

The main transmission network will provide the functions of transmission, interconnection and high power distribution within a utility, constructed for the highest voltage used by the utility, although there may in some be a still higher voltage overlay used for point-to-point transmission of large powers over long distances, or for interconnections between countries on a continental scale. Currently, the alternatives for the whole range of applications are d.c. or three phase a.c. Alternatives of multiphase (above 3) and half wave length a.c. have not, to the author's knowledge, been taken beyond early consideration.

Without attempting any detailed comparison of a.c. and d.c. technologies, some of their attributes relevant to providing the functions above are:

1. A.C. systems can provide any function, although with technical constraints on transmission distances. There is no limit on the geographical extent of an a.c. system, provided reactive and active power sources are distributed through it.
2. D.C. circuits or links have a number of well defined applications (see Section 4.3.1) and are embedded in or superimposed on a.c. systems to provide these.
3. The power transfer through an a.c. circuit is determined by the state variables (voltage magnitudes and phase angles) at the terminal nodes. The power transfer through a d.c. circuit is determined by the settings on the d.c. link, subject to the power and reactive power requirements demanded by the link being available at its terminals.
4. As a consequence of (3), power changes across a d.c. link will be virtually instantaneous. Those across an a.c. link will depend upon the busbar angles and upon inertia and control mechanisms in the terminal a.c. systems. However, this means that an a.c. interconnection provides rapid reciprocal assistance between all the generating units participating in primary reserve. At the same time, it will feed short circuit current from all the interconnected systems into any fault.

5. The need to constrain fault levels may impose restrictions on the planning and operation of an a.c. network.

6. For a given power transmission, d.c. circuit losses will be lower than a.c. ones. Although d.c. terminal losses will offset this advantage for long distances and high powers, the overall losses for the d.c. case will be less than for the a.c. case. Using d.c., efficiencies (in financial terms) of some 95% and 85% have been indicated for the transmission of 3000 MW over 1000 km and 4000 km, respectively [4.3]. In a 4000 MW, 800 km transmission proposal, the losses at the rated power with a d.c. system were about 87% of those on an a.c. system [4.4].

7. Even based on thermal loadings, the transmission capacity obtainable over a wayleave can be much higher with d.c. rather than a.c. Baker [4.5] includes information on the conversion possibilities of a.c. lines to d.c. from which the comparisons in Table 4.2 are taken.

8. Conversely, because a.c. networks obey Kirchoff's second law, some of the current flows do not contribute to useful transmission capacity. These have been called 'loop flows' or 'parasitic flows', and have in part prompted the development of FACTS devices.

9. A d.c. link can be sized to the operational requirement and the economic capacity. Its size will not be dictated by technical issues of maintaining stability, and it will not be necessary to harmonize the frequency and voltage control mechanisms or settings in the terminal systems.

10. In general, it will be cheaper to tap an a.c. line to supply demands along its route than would be the case with a d.c. line. However, this could be a mixed blessing in some societies, since theft of power and energy would be easier in the a.c. case. Very small powers (up to tens of kVA) can be obtained by inductive coupling from antennae located adjacent to the earth wire/s of UHV lines, and have been used to supply small villages along the routes of such lines, for instance in the Brazilian forests.

11. In spite of the advantages of d.c. for various specific applications, there is virtually no alternative to a.c. for routine transmission and distribution tasks within a utility. The task of the system planner will be to adopt and use appropriate security criteria, network configurations, voltage levels, plant ratings and types of construction (overhead, underground, single or multi-circuit, etc.).

12. There is thought to have been limited application as yet of d.c. switching stations, that is a d.c. busbar to which several d.c. links are connected. This is perhaps because the need has not existed although some of the major proposals now being considered may bring this need.

Table 4.2 Line capacities with change from a.c. configuration/operation to d.c. configuration/operation (Reproduced by permission of Cigre from [4.5])

Existing a.c. line	Voltage and current (thermal) limit	Apparent power MVA	Conversion to d.c. line	d.c. voltage and current	d.c. power
Double circuit, single conductor	220 kV 630 amp	480	Bipole, three conductors per pole	±380 kV 1890 amp	1440 MW
Double circuit, twin conductor	400 kV 1440 amp	1000	Bipole, three conductors per pole	±500 kV 2160 amp	2160 MW
Single circuit, twin conductor	500 kV 1440 amp	1250	Bipole, three conductors per pole	±500 kV 2160 amp	2160 MW

13. Unlike a.c. cables, capacitance and charging currents do not limit the length of d.c. cables, or require reactive compensation at intermediate points along the length of the cable.

14. D.C. links pose some technical problems:

    (i) the need to absorb harmonic currents produced by the converter equipment,
    (ii) the provision of reactive power at the a.c. terminals of the link,
    (iii) interactions with the a.c. systems especially if the a.c. short circuit levels are low, and
    (iv) co-ordinating the control of several conversion stations.

15. A.C. links do not 'isolate' the effect of changes in one system from the other to the same extent as do d.c. links. Hence, more detailed joint studies are required from the partners essentially covering all aspects of planning, operational planning and control. Not least, this means there must be a greater exchange of technical, although not economic, information.

4.3.1 The Roles of Direct Current Interconnection and Transmission

The technical characteristics and costs of d.c. links compared with a.c. links determine the d.c. applications as follows (it is found that the two criteria often coincide):

(1) *Long distance high power overhead transmission*
The distances and powers involved will typically be 500 km plus, and some hundreds of MW, requirements found for instance in South America, Africa, Russia and western through to eastern Europe. Over the years, the break-even point between d.c. and a.c. transmission has tended to fall, both for distance and power transmitted.

(2) *Long distance submarine and underground cables*
The need to underground circuits overland will usually result from environmental factors and the lengths involved will rarely exceed, say, 50 km. At this length, a.c. is technically feasible and competitive in cost. In contrast, submarine cables will tend to be longer (although nowhere near the hundreds of kilometres of point one above) and sometimes will be used to interconnect different utilities, even countries, situations which in themselves may favour d.c. solutions.

(3) *Interconnection between power systems*
'Back to back' links have been constructed to provide interconnection, sometimes transmission, facilities between large systems. Even when the

nominal frequencies are the same, stability and control problems are eliminated or much reduced. Such links will typically be rated at a few hundred megawatts.

These d.c. and VHV/UHV a.c. applications are unlike the systems with which most planners and operators will be familiar. Because of this, some of the already constructed or proposed developments across the world are described in Appendix 1.

4.4 SECURITY AND QUALITY OF SUPPLY IN PLANNING AND OPERATION

Security and *quality* describe different aspects of the performance of a power system:

- *Security*: its robustness in terms of its ability to withstand faults and other losses of equipment; it will usually be quantified in terms of security standards, i.e. the relationship between outages of generation and transmission plant and the level of any acceptable loss of demand.
- *Quality*: usually taken to include factors such as the constancy of voltage and frequency (measured, say, by average values and standard deviations); other factors defining voltage quality will be its freedom from sudden steps, transients and purity of waveform.

The system planner and operator will always have to consider security. As regards quality, utilities will often require consumers to control their demands so as to limit the frequency and magnitude of sudden changes or distortion of waveform. This will mean that the planner and operator will, in general, only need to consider average values, standard deviations and deviations under abnormal conditions of frequency and voltage. Planning and operational standards must be compatible, but as the lead time decreases in moving from the planning to the operational timescale, the uncertainties decrease. Hence, it is quite usual to find that the planning standards are more rigorous, for example including an allowance for uncertainty in demand not found in operational standards.

Many comparisons have been made between standards around the world, often by the international and national bodies (CIGRE, IFAC, IEEE, IEE), which have appeared in their publications, but also by utilities. The form of the criteria as well as the numerical targets vary, not least because some standards are framed in probabilistic and some in deterministic terms. A general form will, however, be 'if any one of event *a* and/or event *b* and/or ... occur, then condition *m* and/or condition *n* and/or condition ... should obtain'.

4.4.1 Standards of Security in Planning [4.6–4.10]

The impact of a loss of supply will depend upon its location, magnitude and duration. Hence, it is usual to find that security standards take account of these factors, for instance distinguishing between transmission and distribution standards, or between 100 MW and 10 MW demand levels.

Generation Standards

Generation standards, often expressed as margins between the minimum generation and the maximum demand expected in the planning years, are usually specified in terms of impact on the supply of power to the whole system.

- *loss of load probability* – the probability that over a given time period, there will be insufficient generation to meet the demand. This is one of the most widely used criteria.
- *frequency and duration of generation capacity outages/loss of demand* – the frequency and duration of specific levels of generation capacity over a period of time are calculated. These may be combined with a demand distribution to estimate the frequency and duration of losses of demand.
- *failure probability* (sometimes called *loss of load expectation*) – the number of times a loss of supply is expected over a given period of time, for example days per year, times per year, failure days per 100 years, times per 100 years.
- *failure duration* – hours per year.
- *undelivered energy* – proportion of energy expected to be consumed by the demand but not delivered because of shortage of generation capacity.

Generation capacity margins (for example, (total sent out generation–expected demand)/total sent out generation) are calculated to achieve the criterion adopted. There is a considerable variation in both target capacity margins (from, say, as low as 10% to 30% on thermal systems), and in expected failure rates (variously a shortage of capacity every two to three years, or 50 failure days every 100 years, etc.). The author's experience in a large, fully integrated generation/transmission utility with frequent checks on the generation-demand balance prior to the event, was that various *ad hoc* measures would be taken if a generation deficit appeared likely. These would include returning plant to service early, deferring outages, and finally, allowing plant to be slightly overloaded.

A fairly recent survey [4.6] covering some 30 countries/utilities worldwide showed that the majority (68%) used probabilistic methods only, 12% deterministic (plant margin) methods, and 16% combined methods. The failure duration method seemed the most popular. The author has been surprised that the frequency and duration method of Ringlee *et al.* [4.7] which produces estimates of the frequency of outages of various durations has not found wider support. Reference [4.8] includes statistics (probably for about 1985) on plant margins and failure duration. The countries/utilities included are mostly industrialized. The margins for 20 countries/utilities varied between 10% and 33% (average 18.5%, and most frequent 20%). Failure durations varied between 2 and 35 hours per year for 10 countries/utilities, average 12 hours per year. A survey made some five years earlier suggested a somewhat higher average margin and a lower duration.

The target margins will be affected by several factors:

(1) The plant mix;

  (a) systems including hydro capacity will often have a higher margin, because the capacity may have to be sized to meet energy as well as power requirements ;
  (b) compared to conventional oil/coal stations, gas turbine based stations will tend to be of lower capacity (say, up to 1000 MW, rather than 2000–4000 MW), require less area per kW installed, require smaller site services, shorter times to acquire planning permission and for construction, present fewer problems for connection into the networks, and finally, need a lower capital commitment and offer a quicker return on this; these factors are all conducive to less uncertainty in planning, and hence lower margins.

(2) Interconnection and agreements with neighbours

(3) The supply standard.

(4) The size of the largest unit in absolute terms and its ratio to the system size in percentage terms.

(5) The structure of the industry – in general, one can expect countries with only one organization responsible for generation to have lower margins than others.

Following privatization, there has been no formal planning margin in England and Wales, although NGC provides yearly statements covering the next seven years on the expected supply situation overall and in different parts of the country. It attempts to influence the siting of new plant through its transmission tariffs.

Transmission Standards

Most utilities use some form of probability method in determining generation margins. In contrast, transmission planning both past and present has often used an empirical approach. Typically, the transmission system is designed to meet expected transfers at specified levels of demand and corresponding generation with a specified degradation of the transmission network, frequently any one circuit not available.

There is, nevertheless, a wide range of options available. In one survey, virtually all 24 countries/utilities reporting studied the system conditions at peak demand and the majority at off peak and at minimum demand levels. Some two-thirds used a merit order to select the generation to run, often in conjunction with some form of scaling down for availability of generation. One of the most important factors will be the circuit outage conditions used. Some half of the utilities used single and half double circuit outage conditions. In most cases, the pre-outage system condition was assumed to be with all circuits in commission. Some utilities will also study the loss of a busbar and its associated generation.

The standards just outlined specify the steady state conditions, for example pre- and post-disturbance, for which the system must be able to operate stably and with acceptable circuit thermal loadings. It is also necessary in stability studies to specify the fault conditions which the system must survive, that is demand level, fault location, fault type and duration, and reclosure or not. The usual approach is to specify fault type, clearance time and reclosure condition, the planning engineer being responsible for selecting the worst fault location. Criteria found in practice are:

(1) A three-phase fault affecting both circuits of a double circuit line, correctly cleared, with 'permanent' loss of both circuits.

(2) A single circuit three-phase fault, correctly cleared, with 'permanent' loss of the circuit.

(3) A single circuit two-phase to earth fault, correctly cleared, with 'permanent' loss of the circuit.

Of the 24 countries mentioned above, 21 used a three-phase fault criterion, with several adding two-phase and single-phase faults to check, for instance, torsional effects or voltage transients.

The network configuration will strongly influence the criticality of the location of a fault. The locations will be selected by the system planner based on experience, but will often be close-up faults at major power stations. The planning criteria used in North America differs in detail between the Regional

Reliability Councils. As an example, those used by the North East Council, as reported in Ringlee *et al.* [4.7] were:

Normal conditions

- pre-contingency voltages: line and equipment loadings shall be within normal limits;
- stability of bulk power system shall be maintained during and following the most severe of the following contingencies:
  - permanent three-phase fault on any element (except a circuit breaker) normally cleared with due regard to reclosure,
  - permanent phase-to-phase-to-ground fault on different phases of two adjacent circuits on a multiple circuit tower, normally cleared with due regard to reclosure,
  - permanent phase-to-ground fault on any transmission circuit, transformer, or bus section with delayed fault clearing and with due regard to reclosure,
  - loss of any element without a fault, and
  - permanent phase-to-ground fault on a circuit breaker normally cleared, and with due regard to reclosure;
- post-contingency voltages, line loadings and equipment loadings shall be within applicable emergency limits (loadings shall not exceed 5-minute time ratings).

'Extreme contingencies' may also be considered, for instance loss of an entire generating station, or an entire substation or all circuits on a common right of way. Typically, widespread power interruptions or conditions which might lead to further tripping should not then occur.

4.4.2 Standards of Security in Operation

Operational criteria are often deterministic. General practice to ensure acceptable quality and continuity of supply in operation is:

(1) to provide a margin of capacity to meet:

  (a) changes in demand, either random or resulting from its trend,
  (b) changes in generation capacity as a result of a controlled, but necessary, or uncontrolled, reduction of generation, and
  (c) if applicable, loss of import to the system.

(2) to ensure that there will be no loss of supply or any uncontrollable situation in the event of loss of one, and in some cases, two, transmission circuits or nominated other parts of the transmission network.

The picture which emerges from the various reviews of operational standards of security [4.11, 4.12] is that the loss of the largest single infeed should not lead to any loss of supply or other significant effect. Some countries have adopted a single circuit loss, some a double circuit loss, for the transmission contingency. Quite a few countries have also stipulated a combined contingency–the coincident loss of the largest generator and one circuit.

The operational criteria which have been used in three major interconnections are summarized below.

UCPTE

UCPTE established basic recommendation for the interconnected countries of Western Europe as follows [4.13, 4.14]:

(1) each member to hold a spinning reserve of at least 2.5% of actual generation to be available in a few seconds;

(2) under frequency demand disconnection schemes to be implemented nationally (the first step, of at least 10%, to act at a frequency of 49 Hz or above). Members may separate from the total system at 49.5 Hz if desired;

(3) the $(n - 1)$ criterion should be met at all times on the main interconnected system. The most important switching stations should be operated with double busbars.

North American Electricity Reliability Council

The North American standards are framed by the North America Electricity Reliability Council (NERC) as general objectives, responsibilities and policies which then form the basis for a number of operating guides [4.15]. The NERC regions (Northeastern Power Coordinating Council (NPCC)), etc.) are responsible for establishing regional operating policies based on the NERC criteria, and guides.

The guides are statements of operating policies, procedures and practices designed to promote co-ordinated operation among interconnected systems, and to ensure that high levels of reliability and control are efficiently and continuously achieved. The preamble to the document notes that:

All systems share the benefits of interconnected systems operation and, by their voluntary association with NERC, they recognise the need and accept the responsibility to operate in a manner that will enhance interconnection operation and not burden other interconnected systems. Responsibility to observe the Criteria and Guides ... extends to the member system operators.

The basic operating policies which each control area (either a Power Pool or an individual utility) should follow embodied *inter alia* the following principles:

(1) operation of sufficient generation under automatic control to meet its demand, transfer and frequency obligations;

(2) maintenance of voltages between established limits;

(3) maintenance of frequency and time error within limits;

(4) ensuring that power interchanges are within agreed transmission capacities (these may be contracted for rather than actual figures);

(5) install control equipment to meet system and interconnection requirements and maximising control performance;

(6) monitoring and preventing inadvertent power interchanges.

The operational criteria which members of an interconnected system should follow to implement these policies are:

- a generation power reserve should be provided to cover *inter alia* forecasting errors, plant unavailability, maintenance and regulation. The control error, $T_a - T_s + k(f_a - f_s)$, should be reduced to zero within 10 minutes after a disturbance and prompt steps should be taken to protect against the next contingency; T_a, T_s are the actual and scheduled external power transfers and f_a, f_s are the actual and scheduled frequencies;
- reactive power resources should be held and located for timely correction of voltage levels during contingencies;
- transmission operation should be co-ordinated across and between control areas, and should cover equipment outages, voltage levels, switching and establishment of inspection and preventive maintenance schedules.

England and Wales

The security criteria which have been used in England and Wales are shown in Table 4.3. These, covering virtually the whole range of demand, are deterministic

Table 4.3 Capability of system to meet the demand in groups of various sizes[1]

	Minimum demand to be met after		
Range of group demand	First circuit outage	Second circuit outage	Notes
Uo to 1 MW	In repair time: Group demand	Nil	Where demand is supplied by a single 1000 kA transformer the "Range of group demand" may be extended to cover the overload capacity of that transformer.
Over 1 to 12 MW	(a) Within 3 hours Group demand minus 1 MS (b) In repair time Group demand	Nil	
Over 12 to 60 MW	(a) Within 15 minutes: Smaller of (Group demand minus 12 MW) and two thirds Group demand (b) Within 3 hours Group demand	Nil	Group demand will normally be supplied by at leat two normally closed circuits or by one circuit with supervisory or automatic switching of alternative circuits
Over 60 to 300 MW	(a) Immediately Group demand minus up to 20 MW (automatically disconnected)	(c) Within 3 hours For group demands greater than 100 MW, smaller of (Group demand minus 100 MW) and one-third Group demand	A loss of supply not exceeding 60 s is considered as an immediate restoration. The recommendation is based on the assumption that the time for restoration of full-group demand after a second circuit outage will be minimised by the scheduling

	(b) Within 3 hours Group demand	(d) Within time to restore arranged outage: Group demand	and control of planned outages, and that consideration will be given to the use of rota load-shedding to reduce the effect of prolonged outages on consumers.
Over 300 to 1500 MW	(a) Immediately Group demand	(b) Immediately All consumers at two-thirds Group demand (c) Within time to restore arranged outage Group demand	The provisions of Clas E apply to infeeds but not to systems regarded as part of the interconnected Supergrid to which the provision of Class F apply. For the ystem covered by Class E, consideration can be given to the feasibility of providing for up to 60 MW to be lost for up to 60 seconds on the first circuit outage if this leads to significant economies. This provision is not intended to restrict the period during which maintenance can be scheduled. The provision for a second circuit outage assumes that normal maintenance can be undertaken when demand is below 67%.
Over 1500 MW	In accordance with CEGB planning memorandum PLM-SP2 or Scottish Board Security Standard NSP 366		

[1]This information is taken from MPSP-L

and embody both operational experience and a measure of the impact of losses of supply of various magnitudes to the community. The degree to which normal system conditions should be maintained after the outage/s will depend upon the demand. Typically, the loss of supply should be zero (or at the lowest demand levels restorable in a specified time), the voltages should be acceptable, there should be no system instability, and no unacceptable overloading of plant.

Nordel

The security criteria of Nordel (called *dimensioning principles*) have stipulated that the power system should withstand certain defined contingencies without loss of load, static or dynamic instability, or unacceptable loading changes in voltage or frequency [4.16].

Nordel is a 'loose pool' operating to broad criteria which the national operations managements are primarily responsible for implementing. The criteria are that the network frequency should be 50.0 ± 0.1 Hz, the time variation should be less than ± 10 seconds, and transmission power flow limits should not be exceeded. Amongst measures taken to meet these requirements are that a large number of units are equipped with turbine regulators which change the steam or water flow in proportion to the system frequency (primary regulation). The frequency regulating reserve held on these units was some 600 MW (1992), providing a regulating power of 6000 MW/Hz. A reserve to meet the 'dimensioning fault' (the largest credible fault) is also provided. This reserve is based on the largest of the dimensioning faults for the different countries and is decided on a weekly or more frequent basis. Response times are specified for components of both these reserves as instantaneous (within 30 seconds), fast reserve (within 15 minutes) and slow reserve (within four hours). Lost generation within a country must be replaced within 15 minutes, either directly or by purchase. Some of this response will be provided by secondary regulation, involving manual adjustment of set points or start-up/shut down of units.

Other Interconnections

Some widely used operational security criteria are given in Table 4.4.

4.4.3 Standards of Quality

In basic terms, the quality of a product is its capacity to meet the consumers' needs, which translates for electricity that the supply should be continuous and free of any disturbance or condition which would result in incorrect operation or

Table 4.4 Typical Security Criteria

Timescale	Parameter	Criterion
Real time	Each utility—generation reserve	(a) capacity to meet loss of largest infeed from running generators connected via single circuit breaker or transmission path
	Interconnection—generation reserve	(b) frequently each utility will provide its own reserve as in (a)
	Each utility—transmission reserve	(c) capacity to meet loss of any single circuit
		(d) occasionally capacity to meet loss of a double circuit line
		(e) occasionally capacity to meet coincident loss of any two circuits
	Interconnection transmisison reserve	(f) frequently each utility will provide its own reserve, with utilities agreeing standards to be met
	Each utility—generation and transmission reserve	(g) sometimes the simultaneous loss of generation and transmission is specified
	Interconnection—generation and transmission reserve	(h) frequently each utility will provide its own reserve as in (g)
Operational planning (short term)	Each utility—generation reserve	(j) as (a) for expected demand condition
	Interconnection—generation reserve	(k) as (b) for expected demand condition
	Each utility—transmission reserve	(l) as (c), (d), (e) for expected demand condition
	Interconnection—transmission reserve	(m) as (f) for expected demand condition
	Each utility—generation and transmission reserve	(n) as (g) for expected demand condition
	Interconnection—generation and transmission reserve	(o) as (h) for expected demand condition

failure of connected equipment. The important attributes of quality for most consumers will be continuity, the voltage level and constancy and frequency level and constancy in relation to the nominal values declared by the utility, and accuracy of time keeping. Consumers' actions in recent years have increased the problems of providing a supply of acceptable quality, because:

- devices and equipment now being installed by consumers are more sensitive to minor changes and disturbances than in the past, effectively requiring a better quality;
- increased use of power electronics equipment is creating more harmonic distortion, producing harmonic currents which will result in additional heating, voltage distortion and interference;
- the search for higher system efficiencies may magnify these effects; for instance, shunt capacitors installed to raise system voltages and compensate for reactive demands will reduce system losses, but may lead to resonance effects, amplifying harmonic voltages.

Statistically, disturbances will be classified on the basis of magnitude and duration, from transient (a few milliseconds), through momentary (milliseconds to a few seconds), to interruption (a few seconds upwards).

Continuity

Statistics on continuity seem not to be routinely published by utilities. However, if available, they are likely to be found in one of three forms:

- as a number of consumer disconnections per year or number of hours consumers are disconnected per year, in annual reports;
- as a number of disturbances per year of specified degree of severity in terms of system minutes lost, as used in some CIGRE reports;
- from *ad hoc* reports on major disturbances.

In practice, very high continuity is achieved by many utilities, say an average disconnection time per consumer of about two hours per year, nearly all attributable to distribution and supply.

System minutes (of outage) per year has been proposed by a Cigre Working Group [4.17] as an indication of disturbance severity, as shown in Table 4.5.

Table 4.5 Classification of severities of outage (as used in CIGRE papers) (Reproduced by permission of Cigre from [4.17])

Degree of severity	System minutes	Comment
0	Under 1	a level of unreliability normally considered acceptable
1	1–9	a level of unreliability which may have significant impact to consumers
2	10–99	a level of unreliability which will have serious impact on consumers
3	100–999	a level of unreliability which will have a very serious impact on consumers

Note: system minutes = energy lost in disturbance (MWh) × 60/peak demand (MW) e.g. a loss of 600 MW for 20 minutes on a system with a peak demand of 5000 MW would be (600 × 20/3 × 60)/5000 = 2.4 system minutes.

Voltage

There are two aspects of voltage quality important to most consumers: the range of steady state values; and the 'spikiness'. The former will determine what corrective measures should be provided between the point of supply and the consumer and the efficiency at which the consumers' apparatus performs. The spikiness, including the regularity (if any) and frequency of the spikes, will determine whether it will be suitable for some types of lighting and electronic equipment. A third aspect, waveform, is less important to most consumers, although the situation is changing, as, for instance, more capacitor based reactive compensation is installed.

(1) *Range of steady state voltages*
It is usual to include acceptable minimum and maximum deviations about nominal voltages in design and operating standards. Extreme voltages under normal and circuit outage conditions and step voltages for various changes, for instance in power flow or on switching, will be specified, often for various configurations, voltage levels, and perhaps demands. The structure and detail of voltage criteria will vary, but typically the upper bound at the ehv and intermediate voltage levels will be +10% and the lower bounds from zero to −5% under normal conditions. During outages, the lower bounds may be zero to −10%. If specified, step voltage changes on switching or fault clearance may range up to +5% on increase and down to −10% on decrease.

(2) *Fluctuating voltages*
Frequent variations in reactive and active power requirements imposed by loads such as steel rolling mills and arc furnaces cause variations in network voltages which will adversely affect other consumers if they are too large and too frequent. There is thought to be no international standard or recom-

mendation on acceptable levels of flicker, although some countries have their own figures. A frequently accepted criterion is that an arc furnace load, when short circuiting, should not exceed some 2–3% of the network short circuit power.

Frequency

Wide ranging frequency standards and/or performance are reported from across the world (see Table 4.6). It is often said that system size has a major effect on the frequency performance achieved, but the author suggests that at least equally important will be the generation control mechanism.

Time

Less attention is paid to absolute time than frequency; it has been said that utilities do not sell time. Nevertheless, 'electric time' is adequate for many purposes; it is often within ±10 secs. Subject to there being a plant margin at peak and flexibility in operation of the generating plant, it is not difficult to keep the time error small, although doing this may slightly increase operating costs.

The Measurement of Quality

Many utilities have installed equipment to measure the effects of disturbances and other aspects of quality of supply. EPRI initiated a major project in the early 1990s with the objective of assessing utility power quality at the distribution voltage level in the USA. An instrument, the 'PQ node', was developed which provided simultaneous three-phase measurement of steady state quantities and disturbances, including subcycle transients (impulse and oscillatory), short duration rms variations, long duration rms variations and waveform duration.

Table 4.6 International frequency standards

System/s	Target
North America	The frequency stability is specified in terms of the number of times the frequency may transgress specified limits in specified times. A maximum time error is also specified
UCPTE	Nominal ± 10 mHz
CENTREL	±20 mHz (steady state conditions)
NORDEL	±100 mHz
Great Britain	Nominal ± 1%: the standard deviation over many years was 0.12%

There were 300 sites selected for monitoring, covering the range of load densities, rural to urban, and load types [4.18].

Several utilities have developed equipment to record disturbances [4.19, 4.20]. Some of these rely on the detection of an abnormal parameter or state change to trigger a recording. Others have recorded continuously, and still others provide a continuous record which is overwritten after, say, several days unless an operator wishes to retain it. Recent recorders are multi-channel (e.g. 32 analogue and 32 digital channels). Sampling rates have varied widely from, say, 10 per second to some hundreds per second.

Transient recorders are available 'off the shelf'. The IEEE Task Force Report [4.21] lists 19 manufacturers worldwide marketing some 40 recorders, all microprocessor based. Sampling frequencies are mainly in the 500–5000 samples per second range. Multiple analogue and digital channels, with triggering, may be provided. The recorders often have built in communication facilities, some with printers; event listing and fault scanning data is often provided. Even in the absence of such advanced recorders as those mentioned above, one has to remember that displays in control rooms will invariably include frequency recorders, transfer error recorders, and probably some voltage recorders.

Reporting Supply Quality

Some regulatory authorities will require utilities to report on the quality of supply achieved, but often the information published will be basic, for instance:

- average number of interruptions per customer per year;
- average length of interruptions;
- percentage of interrupted supplies not restored within fixed periods (e.g. 3 hours, 6 hours, 24 hours);
- number of verified voltage complaints per year.

The author has found that such data may only be patchily available. Its main use is felt to be to judge the performance of a utility against its peers, to determine any trends in the figures, and perhaps to establish norms.

4.5 TIMESCALES IN SYSTEM OPERATION AND CONTROL

The work of system planning (the application of capital resources to the extension of the power system) is often taken as ending some two–five years ahead of the event and system operation (the application of revenue resources to the running of the power system) extends from then to completion of the post-event tasks. Four phases of work can be distinguished in system operation:

- *operational planning*, extending between one or two days to several years (say two to five) ahead;
- *extended real time*, between one or two hours and one or two days ahead. Sometimes the term 'operational programming' is used to describe the extended real-time and the short-term component of the operational planning phases;
- *real-time*, between one or two hours ahead to an hour or two after the event;
- *post-event*, covering collection and analysis of data from the previous phases; this may extend from a few hours to a few months after the event.

Each utility will have its own organizational structures for accomplishing this work. For instance, in some the system operation function will cover the whole timescale, whereas in others, system planning will look after the operational planning work down to quite a short time ahead. Nevertheless, the phases and tasks exist, in whichever part of the organization they are placed. Some of the work which has to be done during these phases is outlined here.

4.5.1 Operational Planning

Although rarely framed in this way, the objective of the operational planning work is to empower the necessary expenditures, and to programme all the necessary work so that over an agreed period of time and with minimum use of resources:

(1) Essential and worthwhile operational maintenance will be done.

(2) New construction will be integrated into the system.

(3) Sufficient generating and transmission plant will be available in the event to meet the agreed standards of security.

(4) Sufficient supplies of raw materials will be available; these will cover fuel (including hydro), spare parts, lighting-up and lubricating oils, hydrogen, nitrogen, rare gases, etc.

(5) Trading contracts are agreed.

Additionally, it will be possible to finalize those budgetary estimates needing operational forecasts. Consultation with interconnected utilities will be valuable, if not essential, during this work.

The main tasks of operational planning are summarized in the following sections.

Demand Forecasting

Demand forecasting is fundamental to all aspects of predictive work. Together with the forecasting of plant and transmission availability, the estimates provide the basis for all decisions on resource needs, and are required for the whole of the operational planning timescale. Power and energy estimates will be needed for the whole utility, with a geographical breakdown of the power estimates (probably to individual bulk supply points) and time profiles, or at least demand duration histograms. Estimates of reactive power (or demand power factor), and their geographical distribution, will also be needed.

The techniques used are:

- Years ahead – a judgmental approach using information on recent demand levels, trend and economic factors, as used for planning timescales.
- Months and days ahead – trend- and regression-based methods, incorporating weather factors and date effects such as season, month, week and day of week.
- Minutes and hours ahead–regression and trend methods incorporating the immediate past history of the demand.
- On-line – trend based methods using immediate past and longer term history of demand.

Judgement, including a (generous) allowance for uncertainty, will be used for the occasional 'one-off' event.

Plant Availability Forecasting

This is complementary to demand forecasting, and together these provide the basis for such decisions as the size of maintenance programmes, plant scrapping, fuel requirements, and trading programmes. Forecasts are required for the whole operational planning timespan. Estimates are usually based on judgement from past operating results, levels of maintenance expenditure, experience on similar plant, and the age of the plant.

Generation and Transmission Outage Planning

Generation and transmission outage plans are required to programme manufacturers' and utility maintenance resources, as input to loading simulation and network security studies, as input to studies with shorter lead times, and finally, for the preparation of the real-time outage and switching schedules. The plans will be required for the whole of the operational planning period, in outline for the longer lead times and in detail for, say, lead times of one year or less. Techniques used will include loading simulation, network analysis, possibly mathematical optimization for programming and sequencing the outages to achieve minimum use of resources over the whole operational planning period. In general, the generation programme will be dominant, with the transmission outages placed so as to have minimum, preferably zero, impact on the optimum operation of the remaining generation.

This is a combinatorial problem and 'exact' solutions have been formulated using integer linear programming and dynamic programming techniques. The computation can be massive, and empirical formulations have been used. Nuclear refuelling is a special case for which a two stage approach has been described.

Overall Fuel and Fuel Transport Requirements

Systems using fossil and nuclear fuels—This information is required on a system basis for budgeting purposes, setting tariffs, negotiating contracts with fuel suppliers, trading, etc. Individual station fuel requirements will be needed to determine station fuel purchase and transport and station operating regimes, to estimate staffing and maintenance needs, and to predict station production costs, and hence incremental (merit order) costs for use in all types of predictive, real-time and retrospective unit commitment and dispatch studies.

Sometimes called *energy modelling*, this task will require loading simulation and network analysis, including modelling of transport needs and costs. Each year of the operational planning timespan will be covered, with the model detail increasing as the lead time decreases and the lumping together of different time intervals decreasing as the lead time shortens. This is one of the most computer intensive tasks of operational planning.

Systems with hydro capability—In the case of systems including hydro capacity, records to enable estimates of hydraulicity will be needed. These will enable the need and quantities for other types of fuel and trading possibilities to be predicted and, in the short term, the run-off of stored water to be assessed.

Advice to Control Staff on Operating Arrangements

An important end product of operational planning is to provide advice and supporting information to the control staff on expected operating conditions. Typically, this will include for the day/s ahead:

- generation incremental costs or merit order,
- outage programmes,
- expected available generation,
- expected peak and minimum demands,
- preferred network configuration including constraints on power flows and remedial switching in the event of faults,
- preferred voltage profile and reactive sources,
- any special situations (e.g. a local insecurity due to transmission outages); abnormal demands (both high and low).

Demand forecasting, loading simulation and network analysis models will all be used.

Protection Settings

Although types of protection are likely to have been settled at the planning stage, the settings will need to be calculated, and also any changes necessary to cater for outages and the phases of the construction programme. The main requirement will be for network analysis, but in more detail than that for other tasks (for instance including phase-to-phase as well as three-phase and phase-to-earth faults, coupling between circuits, and faults at more points on the system).

Automatic Protection Systems

This term is used to cover such facilities as automatic rejection of generation or circuit switching to reduce network loadings, and disconnection of demand by under-frequency and/or frequency trend relays. The need for some of these may have been determined at the system planning stage, but others may have to be introduced to accommodate stages in construction programmes. In some cases, the times available for remedial actions, such as reduction of generation in the event of loss of transmission out of an exporting area, may allow these to be left to the operators, in which case operational instructions will have to be prepared.

Network analysis will mainly be used, and for the under-frequency studies dynamic analysis extending over several seconds.

Preparation for Abnormal Situations

It is sometimes necessary to prepare contingency plans for possible periods of abnormal operation, for example interruption in the supplies of fuel, loss of communications, exceptionally severe weather, or restoration of supply following a large scale loss of supply.

The emphasis in such studies will be on maintenance of supply when faced by shortages of resources. Demand prediction, loading simulation (perhaps with a different objective to the usual one of minimum cost of operation) and network analysis will all be needed. Abnormal operating states may need to be analysed, including possibly curtailment of demand.

Operational Standards

One of the functions of operational planning is to review past operating experience and, in the light of this, the expected development of the system and any inter-utility agreements, to decide whether any changes are needed in the operational standards of security and quality of supply. Such studies are likely to require an exhaustive analysis of the possible changes and simulation of their effects, covering ranges of system conditions over a number of years. The methods and programs of the sections above will be used. Probability analysis may also be used, for instance, in assessing running spare requirements.

Operational Memoranda and Procedures

It is essential that operational memoranda and procedures are updated in line with system and organizational developments. This may be done jointly by operational planning and control centre support staffs, and should include liaison with planning staff. It is unlikely to involve much computational work.

Facilities for Operational Planning and Real-Time Control

Experience shows that EMS and SCADA systems have been updated, sometimes replaced, at intervals of, say, 5–15 years. The lead time from first thoughts to commissioning is likely to be between some three and seven years. Whether system operation has the prime responsibility or not, it will provide a major input to the user and functional specifications.

Facilities for operational planning are also likely to need enhancement, in the author's experience more frequently than the real-time facilities. System operation will provide a main input to the specification, and may be responsible for the implementation.

Computational Tasks

It is evident from the above that, in addition to demand prediction, and in systems with hydro capacity water storage and flow estimates, the main computational tasks are loading simulation and network analysis. The former is, in one form or another, an essential component of all predictive studies, e.g. for economic studies to estimate operating costs and for network security studies to estimate plant outputs and together with demand estimates, power transfers at substations. The information required will be the operating state and output of each generating unit at specified times of selected (typical) day/s, or at specified load levels. Invariably, some form of optimization will be involved, even if this is only summation down a merit order to meet a given demand at lowest cost. The problem will usually involve constraints, for instance from network capability, system reserve needs, and generation response limits (in increasing order of detail and, inversely, importance).

Network analysis will in one form or another be an essential component of most predictive studies. There will be three main purposes: to confirm that the proposed operating conditions will provide adequate security of supply: to determine the capability of the network and hence any constraints on system operation imposed by the network; and finally, to assess transmission losses and hence any adjustments to merit order selection of generation to achieve minimum cost operation. All the standard analyses will be needed–load flow (full a.c. and active power only, and preferably also an optimum power flow program), transient and dynamic stability, including extension of the study period to several seconds, fault levels, and voltage studies.

4.5.2 Extended Real-Time Analysis

This term has come into use in recent years to cover the aids provided to the control engineer for the period of up to, say, 24-hours ahead. It therefore complements the advice received from the operational planners, refining this to take account of changes, for instance, in demand and plant availability, or the outage programme, and detailing it. The main tasks will be demand forecasting, scheduling of generation (unit commitment), trading, and network security analysis. In each case, more detail will be necessary than may have been provided in the earlier studies, for instance:

- demand forecasting – demand profile at half hourly or less time intervals;
- scheduling – on/off times of generators, outputs over each time interval (say half hour or hour);

- network analysis – a wider search of fault contingencies for more system conditions.

In addition to program modifications to provide the extra detail, there may be advantages in automatically picking up real-time data as a basis for some of these studies. There will be increased emphasis on making manual data input as small and simple as possible.

Generation Scheduling/Unit Commitment

The generation schedule and commitment are the final and most important outcome of the operational planning phase. The terms are sometimes used interchangeably, to mean the determination of the operating states of each generator on the system for a specific time period ahead. Where a distinction is made, as in North America, scheduling is taken as the intention some hours or days before the event, and unit commitment as the immediate pre-event decision. A comprehensive schedule will give, for each specified time interval, the output and spinning reserve of each generator, the system demand, conditions in each section of the system on which transmission constraints exist, trading information and transmission loss factors. The information on transmission constraints will depend upon the way in which these constraints are modelled. The simplest, although with some approximations, is the group transfer technique in which the summated power flow over all circuits into the group is compared with an estimate of the maximum firm transfer into the group. The actual schedules can be presented in tabular form, for instance showing running/not running states (Figure 4.2(a)), or also including the generation outputs when running (Figure 4.2(b)).

Responsibility for the technical, economic and safety management of the power system will pass to the control staff in the event. In addition to liaising with the control centres of neighbouring utilities, making the necessary decisions and issuing instructions to field staff, they may also have some responsibility for studying conditions for the next shift (this will apply particularly during the overnight shift).

4.5.3 Real-Time Operation

The objectives of real-time operation are to ensure that a supply of agreed security and quality at minimum cost is provided to consumers, to provide necessary access to plant and system for maintenance, repair and new construction, to minimize the effects of disturbances and, if these occur, to restore

Date..............................

Generator \ Time of day (hours)	1	2	3	4	5	6	7	8	9	10	11	12	13	14	15	16	17	18	19	20	21	22	23	24
8.1	1	1	1	1	1	1	1	1	1	1	1	1	1	1	1	1	1	1	1	1	1	1	1	1
2.1	1	1	1	1	1	1	1	1	1	1	1	1	1	1	1	1	1	1	1	1	1	1	1	1
10.1	1	1	1	1	1	1	1	1	1	1	1	1	1	1	1	1	1	1	1	1	1	1	1	1
1.1	1	1	1	1	1	1	1	1	1	1	1	1	1	1	1	1	1	1	1	1	1	1	1	1
4.1	1	1	1	1	1	1	1	1	1	1	1	1	1	1	1	1	1	1	1	1	1	1	1	1
2.2	0	0	0	0	0	1	1	1	1	1	1	1	1	1	1	1	1	1	0	0	0	0	0	0
4.2	1	1	1	1	1	1	1	1	1	1	1	1	1	1	1	1	1	1	1	1	1	1	1	1
1.2	0	0	0	0	0	0	0	0	0	0	0	0	0	0	0	0	0	0	0	0	0	0	0	0
10.4	0	0	0	0	0	0	0	0	0	1	1	1	1	1	1	1	1	1	1	1	1	1	1	1

(a) Schedule in time of day/running -not running format

Date..............................

Generator \ Time of day	1	2	3	4	5	6	7	8	9	10	11	12	13	14	15	16	17	18	19	20	21	22	23	24
1.1	165	165	165	165	168	172	179	205	206	240	240	240	212	213	214	216	217	218	218	213	177	176	175	175
1.2															95	108	42							
1.3																								
1.4																								
2.1	165	165	165	165	168	206	210	212	185	240	240	240	212	213	214	216	217	218	218	216	212	207	176	175
2.2								69	192	41														
4.1	165	165	165	165	168	172	175	177	240	240	241	240	212	203	214	216	183	215	204	180	177	176	175	175
4.2					109	172	175	177	240	240	240	221	179	167	214	216	181	182	182	180	177	175	54	
4.3							40	89	96	115	41							91	91	90				
4.4						40	78	89	96	120	120	40	40	40	40	108	91	91	91	90	86			
6.1							40	70	96	120	120	120	89	89	90	108	91	91	91	55	40			
6.2									96	120	120	120	89	89	90	108	91	91	91	40				
6.3									50	31			41	42	42	42		18						
6.4									50	50	25		42	42	42	42		42	4					
6.5									40	50	50	50	42	42	42	46	32	42	42					
8.1	192	192	192	192	195	206	210	212	240	240	240	240	212	213	214	216	217	218	218	216	212	211	210	208

(b) Schedule in time of day / output format

Figure 4.2 Alternative presentations of generation schedules. (a) Schedule in time of day running–not running format; (b) schedule in time of day/output format. (Note that, although the same system, the two schedules are for different configurations)

conditions to normal as quickly as possible. Some key functions and tasks are itemized below, with comment on their relative importance.

It has to be said that the level of responsibility given to control staff varies between utilities. In some they are effectively autonomous whilst on duty; in others they will seek instructions from management or day staff in the event of any minor abnormality.

SCADA Functions

The essential SCADA (supervisory control and data acquisition) functions will be the acquisition and display of current system information, normally with a cycle time of a few seconds. The essential telemetered data will include equipment states (on-line, off-line), flows, voltages, frequencies, alarms for status changes, protective gear operations and possibly operating variables outside limits. A hierarchy of displays will be used, from block diagrams of the system showing basic operating quantities in geographical areas, to system diagrams, substation and circuit operational diagrams, and sometimes substation and circuit safety diagrams showing the isolation and earthing states of equipment. Alphanumeric displays of status changes and alarms will be provided (with provision for acceptance of alarms), supply conditions in the whole and sections of the system, settings and condition of automatic control equipment, etc. Monitoring the values of the telemetered quantities is essential to check whether any operating quantities are outside limits. This will be additional to, and more comprehensive than, alarms generated in the substations.

Quite often, facilities will be provided to telecommand the operation of equipment from the control centre. This may include adjustment of governor speeder-motor settings (i.e. automatic generation control (a.g.c.)), circuit breaker operation, tap changing, changing of protective gear status, RTU (Remote Terminal Unit) status and signalling path, demand disconnection and reconnection. Hydro and pumped storage stations in particular are likely to be remote controlled. Transmission telecommand may include facilities for sequence switching, that is sequences of circuit breaker and power operated isolator operations to achieve, say, the transfer of a circuit from main to reserve busbar will be initiated on one command. Decisions regarding, say, a.g.c. and the telecommand of transmission plant will often by taken independently.

An increasingly discussed topic nowadays is alarm analysis, that is to determine from the alarms available what system event has occurred, noting that there will be redundancy in the alarms, that some may be incorrect, and that some of the secondary equipment operations such as those of protective gear initiating these, may not have been correct. With the increasing use of remote control and demanning of substations, the control engineer may no longer be able to complement his telemetry by discussion with the substation attendant. This human interaction, though time consuming, may provide very useful information to the control room when establishing the cause and evolution of complex incidents.

Frequency Bias Tie Line Control

Done automatically in many systems and essential in any event, this is the procedure to maintain the frequency of the interconnection system sensibly constant and flows between neighbours at agreed values. When done automatically, the generation change instructions are issued about every two seconds.

Economic Dispatch

Done automatically on quite a few systems, but essential in any event, this is the determination of output from running and quick start plant and the pump or generate duty on pumped storage plant which will minimize the total cost of generation in accordance with the current merit order. Strictly, it is a multiply constrained minimization problem, although many systems even now only minimize the cost of generation subject to meeting the total system demand, transmission constraints being included, often with some approximation, by placing limits on generation outputs. In the case of mixed hydro-thermal systems, it has been common practice to assign to the hydro plant whose running is optional (that is, plant with storage) an incremental cost, which causes this plant to be operated so as to use the water run-off determined from longer term studies.

State Estimation

This is the process of determining the complete and consistent set of variables which best fit, usually by a least squares criterion, the telemetered data. It will include as a first stage some form of network validation which will attempt to determine a network configuration most likely to be that of the actual system. Some form of network configurator and state estimation are essential if contingency evaluation or other computation including system analysis are to be implemented.

Contingency Evaluation

This is a desirable function in which, hopefully, the effects of all potentially critical outages classified as credible contingencies are evaluated. It will comprise at least a series of load flows (preferably a.c.), one for each of the outage cases, with the resulting flows and voltages checked against limits and, if outside these, alarmed. A scan using an active power approximation is sometimes used to determine the contingency cases needing the more accurate and time consuming a.c. analysis. Increasingly, the contingency analysis will include a procedure to check short circuit levels. Either within the contingency analysis or as part of the real-time load flow (i.e. a load flow using telemetered data to define nodal *PQ* conditions), there may be a facility to estimate conditions in the near future.

Real-Time Power Flow

Facilities are often provided for the operator to initiate a power flow, usually a.c., based on current telemetered conditions but including a facility to change the switching and generation conditions on demand; this enables a system state expected in the near future to be studied with minimum work from the operator.

Trading and Accounting

The operator must operate the system so as to fulfil longer term contracts and take advantage of current differences in generation costs between his and neighbouring systems (opportunity trading). This will require knowledge of marginal cost levels derived from economic dispatch, and on the system status and capability. This work area has become increasingly important with privatization.

Load Managment/Demand Side Management

This is the ability to control demand, usually by prior agreement with the consumer, so as to reduce the operation of expensive generation, or to avoid low frequency operation or forced disconnection of demand. The main decisions are when and how much demand to manage, essentially simple decisions in computational terms.

Automatic Voltage/Var Control

Automatic excitation systems are always fitted to generators, whilst step down transformers in the chain between transmission and consumer are often fitted with automatic voltage regulators which control the operation of tap changers.

However, a few utilities, notably EdeF and ENEL, have installed equipment to control voltages across a section of the system in accordance with a voltage specified at one point in the area.

Containment of Disturbance and Restoration of Normal Conditions

An essential part of an operator's work is to restore the system to normal after a fault. In rare cases, the disturbance may have developed to the extent that demand is disconnected, or the system split or large amounts of generation lost. In this event, the operator's first priority will be to stabilize and restore the situation (see Chapter 7).

The operator's task will be significantly eased by auto-reclose of individual circuits and, in a few utilities, by automatic restoration of circuit paths through a network. Otherwise, the operator must often rely on previously evaluated strategies and procedures. Studies have been made into the use of expert systems to aid restoration.

4.5.4 Facilities

Essential facilities to enable the control staff to carry out the tasks outlined above are:

- a telemetry network and computer system which will ensure that power flows, equipment states (closed/open), voltages, etc., are available for display at the control centre within seconds of occurrence (a);
- a communications network for voice and other communication with other control centres, field staff, etc.;
- a human-computer interface at the control centre which presents the essential information (mainly from (a)) clearly and without ambiguity;
- facilities to telecommand transmission plant as required;
- facilities for generation control (e.g. automatic generation control) as required;
- back-up facilities as required (e.g. communications, stand-by computers, secure power supplies, etc.): a stand-by control room may be provided;
- access to computer aids, including results of operational planning studies.

Further information is provided in Section 4.6.

4.5.5 Post-Event Tasks

The statistics of past operation are a most important source of technical data for estimating future commitments and requirements. These also provide the raw data by which operational and general management can monitor the efficiency with which many aspects of the utility's work is being done. A further area of work will be the analysis of abnormal events such as major losses of supply and making recommendations to avoid recurrences. The data collected and in particular the analysis done will be particular to each utility, but a representative sample is given below:

(1) *Routine on-line and real time data*
Typical data collected routinely is indicated in Table 4.7 (most of this will be logged automatically).

(2) *Performance analysis*
This raw system data of Table 4.7, plus input from predictive studies, will provide information for analysis of the performance of the system and its operational and control procedures. Some of the analyses of performance possible on the raw data are suggested in Table 4.8.

(3) *Analysis of abnormal occurrences*
The data for the monitoring and analysis of abnormal situations will come from on-line sources and staff reports, the main purposes being to obtain data on system performance in such conditions for future design work, and to assess what changes to plant, control facilities or organization would be valuable. Table 4.9 lists some of the analyses possible. A special case will be the analysis of conditions leading up to and during a major disturbance. This

Table 4.7 Routine on-line and real time data

Quantity	Main source of data
Frequency	SCADA
Tie line flows	SCADA
Flows in other circuits	SCADA
Voltages	SCADA
Generator outputs	SCADA
Demands met	Raw or processed SCADA, Printometer
Power and energy traded	SCADA, printometer
Generator availabilities and operation	Station records, SCADA, printometer
Fuel consumed	Station records
Fuel delivered	Station records

Table 4.8 (a) Routine on-line and real time data and analysis; (b) analysis of performance – normal operation

(a)

Quantity		Routine analysis
(1)	Frequency	Number of times and durations outside limits
(2)	Tie line flows	In conjunction with (1), analysis of agc performance
(3)	Flows in other circuits	Flows outside limits, histograms of loadings
(4)	Voltages	Voltages outside limits
(5)	Generator outputs	Deviations from instructed values, plant flexibility
(6)	Demands met	Maximum and minimum demands, geographical distribution, time profile, sensitivity to weather and frequency and voltage. Accuracy of forecasts
(7)	Power and energy traded	Billing, settlement
(8)	Generator availabilities and operation	Trends; response to instructions, predictability of operation, overload capacities
(9)	Fuel consumed	Generation efficiency
(10)	Fuel delivered	With (9), fuel in stock

(b)

Quantity		Performance analysis
(1)	Demand forecast	Errors in forecasts; errors in predicted margins
(2)	Generation forecast	
(3)	Assessed total system cost	Computed from generator outputs and merit order; preliminary estimate of costs of operation
(4)	Generation schedule versus actual operation	Accuracy of predictive modelling of operation
(5)	Transmission constraints and costs	The cost of out of merit operation due to transmission constraints. Accuracy of predicting transmission constraints
(6)	Operation of generation and transmission	Comparison of actual with programmed outages
(7)	Various forms of ideal operation	Efficiency of the scheduling and dispatch process, including both control centre and station performances
(8)	Pumped storage operation	Loading cycles on plant, savings from use of pumped storage
(9)	Two shifting of plant	Plant flexibility
(10)	Demand management instructions	Comparison with contracts; viability of planning and operational standards
(11)	Frequency and voltage records	Quality and reliability of supply; viability of planning and operational standards
(12)	Interruptions of supply	

Table 4.9 Analysis of performance–abnormal operation

Quantity	Performance analysis
(1) Protective gear performance	Reliability, maintenance needs, type and installation problems, application problems
(2) System response to generation losses	Time response of the system to sudden generation-demand imbalances; support from neighbours
(3) Under-frequency relay operations	Reliability of under-frequency relay protection
(4) Instructed reductions in demand	Reliability of supply, adequacy of planning and operational planning margins

will require the collection of system wide data – telemetered, manually logged and from disturbance recorders – followed by detailed analysis of all aspects – load flows, transient and dynamic stability, protective gear operations. Time tagging of telemetered data is particularly valuable in these circumstances.

There will clearly be overlaps between the analysis included in this and the previous two sections.

(4) *System tests*

The larger utilities may mount tests to assess system and plant performance and the validity of models used to predict these, usually in the transient and dynamic stability areas. Such tests will require additional instrumentation, in particular for transient recording. Often the objective will be to operate the system normally during the test period, but as the tests may well incur throwing faults on the system or subjecting it to significant generation-demand imbalances, this may be difficult to achieve. Exhaustive analysis will be done on the test results, and when models are being studied, utilities have in the past sometimes made these available to other utilities and manufacturers to pool developments in knowledge of plant performance and modelling.

It is judged that privatization and restructuring have made both conducting tests and subsequent release of information more difficult.

4.5.6 Operator Training

Some form of operator training is essential, and this ranges from *in situ* (*father-son*) tuition, through discussion on faults, seminars and short courses to switching simulators, loading simulators, and now real-time dynamic simulators. The latter are often implemented by arranging for the EMS and SCADA software to be run in a training mode on suitably extended operational hardware.

Further information is provided in Chapter 8.

4.5.7 Models Used in Post-Event Tasks

The same range of loading simulation and network analysis models will be used as in the other timescales, plus models to cost a given operating state. Statistical analysis (means, standard deviations and histograms, etc.), particularly of on-line data, will be needed. The models used will fall into these main areas:

- demand prediction
- generation scheduling (including dispatch)
- economic dispatch
- fuel and energy modelling, voltage stability
- assessment of operating costs for given generation schedule or dispatch using normal merit order costs
- assessment of alternative operation
- load flows and short circuit levels
- network configurator and state estimation
- transient stability, voltage stability
- protection modelling
- longer term dynamics
- real time system modelling
- assessment of operating 'costs' using abnormal 'merit orders', based say on fuel stock-days or plant efficiencies.

4.6 SCADA

SCADA (Supervisory Control And Data Acquisition) is commonly taken to mean the chain of equipment which:

- collects the status and measurand system and other data at substations and power stations, codes and transmits this to the control centre, processes and displays the results (automatically or on request) to the operator, and logs selected items;
- enables instructions on status and output (that is, where this is not done by telephone or telegraph) of plant to be sent to substations and power stations for implementation automatically or by the local operator. Tie-line frequency

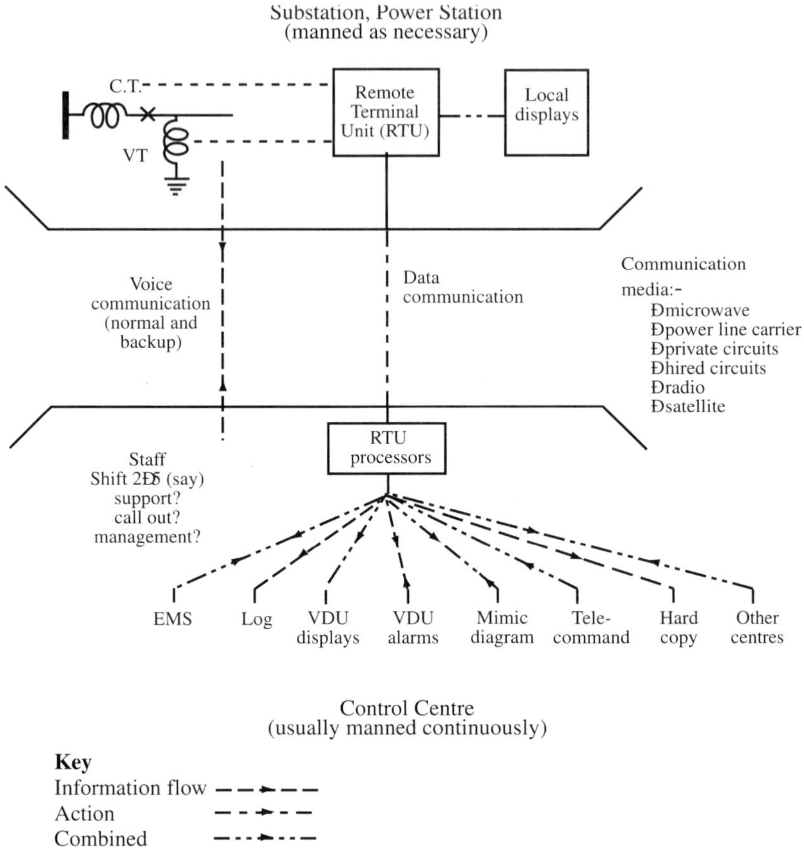

Figure 4.3 Main elements of SCADA

control is fundamental to operation in many utilities, and SCADA is often taken to include this.

The EMS (Energy Management System) comprises the hardware and software provided for computational support – mainly economic dispatch, contingency analysis, on-line load flow, optimal load flow and voltage profile, generation scheduling, unit commitment, load prediction, interchange scheduling, var dispatch, etc.

The basic structure is shown in Figure 4.3. At one time, it might have been said with little exaggeration that there were as many major variations in implementation as there were control centres, but some standardization is now being introduced. Some of the factors which must be considered when providing these system are discussed below.

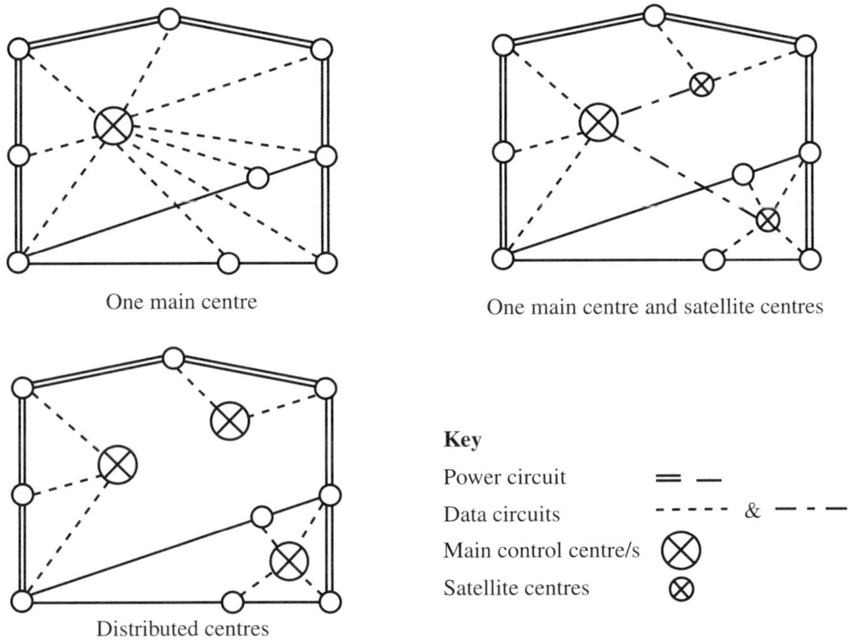

Figure 4.4 Some possible control structures

4.6.1 Questions on Functions and Structure

The fundamental issues listed below must be reviewed if the SCADA and EMS facilities of a large utility, possibly having several control centres, are to be replaced. Brief comments follow each.

(1) *What should the control structure be?*

- one system control centre*
- one system control centre and satellite centres (hierarchical)
- several centres (distributed)
- other.

These are illustrated in Figure 4.4.

The tendency in various countries for a period of some years was to disperse control into several centres, each monitoring its own geographical part of the system. It was argued that this provided better security against catastrophic loss of a centre, and also more manageable tasks in both human and

* The terms 'system control centre' or 'national control centre' are often used to denote the top-level control centre within a utility, interconnection or country

engineering terms. This dogma now seems to have reversed. In the author's view, if there is potentially substantial interaction between different parts of a system, there needs to be one control person who comprehends the overall current and short-term status of the system, and is provided with sufficient information to do this.

(2) *What control functions will be exercised at the different hierarchical control levels for the different system voltage levels?*
The control functions will be broadly supervision, followed by decision making and implementation for each of the principal tasks, these being scheduling and unit commitment, monitoring, system status (including alarms), switching, dispatch, recovery from fault conditions, trading and review of near future conditions. The decision making and implementation actions may be separated, the first into a 'main' control room and the second into a switching centre.
Referring to Figure 4.5 as an example of hierarchical control, the main transmission circuits *T* would typically be controlled from the national (system) control room; the instructions for switching might be given from this room, an associated switching centre or a second level (e.g. Regional) room. Loop *B* providing the main transmission function in that part of the system would be controlled from the national room or a second level room for that part of the system. The 'high power distribution' loop *A* would be controlled from the appropriate Regional room, or possibly the national room as would the local transmission circuits *T*(*L*).
The actual allocation of duties will vary between utilities, but a common arrangement will be as shown in Table 4.10.

(3) *How many control centres should be provided at each hierarchical control level for the different voltage levels?*
Usually, the number of centres per voltage level will increase as the voltage level decreases, not least because the numbers of substations will increase. The numbers are probably set more by organizational needs and judgement than technical criteria. System size (power and geographical area) and perceptions of the impact of the control centre structure on system security will influence the judgement. Typically, there will be one top level centre for the whole national or utility system, and between two and ten regional centres if these are used. Each sub-transmission/distribution utility may have its own control structure, with one or more centres controlling the subtransmission network and step down substations to the distribution network. The owners of generation may establish control rooms for their own generation.

(4) *Which control functions will be performed by remote control from the control centres, and which by local operators on instruction from the centres?*

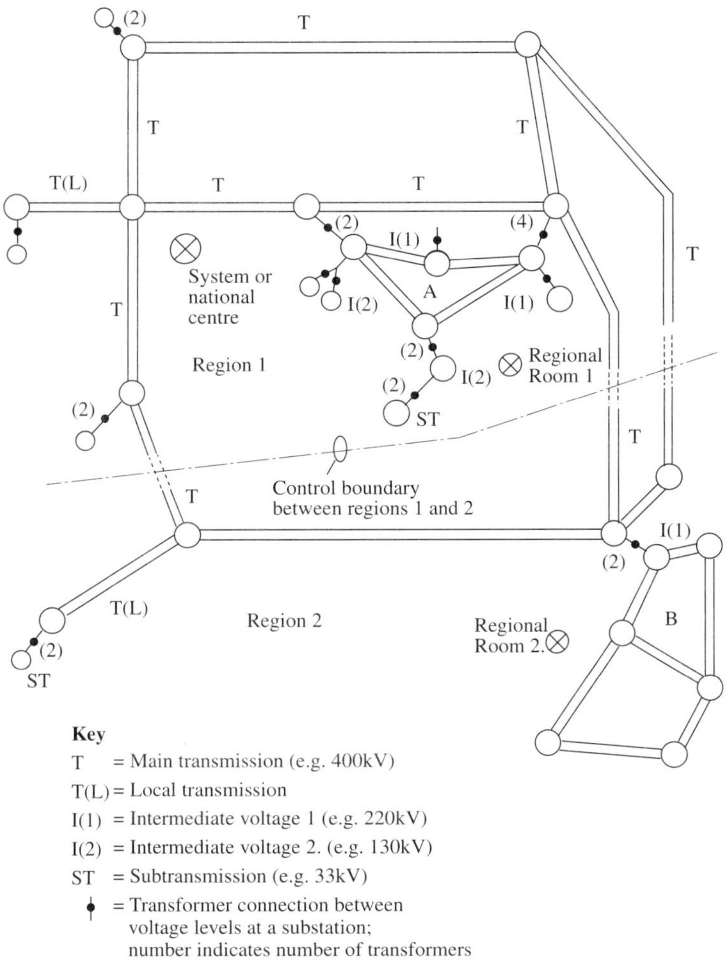

Figure 4.5 An illustration of hierarchical control

The tendency is towards wider application of remote control, since this should decrease the annual operating costs as well as the number of trained staff needed. Shift rotas and staff absences will present fewer problems. On the debit side there will be no 'double check' by the substation operators on instructions from the control centre and the pool of trained operators to supplement normal staff during emergency conditions will be much smaller.

(5) *What provision should be made for system growth?*
This issue is less important with the advent of distributed systems. These enable the functions and capacity of the control system to be expanded in line with the power system needs in a modular fashion avoiding the need, as

Table 4.10 Allocation of duties to Control rooms

Task	Control room/s used
Scheduling/unit commitment (whole system)	National room. The number of small generators (up to a few MW) is growing, and discussion on their operating regimes and outputs may be left to the local sub transmission or distribution utility and/or the owner of the plant
Automatic generation control (whole system, but the smallest units may not be included in the agc)	National room
Generation dispatch (centralized, the smallest units may not be included)	National room
Generation dispatch (transfer system)	Transfers set by the national room, dispatch to units either from regional rooms, national room or generation owners
System status monitoring, including alarm analysis and data logging	
For the main transmission system	National room
For the regional systems	Regional rooms
Switching and remote control if used (main transmission)	National room and associated switching centre if there is one
Switching and remote control if used (high power distribution)	Appropriate regional room or national room
Switching and remote control if used (sub-transmission)	Appropriate regional room or distribution utility
Emergency control and restoration	These topics are dealt with at some length in Chapters 5 and 7

happened with the earlier monolithic systems, for a massive replacement exercise every few years.

(6) *What performance criteria are required for the SCADA facilities?*
These are discussed below.

(7) *What emergency/back up control facilities will be provided?*

These are discussed in Section 5.8.

4.6.2 Questions on Performance Criteria

The acceptability of the SCADA system to the control operators will depend in part upon the usefulness of results provided and in part on the ease of use, that is the ergonomic features. Some of these are listed below [4.22].

(1) The loading conditions, for instance normal, high, emergency, for which the various response times are specified.

(2) Elapsed times from system event to screen display and other operator interfaces; maximum times between presentation of the same information on the different interfaces.

(3) Elapsed times from operator input to screen display or other change; maximum times between these changes appearing on different displays.

(4) Cycle times (repetition rates) for measurand data; status changes will probably be sent on occurrence.

(5) Reliability criteria, for instance frequency and duration of total and partial down times of displays.

(6) Measurement plus telemetry accuracy; for measurements say 1% for critical ones, up to 2% for others.

(7) Acceptable error rates in data transmission.

These parameters will determine the technical requirements of the installation, such as the configuration and power of processors in the remote terminal units and at the centre, and the data transmission rates.

4.6.3 Information Required at Control Centres

The most important on-line data acquired and displayed in a comprehensive SCADA system are shown in Table 4.11 for a national (system) and regional rooms (based mainly on MPSP Vol L).

The content and amount of off-line data will depend mainly upon the computational aids provided. Essentially, it will be the additional status information needed to define the network configuration but which is not telemetered, for instance status of non-powered isolators, and measurand data missing because of equipment problems or by design. Usually less and less data is collected as the network voltage decreases; on subtransmission networks, for instance, active power might only be telemetered for outgoing circuits at the infeeding points. Disturbance recorders are increasingly popular, and some utilities make arrangements to collect data from these via the telemetry system.

4.6.4 Information Sent Out from Control Centres

Depending on the amount of local automatic control, the control centre may need to issue information on operating status by telecommand, or by telephone/

Table 4.11 Information telemetered in a hierarchical (National and Regional) control structure

Equipment	Active power, MW (one end)	Reactive power, MVAr	Circuit-breaker and auto isolator states	Current, A (one end)	Overload alarm	Tempera-ture	Voltage (kV)	Tap position	Other[3]	Operational isolator states	Line end open
400 kV and 275 kV O/H circuits											
internal to Area	✓X	✓X[1]	✓X	✓[2]	✓X					X[2]	X
inter-Area	✓X•	✓X•[1]	✓X•	✓[2]	✓X				✓	X[2]	X
400 kV and 275 kV cable circuits											
internal to Area	✓X	✓X	✓X	✓[2]	✓X					X[2]	X
inter-Area	✓X•	✓X	✓X•	✓[2]	✓X				✓	X[2]	X
400/275 kV transformers											
internal to Area	✓X	✓X	✓X	✓[2]	✓X	✓[2]	✓[2]			X[2]	X
inter-Area	✓X•	✓X•	✓X•	✓[2]	✓X	✓[2]	✓[2]		✓	X[2]	X
400/275 kV to lower voltage transformers	✓X	✓X	✓X	✓[2]	✓	✓[2]	✓[2]	✓X[2]		X[2]	X
400 kV and 275 kV reactors	✓X	✓X	✓X	✓[2]	✓X					X[2]	X
400 kV and 275 kV quad boosters	✓X	✓X	✓X	✓[2]	✓X			✓X		X[2]	X
400 kV and 275 kV bus-section and bus-coupler			✓X	✓[2]						X[2]	X

Generators–120 MW (and larger) plus main gas-turbines	Individually ✓X	Individually ✓X	✓X			✓X		X^2	
All generating stations (GTs are excluded)	Station Total ✓X	Station Total ✓X	✓X				✓	X^2	
Main gas-turbines	Total ✓X	Total ✓X	✓		✓		✓	X^2	
Auxiliary gas-turbine	Total ✓X	Total ✓X	✓				✓		
400 kV and 275 kV substations					✓x		✓[4]		
132 kV and lower voltage									
internal to Area	✓	✓	✓	✓			✓		
inter-Area	✓X•	✓X•	✓•	✓			✓		✓X
132 kV substations					✓		✓		

Notes: ✓ = data transmitted from substation to regional control centre
X = data transmitted from Regional to National control centre
• = data transmitted between Regional control centres
(1) Reactive power metering (one or both ends) will depend upon circuit length; both ends on cable circuits
(2) For plant telecommanded from control centre
(3) Other alarms will often relate to loading or temperature conditions
(4) Frequency may be transmitted to control centres.

teleprinter/fax to the local operator in manned substations. Even if telecommand is provided, it would seem prudent to provide back up telephone links for possible operator action at the substation. This information may include:

- for transformers – on/off line, tap setting or AVR setting;
- for overhead lines and cables – on/off line;
- for reactive compensation plant – on/off line, settings including type of control;
- for generators – on/off line, power and reactive power outputs, loading profiles, AGC setting, AVR setting; generator transformer tap setting;
- for pumped storage – on/off line, operating mode (pumping, spinning, generating), generation or motoring powers and reactive powers, AVR setting, tap setting or unit transformer;
- for quadrature boosters – on/off line, tap setting;
- for busbars – sectioning arrangements, target voltages.

Each plant item will also be equipped with protection and alarm relays. Provision is sometimes made to change the status of such relays (e.g. to reset trip relays) or their settings from the control centre.

A substantial amount of information will also be sent to the control centres of neighbouring utilities. This will relate to the status and flows of tie lines between the utilities, possibly overall utility conditions and foreseen problems such as major changes in security levels, generation availability and transmission availability. Each utility will wish to balance the value to total system integrity of releasing such information against its commercial value to a competing utility. Telephone, teleprinter and fax will be used, and possibly a link between the SCADA/EMS computers of the utilities.

4.6.5 The Human–Computer Interface

The human–computer interface (in less politically correct days, the man–machine interface) is the term used to describe the equipment presenting the SCADA and EMS information to the operator. The importance of the subject is evident from the results of investigations into major incidents where part of the blame has sometimes been attributed to a confusing presentation of information. The word 'ergonomics' is frequently found in this connection. It is used to describe the characteristics of the HCI which make it easy and comfortable to use. A similar concept used in North America is 'human factors engineering'.

Any or all of the following may be provided:

- VDUs (alpha-numeric and graphical displays);
- animated or static, wall or desk mounted mimic diagrams;
- meters (analogue and digital) and chart recorders;
- display panels (plasma, liquid crystal);
- audible indications (chime, bell);
- means to control the displays.

VDUs are judged to be the most important medium for large systems. They can present information in all forms (alpha-numerics, schematic, picture, graph, histogram), in monochrome or colour, with screen and character dimensions to suit workstation dimensions, and considerable manipulation of the screen image is possible (zoom in, zoom out, rotate, invert (black↔white), scroll and superimpose). Frequently, a hierarchy of graphical display levels is provided.

VDU graphical displays

- System level 1 – the total system, as in the mimic diagram, sometimes using scrolling if the system is too large to fit on one screen;
- System level 2 – part total system, e.g. a zone or region, again with scrolling if needed;
- System level 3 – substation configuration;
- System level 4 – substation or circuit safety information for switching.

VDU α-numeric displays

- alarms and alarm acknowledgements;
- summaries for instance total and area generation, demands, transfers, circuit states and loading, etc.;
- safety records;
- technical data;
- computational results for instance generation schedules and commitment, economic dispatch, contingency analysis, transfer/trading schedules, demand prediction, etc.

VDU graphs and histograms

- trends and forecasts (e.g. of demand);

- comparisons (e.g. of operating costs);
- generator capability charts;
- reactive capability charts.

Free standing mimic diagrams provided the main control room display before the introduction of the CRT/VDU-based systems, and the value of retaining these has been discussed at length. Mimic diagrams are expensive, and with their requirements for space and viewability, have a major influence on the size and layout of the control room. Nevertheless, they provide an excellent focal point for appreciation and discussion of the status of a power system during disturbances. Typically, circuit status (e.g. 'line-end-open' and busbar selection), nodal status (section and coupler breaker status), voltages, strategic circuit loadings (possibly in quartile form), alarms and abnormalities will be shown. Either on the mimic or elsewhere will be shown such overall system information as total demand, generation, transfer, frequency, system and clock time, and area control error. Sometimes, the mimic diagram will be flanked by geographical diagrams showing, say, zonal conditions.

Considerable attention is paid to the integrity of control room information, for instance, by duplicating or more information paths and displays. Several VDUs will be available to each operator, and each VDU will be able to access all the information relating to specific functions, or even all functions. Large chart recorders are used to record critical information such as total generation, transfer, demand and control error.

4.6.6 Availability Requirements for SCADA Systems and their Structure

Reference has already been made to the very high availability required from SCADA systems overall. However, when there is a question of priorities, for instance in the number of display consoles and access to these, it may be necessary to place priorities on the various control functions. The author's judgement of these priorities, admittedly placing security before economy of supply, is:

Priority	*Control function*
1	System management for security
2	Operational switching for security
3	Safety switching
4	System management for economics
5	Operational switching for economy
6	Monitoring and logging

Each function will require a specific set of data, and considering these in conjunction with the suggested priorities, the following priority is obtained for availability of system information particularly during disturbed conditions: frequency, total system generation and transfer, voltages, system configuration, circuit flows, station outputs, busbar switching details, cost information, logged information.

A comprehensive specification is likely to state different availabilities for the more or less critical functions, for instance 99.9% (some nine hours downtime per year) for critical functions and 96% (350 hours downtime per year) for non-critical functions such as training.

These high availabilities are achieved by replication of processors, data links, etc. Until the late 1980s, a typical SCADA system would consist of duplicate processors, each capable of handling all the functions. Automatic change-over between computers would be provided, and hence each would be kept updated with the current telemetry data. In some utilities, a third processor has been installed to provided 'hot spares'. Front end processors would also be duplicated, and data links to the outstations would either be duplicated or 'triangulated' (Figure 4.6).

More recently, the trend has been towards 'distributed systems' (Figure 4.7). Duplicated processors are provided for several of the individual functions, the processors and all the peripheral units (displays, front end processors, remote terminal units, memory) being connected to duplicate data busbars.

Further security may be provided by a 'backup' control centre. This will be established on a geographically separate site from the main centre, and eliminates risks from that centre being put out of action by hazards such as fire, sabotage, aeroplane crash, earthquake, air conditioning failure, environmental disaster, etc. Computer hardware and software errors have sometimes been included in this list, but unless the backup centre uses different systems hardware and software, there seems no reason why a backup centre should not potentially be prone to the same failures as the main centre.

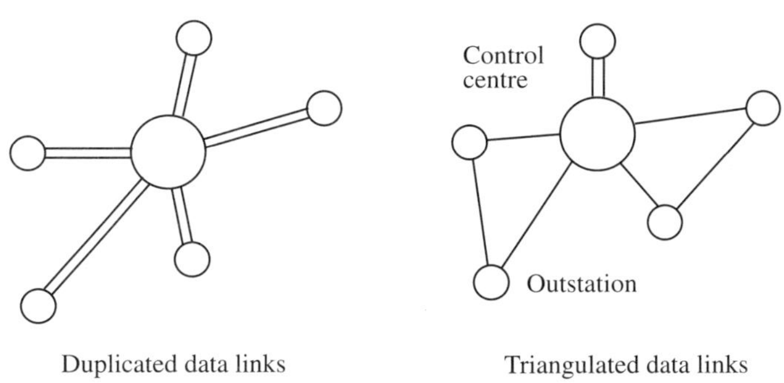

Figure 4.6 Telemetry links between outstations and the control centre

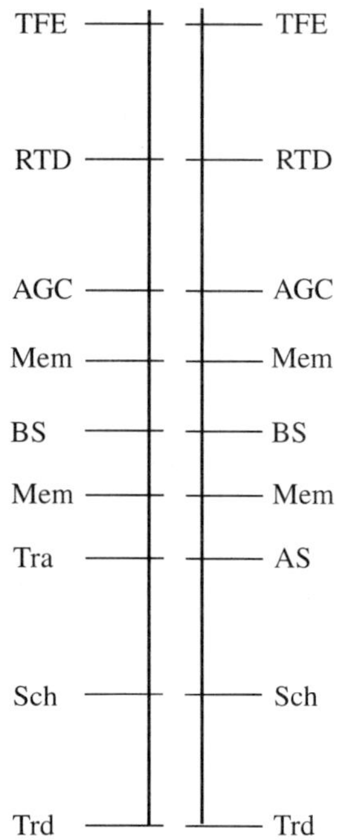

Figure 4.7 Data busbar and SCADA/EMS applications. (TFE: telemetry Front end; AGC: Automatics Generator Control; AS: Advanced Security Application; Sch: Scheduling; RTD: Real time display; BS: Basic Security application; Tra: Training; Trd: Trad; Mem: memory)

These benefits will only be achieved if due thought has been given to the location of the backup centre. Some of the points to be considered are:

(1) It should not be so close to the main centre that the same physical disaster (e.g. flooding) could affect both.

(2) In the event of it being necessary to evacuate the main centre, it should be possible for the control staff to travel to the backup centre within, say, two hours at the most. This necessarily implies that the backup centre will also be located within reasonable travelling distance of the control staffs' homes.

(3) It should be possible to provide data links into the centre to access both private and public communications networks. It is unlikely that this would present difficulties.

(4) Services to the backup site should be continuously available, as should transport and access.

To some extent, the above is a counsel for perfection, and in practice the provisions made will vary widely, being influenced by the utility's appreciation of the security and economic risks to the system if the main centre were lost, the costs of providing and maintaining the backup centre, and the use that can be made of previous installations. Some of the approaches which have been used are, in decreasing order of functionality:

(1) If the utility has a second system control centre, provide the database and communications to enable either centre to control the whole system, although with some degradation of performance.

(2) As (1), but using a regional control centre when the system has a two or more level control structure.

(3) If the spares are being provided as a hot stand-by suite, extend this as necessary, and provide data links to it so that it can function as an operational system.

(4) When a new SCADA system is being installed, retain one of the existing installations with data links and software reconfigured for the stand-by role.

(5) Provide a purpose-built stand-by control room. This can range from a room with a frequency meter and telephones, to public and private telephony networks, up to a complete processor system.

The human-computer interface presents a problem with stand-by centres. Mimic diagrams are expensive, and usually specific to a single power system. Hence if the stand-by centre is to function for several main centres (e.g. one stand-by for several regional centres), it will probably not be possible to provide a mimic diagram in it.

4.7 ENERGY MANAGEMENT SYSTEMS

A wide range of computational aids is available, covering time spans from immediate to months ahead. As far as the author is aware, there is no universally accepted borderline between SCADA and EMS aids, and some utilities have adopted the term 'extended real time' to cover the grey area between real-time and operational planning. The majority of the computational and logical applications found in SCADA and energy management systems are listed in Tables 4.12(a) and (b) in order of very short to long lead times. An indication is included of the frequency at which the various tasks would typically be done.

The logical tasks may include supervising control and sequence switching. Those associated with faults will be largely unpredictable, both as regards timing and content, ranging from the outage/restoration of a single circuit to handling a major disturbance in which tens of circuits may be involved over several hours, with a maximum fault incidence rate of, say, one or two per minute. What is essential is that the SCADA system continues to present system data more or less in real time (one or two early SCADA systems 'froze' when deluged with information from a major system incident).

Table 4.12 (a) Computational applications in EMS and SCADA systems; (b) logical applications in EMS and SCADA systems

(a)

Computational tasks	Frequency of computation
Tie-line frequency control	between 2 and 10 seconds
Automatic generation control	between say 2 and 10 seconds
State estimation	approx. every 30 minutes
Economic dispatch set points	15 to 30 minutes
Automatic V/Q controls	
Contingency (outage) analysis	between 15 mins and 60 mins
Short circuit analysis	between 15 mins and 60 mins
Load prediction (this can spread over virtually the whole time scale)	every 5 mins or so for on-line prediction, several times daily for unit commitment, daily/weekly/monthly/yearly for other tasks
Dynamic security analysis	on demand
On-line power flow	from every 30 mins to on demand
Optimum power flow	on demand
Unit commitment	from weekly to several times daily
Generation scheduling	from weekly to daily
Hydro scheduling	from yearly to weekly to daily depending on the type of hydro
Plant maintenance scheduling	yearly with shorter term revisions

(b)

Logical tasks	Frequency of task
Alarm processing	on receipt of new alarms
Switching for maintenance and construction purposes	as required by work programme/s at substations and power stations
Switching for system restoration after faults	as dictated by incidence of faults
Switching to accommodate demand and generation changes	either as changes require, or to a time programme

4.8 COMMUNICATIONS AND TELEMETRY

Reliable communications and telemetry are a *sine qua non* of efficient system operation. The main information flows, together with suggested data rates and security levels, will be:

- substations to control centres and control centres to substations – real time telemetry (50–75 bits per second (bps)), duplicated or triangulated data links, i.e. routed direct and also via an adjacent substation;

- power stations to control centre and control centre to power stations – real time telemetry (50–600 bps), duplicated data links; predictive (up to Mbps, duplicated links);
- between control centres – real time telemetry and predictive (Mbps, duplicate dedicated circuits or switched packet network);
- substation to substation – real time telemetry (50–600 bps, triangulated configuration). Most types of protection will need data links for intertripping if these are not an integral part of the protection system.

Telephony will always be provided using the utility's own circuits or the public service network, often both.

The features of power system telemetry by comparison with other industries will be its small volume and low data rates, but large geographical cover. Physical integrity must be high; target availabilities of 99.99% are suggested for communication links associated with operation of the main system. Protective gear applications will demand still higher targets (e.g. probability of a command being received 99.999%). Typical volumes of data for a large substation would be, say, 50 measurands and 250 states transmitted to the control centre. Traffic into the system or national centre of a large utility would be, say, 5000 measurands and 10000 states (these figures illustrate the need to filter out and highlight the data really required by the operator). Most forms of communication channels are used as listed below:

- *power line carrier* (plc) – a high frequency carrier, 30–500 kHz, is injected into the power circuit/s via coupling capacitors. The carrier is modulated for state and analogue information, including an audio channel. PLC offers long range (several hundred kilometres without a repeater) and high availability. Some limitations are that channel frequencies must be organized to prevent breakthrough between contiguous channels, and hence the number of channels will be limited. Transmission capacity is limited and noise level high, but no infrastructure support external to the utility will be needed. It is a major communication medium in less developed systems.
- *fixed link radio* – this term is used to describe radio links between geographically fixed points. A wide range of frequencies is used, for instance VHF, below 100 MHz, to give coverage up to some tens of kilometres (mainly speech), and microwave links (speech and data) for high capacity links. These will require line of site location, and may be subject to fading depending on weather conditions. The length of individual spans will usually not exceed some 50 km. The advantages of radio links are high quality, high reliability (assuming propagation problems are overcome), easy maintenance/replacement, high capacity and low cost compared to other media. Disadvantages are difficulties

in obtaining suitable frequencies, impact of adverse weather on propagation, and location, cost and power supply for repeater stations.

- *fixed-mobile and mobile-mobile radio* – the fixed-mobile installation will consist of a base station operating in the low and mid band VHF range (70–165 MHz). Mobile-mobile radio may be provided by 'talk-through' a base station (that is, one mobile communicates with the base station which relays the information to the other mobiles) or mobile-mobile direct using the UHF band.
- *optical fibre* [4.23] – an optical fibre generally consists of an inner core of glass with a high refractive index surrounded by an outer glass cladding with lower refractive index. Plastic, which is cheaper, has higher losses, but can be used for short runs. Light beams change direction at the interface between materials with different refractive indices, and hence a beam of light introduced at one end of the cable 'bounces' along the core until it reaches the other end. Attenuation depends upon the light source (LED or laser), mode of transmission and installation details giving feasible distances between repeaters from some 25–85 km. Modulation frequencies (i.e. the frequency of the signal to be transmitted) in the gigahertz range can be used capable of providing 30 000+ channels.

  Three types of optical fibre are used: 'optical conductors' are overhead line conductors which have optical fibres incorporated within them, for instance 24 fibres within a central aluminium tube, surrounded by aluminium and aluminium clad conductor strands. The earth wire will usually be used on 132 kV and higher voltage lines. Phase conductors have been used at 33 kV. A very popular type is the system in which the fibre optic cable, say 24 fibres, is wrapped helically around a ground or phase conductor. Installation may be possible with the line alive. This method offers a potentially inexpensive cable with low installation costs. A third type is the 'all dielectric self supporting cable'. These can usually be strung on existing towers without strengthening and their tensile strength allows spans approaching 1 km.

  Other attractive features of fibre optic cable based communication channels are their high reliability–a life well in excess of 25 years has been suggested – and the complete physical separation between the power equipment and the channel giving immunity to electrical interference. The bandwidth will usually be well above any power system requirement and having installed fibre optic cables, utilities have hired channels to external users. In the UK, the National Grid Company set up a telecommunications subsidiary, Energis, in the early 1990s, and in three years built up a network of over 3000 km of fibre cable together with some 75 equipment sites, providing a full range of network services. Coupled with the use of optical fibre cables laid with distribution circuits, access is available to city and town centres.

- *public communications networks* – public service communications carriers may offer long-term lease of communication channels ranging up to Mbit/sec capacity. As far as is known, the leasor will not know the form of the channels he is using, but sometimes the hirer will tag such circuits, identifying them so that an interruption free service can as far as possible be provided.
- *pilot cables* – utilities may lay pilot cables with power cables, not least for protection schemes on the power cables. In these and other communication channels, the total bandwidth will be multiplexed by high and low pass frequency filters into a number of channels ranging from some 100 Hz for state indications to 2 kHz for speech.

4.9 TELECOMMAND

The excellent surveys of system control centres by Dy Liacco and Rosa [4.24] have provided comprehensive information on facilities. Product guides on SCADA systems, largely hardware have been published in the journal *Modern Power Systems*. In round terms, some 70% of utilities use supervisory control of the network, with a slight bias to a wider application in smaller utilities.

The technical pros and cons and supervisory control of networks have been discussed in Section 4.6. Additional considerations will be that the presence of staff is a deterrent against theft and vandalism, and staff on-site could provide early warning of incipient plant, system or environmental problems.

4.10 DISTRIBUTED GENERATION

It has been suggested, e.g. [4.25], that beginning around the year 2000 in the USA, small distributed generating units will emerge, initially in niche markets, and that new manufacturing firms will begin to appear focused not on large boilers and turbines, but on assembly-line production of micromachines, fuel cells, photovoltaics, and other, yet to be developed, generating options. Industry restructuring could facilitate these developments. Some of the distributed generation would be installed by end users, on their own sites.

The information on some of the options for distributed generation in Table 4.13 has been taken mainly from Moore [4.25].

4.11 FLEXIBLE A.C. TRANSMISSION SYSTEMS (FACTS)

Pressures to minimize capital expenditure, operating costs and the use of wayleaves have led to the development of control equipment to maximize the use of a.c. networks as determined by circuit loadings and nodal voltages. Such

Table 4.13 Some options for distributed generation

Type	Size range	Efficiency (%)	Application
Diesel engines	50 kW–6 MW	35	In utilities – black start, back-up house service supplies in substations, mobile emergency supplies For consumers – stand-by power for commerce and small industry
Combustion turbines	1–100 MW	33–45	In utilities – peaking and emergency power, network support in heavy and abnormal loading conditions For consumers – industrial Combined Heat and Power (CHP)
Fuel cells	From 25 kW to 5 MW depending on type	40–65 depending on type	In utilities – for remote locations For consumers – commercial CHP, high power quality
Photovoltaic arrays	Under 1–1000 kW		For consumers – power for remote locations, high power quality

equipment is often referred to as 'FACTS devices', and in contrast to earlier generations of control means, will be capable of acting sufficiently rapidly to improve all forms of stability, as well as steady state conditions (c.f. the IEEE definition "alternating current transmission systems incorporating power electronic-based and other static controllers to enhance controllability and increase power transfer capability"). These effects are achieved by controlling the three main parameters directly affecting a.c. power transmission: voltage, phase angle and impedance. The devices are not new in basic concept, but the use of power electronics has revolutionized their design and engineering [4.26]. The main objectives in applying them are to increase the power transfer capability of transmission networks, and to provide direct control of power flow over designated transmission routes.

4.11.1 Factors Preventing Full Thermal Loading of Circuits in an A.C. Network

The aim of planners must be for all circuits to be loaded to their thermal capacity under the most adverse system loading conditions just prior to system reinforcement. If this ideal condition obtains, the planner can be sure that the network is not overdesigned. Factors which may prevent this being achieved will be:

- poor power sharing between circuits or unacceptable voltage levels under normal conditions; quick action (i.e. seconds rather than minutes will be unimportant when eliminating these problems);
- the need to avoid system conditions which with credible changes might lead to any form of instability–steady state, transient, dynamic, voltage; subsynchronous resonance; quick action might prevent instabilities;
- the need to ensure that fault levels are within the switchgear rupturing capacity (except possibly under very well defined and carefully controlled conditions); quick action (seconds rather than minutes) will not be important.

4.11.2 Some FACTS Devices

FACTS devices are currently based on the thyristor, and on its development the Gate Turn-Off Thyristor (GTO). The thyristor is a semiconductor device, with a maximum rating at present of some 4 kA and blocking voltage up to some 6 kV. It is ‘turned on’ (made conducting) by applying current at its gate, but only ‘turns off’ (ceases to conduct) when the through current falls below a minimum value for a minimum time period. The GTO can be turned off by the application of a large reverse gate current (e.g. −750 A), with a peak controllable on-start current of some 4 kA and blocking voltage of 4.5 kV. The thyristors or GTOs are built up in series to achieve the required line voltages.

Table 4.14 lists the majority of the FACTS devices and their commercial/development status in the mid-1990s, taken mainly from N.G. Hingorani, ‘FACTS technologies and opportunities’ in [4.27], plus brief comment on their mechanism and function. Some of these are discussed in Chapter 10.

The articles suggest the following benefits from using FACTS devices:

- FACTS technology is often the most economic alternative for solving transmission loading problems. It provides a mechanism to make the best use of existing transmission.
- The siting of some FACTS devices are flexible (e.g. a quadrature booster will have the same effect wherever it is installed between specific nodes in a transmission circuit). Siting requirements are not excessive, and devices can be retrofitted without the widespread system changes the introduction of d.c. links would entail, yet still provide control of flows in individual circuits.
- Faster response speeds are provided than with electromechanical devices.

Table 4.14 FACTS devices

Device	Application Status (mid 1990s)	Mechanism	Function
Static Var Compensator (SVC)	Commercially available	Adjusts reactive load, using thyristor to control shunt reactors and capacitors	Voltage control, reactive compensation, damping of oscillations
Thyristor Controlled Series Capacitors (TCSC)	Commercially available	Adjusts circuit series reactance by switching series capacitors	Series impedance adjustment, giving control of power flow, transient stability and oscillation damping; control can only act to decrease circuit normal impedance
Thyristor Controlled Series Reactor (TCSR)	Close to commercially available	Adjusts circuit series reactance by switching series reactors	As TCSC, but circuit normal impedance can only be increased
NGH Subsynchronous Resonance Damper	Commercially available	Adds circuit series resistance to inhibit low frequency oscillations on long heavily loaded circuit	Series impedance adjustment and hence damping of oscillations and transient stability control
Thyristor Controlled Phase Angle Regulator	Design studies	Basically a quadrature booster with thyristor control of series voltage injected into circuit	Phase angle injection into circuit giving control of power flows, transient stability and oscillation damping
Static condenser (STATCON)	Demonstration	Uses GTO-based converter	Adjustment of shunt capacitance and hence control of voltage, reactive compensation, oscillation damping and transient stability
Thyristor Controlled Dynamic Braking Resistor	Demonstration feasible	Provides adjustable dynamic braking load	Improvement of stability, damping of oscillations
Unified Power Flow Controller (UPFC)	Design studies for demonstration purposes	Injects variable phase voltage into circuit giving control of active and reactive power flows. Basically a quadrature booster plus in phase voltage tap changer	Could provide control of power and reactive power flows, voltage, transient stability and damping of oscillations

A specific illustration of flexibility has been demonstrated by NGC which, in conjunction with manufacturers, has developed conventional quadrature boosters of 2000 MVA rating which can be moved between sites on the NGC system.

The subject is considered further in Chapter 10.

REFERENCES

4.1. Fink, L. and van Son, P. J. M., 1998. 'On system control within a restructured industry'. *IEEE Trans. Power Sys.*, **13** (2).

4.2. CIGRE Working Group 37.12, 1994. 'The extension of synchronous electric systems: advantages and disadvantages'. Paper 37-110, Cigre.

4.3. Clerici, A. et al., 1996. Long distance transmission: the d.c. challenge. IEE Conf. Publication 423.

4.4. Bowles, J. P. et al., 1990. 'AC-DC economics and alternatives – 1987 panel session report'. *IEEE Trans. Power Delivery*, **5** (4).

4.5. Baker, M. H., 1997. 'The technologies for interconnection'. *Cigre Symposium*, Tours.

4.6. CIGRE Working Group 37.02, 1993. 'Review of adequacy standards for generation and transmission planning'. *Electra*, 150.

4.7. Ringlee et al., 'Frequency and duration methods for power system reliability calculations'. *IEEE Trans. PAS*, **8**.

4.8. Jusert, R., 1987. Comparison of the reliability criteria used in various countries. *Cigre paper* (also *Cigre Electra* No. 110, 1987).

4.9. British Electricity International: *Modern Power Station Practice Vol K: EHV Transmission*, 1991.

4.10. Hanbrich, H.-J., and Nick, W. R., 1993. 'Adequacy and security of power systems at the planning stage', *Cigre Electra*, No. 149.

4.11. Buxton, P., 1998. 'Operational measures to alleviate the possibilities of supply failure'. *IEE Colloquium on Measures to Predict Power Blackouts.*

4.12. Hyman, L. S., 1999. 'Transmission, congestion, pricing and incentives', *IEEE Power Engineering Review*, August.

4.13. UCPTE, 1998. 'Ground rules covering primary and secondary control of frequency and active power within UCPTE'.

4.14. UCPTE, 'Measures to take in the event of overloads on lines. UCPTE'.

4.15. North American Reliability Council, 1989. Electricity transfers and reliability.

4.16. Holmberg, D., et al., 1998. 'Reliability standards versus development of electric power industry', *Cigre Electra*, No. 177.

4.17. Winter, W.: Bulk electricity system operational performace: measurement systems and survey results.

4.18. Gunther, E. W., 1996. 'The EPRI distribution system power quality project (1996). EPRI (internet)'.

4.19. Clark, H. K. et al., 1992. 'Experience with dynamic system monitors to enhance system stability analysis'. *IEEE Trans. Power Sys.*, 7 (2).

4.20. IEEE Task Force Report, 1986. 'Instrumentation for monitoring power system dynamic performance'. *IEEE Power Engineering Society Winter Power Meeting*, Paper 86 WM072-3.

4.21. Anon, 1993. 'Disturbance monitors product guide', *Modern Power Systems.*

4.22. Knight, U. G., 1991. Energy management systems, NGC training course.

4.23. Carlton, G. et al., 1995. 'UK power utilities experience with optical telecommunications cabling systems'. *IEE Power Engineering Journal.*

4.24. Dy Liacco, T. E. and Rosa, D. L., 1985. 'Statistics on control centres around the world', e.g. *Electrical World.*

4.25. Moore, T., 1993. 'Emerging methods for distributed resources. *EPRI Journal* (see also Distributed generation', *EPRI Journal*, 1993).

4.26. British Electricity International, 1991. *Modern Power Station Practice, Vol. L Power System Operation.*

4.27. Hingorani, N. G., 1994. 'Facts, technology and opportunities', Colloquium on Flexible a.c. Transmission Systems (FACTS) – The Key to Increased Utilisation of Power Systems, *IEE Digest No. 1994/005.*

FURTHER READING

Webb, M. G. and Carstairs, J., 1996. 'Steps to develop regional trade'. *IEE Conf. Publication*, 423.

Valtorta, M. M., 1983. 'Electric power transmission at voltages of 1000 kV and above'. Cigre Working Group 31.04, *Electra.*

CIGRE Working Group 38.04/Task Force 30.04.04, 1988. 'Electric power transmission at voltages of 1000 kV a.c. or ±600 kV d.c. and above'. Paper 38-12, *Cigre brochure 96 TP 113-0.*

IEEE Power Engineering Society. 'East and central European policy on electricity infrastructure, interconnections and electricity exchanges'.

Wito, A. G. et al., 1994. 'The European supergrid – looking east and west'. *Universities Power Engineering Conf.*

Carlsen, T. H. et al., 'Feasibility study for increased power exchange between Norway and continental Europe by new hvdc links'. *IEE Conf. Publication No 423.*

Schneider, J. et al., 1994. 'Technical requirements and possibilities of an all-European east-west interconnection'. Paper 37-103, *Cigre.*

Dwivedi, P. K. et al., 1996. 'Planning and interconnections for disparate regional grids – a challenge'. *IEE Conf. Publication No. 423.*

Praca, J. C. G. et al., 1992. 'Amazon transmission challenge – comparison of technologies'. Paper 14/37/38-01, *Cigre.*

Sackey, T. and Zakhary, S. Z., 19. 'Power wheeling through the West-African interconnected system'. *IEE Conf. Publication No. 423.*

MPS Review, 1996. 'Is the east-west power bridge economic?' *Modern Power Systems.*

Soderberg, L. and Johnson, T., 1997. 'Swedish and Polish grids to be connected by SwePol link'. *Modern Power Systems.*

Kundar, P. (chairman panel discussion), 1988. 'Power system disturbance monitoring: utility experience'. *IEEE Trans. Power Systems*, 3 (1).

Buxton, P., 1994. 'Transmission operational security standards'. *IEE Discussion Meeting*, May.

5
Measures to Minimize the Impact of Disturbances

One of the core topics of emergency control will be reviewed in this chapter, namely what measures should be taken in the management, planning and operation of power systems to minimize the effects of disturbances on their viable operation. The objectives of the measures should be to reduce both the frequency of the disturbances and their deleterious effects if they do occur. Detailed plant design issues, for instance insulator string creepage and flashover distances, will not be discussed.

The significant factors which affect the impact of disturbances have been described in Chapter 2. Following a brief assessment of the relative importance of these, the measures which can be taken in the various timescales to reduce the risk of a disturbance occurring will be discussed. These will include the use of automatic mechanisms and 'defence plans'. The extent to which a disturbance spreads is very important – are the disconnections restricted to the plant on which fault/s have actually occurred, or has the redistribution of currents and voltages caused protection on other plant items to operate and trip these? The term 'containment' is used to describe this effect, and again, the measures available to minimize the spread are described. The measures can be included in the plans for the system, and provided in that timescale or decided during the operational planning period. Before going into detail, it is worth recalling that measures will often contain three main elements: to detect the possibility of a disturbance and its type; to assess the best way to prevent it, or at worst to minimize its effect; and to restore normal conditions.

It would be wrong to assume that all the operational measures involve hardware and/or software, except in the sense that the system instrumentation is likely to be the main source of information for the decisions. Some will be a codification of the manual information exchange and manual decision making and instruction processes.

5.1 FACTORS IN ONSET, SEVERITY AND PROPAGATION OF A DISTURBANCE

Various factors which determine the risk of a disturbance occurring, its severity and the immediately following events are suggested below (see also Chapter 2). These are listed in a judged order of priority, based on a general appreciation of the performance of power systems. The classification is worthwhile, since it indicates priorities in providing preventive measures, and also their scope. However, to avoid undue fuzziness in interpretation, where necessary the disturbances have been classified as:

- *severe*: the initial cause *considerably* exceeds the plant outages specified in the usual security criteria;
- *moderate*: the initial cause exceeds the outages specified in the usual security criteria. Some quantification is suggested below.

Factors contributing to the risk of a *severe* disturbance:

- Exceptionally severe weather (causing, for example, multiple trippings of transmission circuits).
- Unexpected and sudden bad weather (causing a rapid increase in demand and increased plant failures).
- Sudden and large loss of generation.
- Excessive non-availability of generating plant.
- Failure of anti-disturbance or protection equipment.
- Some sudden changes in ambient conditions, for instance a thaw after a long period of freezing weather leading to insulator flashovers.
- Errors by control staff.
- Errors by operational planning staff.
- Errors in planning.
- Errors by field staff.

Factors contributing to the risk of a *moderate* disturbance:

- In general, any event/s which exceed the standard criteria for secure operation.
- In general, any of the more specific risks listed above but at a reduced severity.

It is difficult to be precise on the boundary between 'severe' and 'moderate'. If one takes the normal security criteria as covering single, sometimes double, circuit outages and generation losses as the largest single infeed, a sensible distinction would be:

- *moderate* – three or four circuit outages; generation loss up to 15 percent.
- *severe* – five or more circuit outages; generation loss above 15 percent.

5.2 MEASURES IN THE PLANNING TIMESCALE TO MINIMIZE THE RISK OF A DISTURBANCE

Within the confines of capital, environmental and political limits, the planning engineer has at his discretion the magnitude and location of all new plant. He is also in a unique position to know what protective and control measures will be needed on a system-wide scale. Some of these (for instance, under-frequency load shedding), can be included in the system plan, and others (say, intertripping schemes with local impact, the requirement for which becomes evident on a shorter timescale) can be included in an operational plan.

Two system models are described below. The first of these examines the instantaneous generation-demand balance. In the second, the system dynamics are modelled. In the simplest spatial model, the system generation and demand will be summated to single totals. At an intermediate level, the system will be considered as a number of discrete regions of generation and demand, the summated values within each region acting at central points within the regions. In the full spatial model, the system is modelled as it would be for a detailed transient stability study.

With the addition of objective functions, these security orientated models can be used to determine minimum cost operating states.

5.2.1 The Basic Formulation

The fundamental requirement is that, over short periods of time, the average value of generation should equal the average values of demand plus losses:

$$\sum_{i} G_i + \sum_{j} T_j - \sum_{k} L_k - P_L = \phi = 0 \tag{5.1}$$

where

G_i is the generation at node i and is summed over all nodes with generation,

T_j is the transfer at node j and is summed over all nodes with transfer to external systems,

L_k is the average demand at node k and is summed over all nodes with demands, and

P_L is the total system loss.

As an aside, the addition of an objective function, usually to minimize the total system operating cost, will enable a unique solution to be obtained, for instance

$$\text{Min } C_o = \sum_i C_i G_i$$

where C_i is the operating cost of the generation at node i at output G_i, and C_o is the total system operating cost. The appropriate method will depend upon the form of the generating cost/output function of the generators. If these are as in Figure 5.1(a), the minimum cost solution can be obtained most simply by listing the operating costs at full outputs in ascending order, and summating down this list until the system demand total is reached. In an extension of this, in which the generation has a two-part cost-output curve (Figure 5.1(b)), this generation can be treated as two units with incremental costs p_{i1} and p_{i2}, and the added constraint that the output of the first part must be fully used before the second part can be loaded.

If the cost output curve is non-linear (Figure 5.1(c)), the 'co-ordination equations' will be solved:

$$\frac{\partial C_o}{\partial G_i} = \frac{dC_i(G_i)}{dG_i} \tag{5.2}$$

and

$$\frac{\partial \phi}{\partial G_i} = 1 - \frac{\partial P_L}{\partial G_i} \tag{5.3}$$

This can be done using the Lagrangian multiplier (λ) method for constrained optimization:†

$$\frac{\partial C_o}{\partial G_i} - \lambda \frac{\partial \phi}{\partial G_i} = \frac{dC_i(g_i)}{dG_i} - \lambda\left(1 - \frac{\partial P_L}{\partial G_i}\right) = 0 \tag{5.4}$$

λ is a parameter changing the value of which will change the operating point of the system to produce more or less generation. A characteristic appropriate to a multi-valve turbine is shown in Figure 5.1(d), those for gas turbine plant in Figure 5.1(e)(1).

† Models involving more constraints, e.g. for transmission, can be solved using Kuhn and Tucker multipliers or, with linearisation, linear programming.

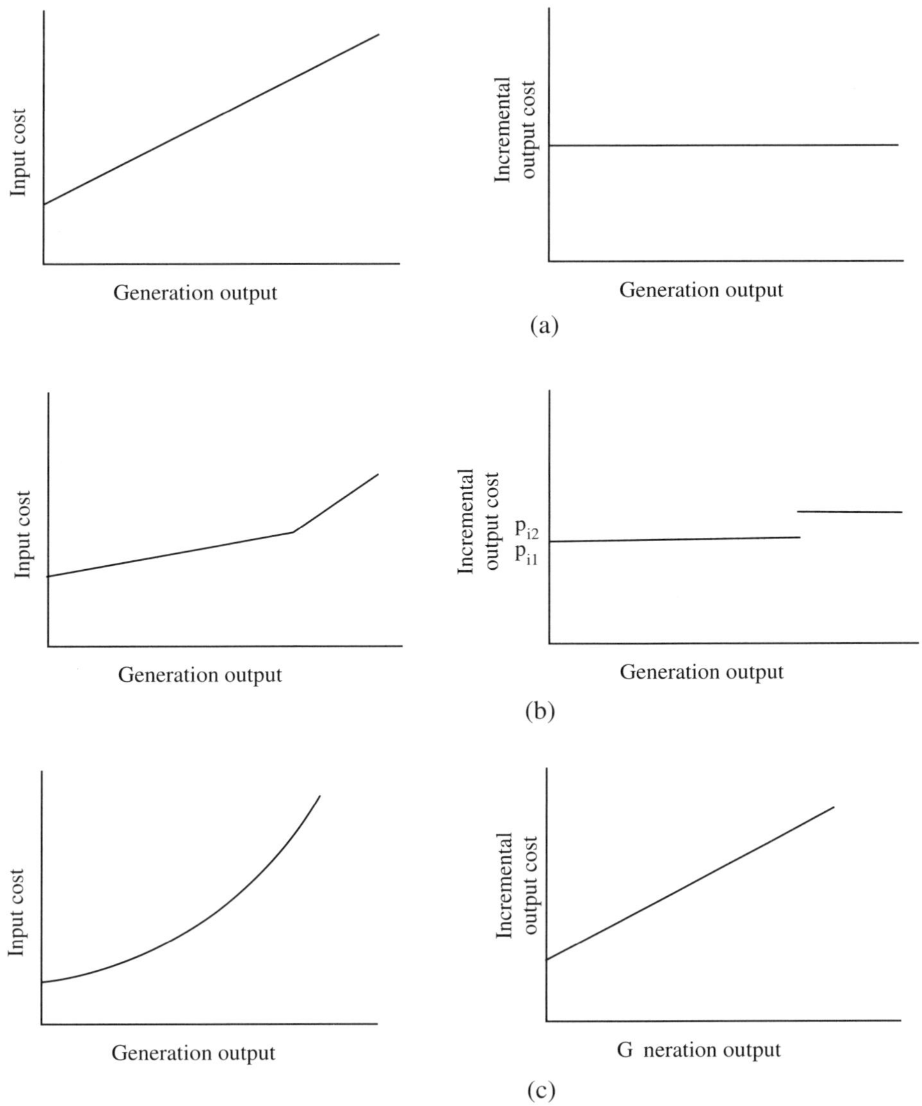

Figure 5.1 Input cost–output power curves for different types of plant. (a) Linear, (b) two-part, (c) non-linear, (d) multi-valve, (e) combined cycle

If some event occurs so that equation (5.1) is no longer satisfied, the operating state will be modified until it is, usually at minimum cost or with minimum delay, as judged appropriate by the planner. He must include facilities in the system plan for this to be done: either spare generation; disconnectable demand (agreed/contracted with consumers); or changes in external transfers (agreed/contracted with neighbouring utilities).

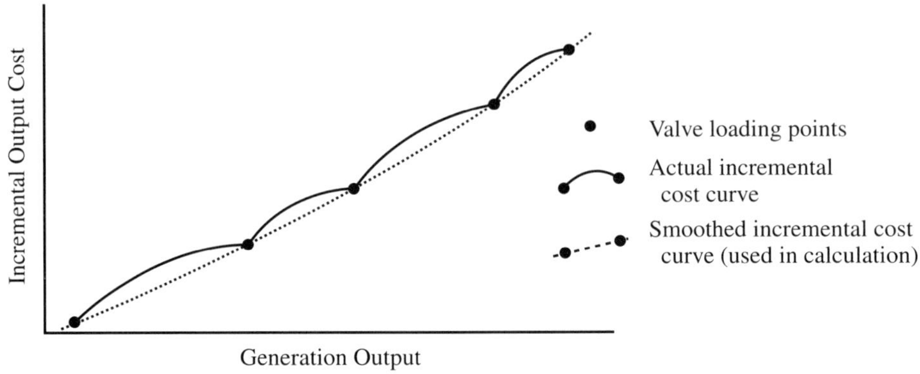

(d)

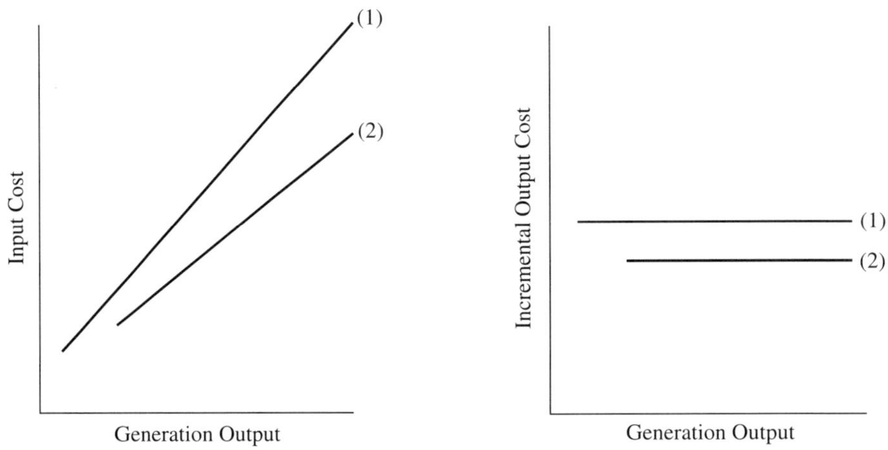

Curves 1 - Open cycle gas turbine
Curves 2 - Combined cycle gas turbine

(e) Combustion turbine based plant

Note - all the curves show shape only, not magnitudes

Figure 5.1 *(continued)*

5.2.2 Generation Provisions in the System Plan

The levels of spare provided include margins for demand forecasting errors, bad weather and delays in commissioning, as well as plant breakdown. The allowance for the latter will typically equal the capacity of one or two of the largest generators. This will also cover the sudden loss of the largest import from neighbours in most cases. The rates at which the generation can be increased (load pick up capability) or decreased (load rejection capability) will also be important, and should have been defined in the plant specifications. The response is also sometimes categorized as 'primary response", that is the response which is sustainable over a period of at least 10–30 seconds after a fall in system frequency, and secondary response, that is the response which is sustainable over a period from, say, 30 seconds to at least 30 minutes after a system frequency fall.

Noting that, on occasion, any part of the system might have to operate in isolation. The pick up capability for a steam unit might be given as

(1) When operating in the range *a–b* percent (e.g. 50–75 percent), a unit should be capable of picking up and sustaining a step increase of *x* percent (e.g. 15 percent).

(2) When operating in the range *b–c* percent (e.g. 75–85 percent), a unit should be capable of picking up and sustaining a step increase in load up to half its nominal spare capacity, that is from $\frac{1}{2}(100 - b)$ percent to $\frac{1}{2}(100 - c)$ percent (e.g. between 12.5 and 7.5 percent).

Minimum loads for gas and coal fired large steam units have been quoted as 40–50 percent, and for oil fired units at 10–30 percent. Sustained (that is, from minimum to 100 percent load) response rate capabilities were 2–4 percent per minute for steam type, oil and gas fired units, and 1.5 percent per minute for coal fired units. Rates over smaller excursions can be higher.

The type of control will also affect the response. 'Turbine following boiler' (the control signals are sent to the boiler, and the turbine output follows the boiler output) will be slower than 'boiler following turbine'. In the 'sliding pressure' operating mode, the turbine speeder gear is wound to its limit, and hence there is no free governor action. Valve throttling losses are minimal, improving unit efficiency. Load changes are initiated by changing the boiler pressure. However, the loading range of units may be limited.

Gas turbines will respond very quickly, within a few seconds, to commands for changes in output, for instance from 75–100 percent load in about five seconds, and will maintain the increased output, although to do this they must be operated at part load, with some reduction in efficiency, in the 'frequency sensitive' mode. The combined cycle gas turbine offers a response between that of the gas turbine and steam turbine, depending on its configuration and operating state. Its minimum stable generation may be 30–40 percent of base load (defined as its most efficient operating point). In a typical unit configuration of three gas turbines and one steam turbine, one third of the output will be provided by the steam turbine, and two thirds by the gas turbines. Of a total plant reserve of 15 percent say, 23 percent would be allocated to the gas turbines [5.2]. Response levels will vary significantly between plants but indicatively could be zero to base load in about two minutes.

The operating flexibility of a pumped storage station provides dynamic adjustment of the generation-demand balance over and above the direct benefits of dependable peak power and the associated economical energy transfer from off peak to peak periods. The plant design will be dictated by the availability of sites with suitable upper and lower storage volumes, and the relative values attached to energy storage and peaking capacity *viz-a-viz* power regulation needs. Sometimes the latter is added as a fairly minor item in benefit assessments of pumped storage. Some of the installations reported in the past had the characteristics shown in Table 5.1.

Environmental factors may place constraints on the operation of hydro plants, for instance the maximum changes in storage water level permitted in one day. One survey paper [5.3] covering 18 installations summarized minimum times for mode changes for single stage reversible Francis type pump turbines as

standstill to no load generation	60–140 secs
no load generation to full load generation	10 secs
no load generation to standstill	120 secs
standstill to full load pump	105 secs
full load pump to no load generation	140 secs

Low head hydro units as used in run of river plants have excellent response capabilities. Many can be cycled over their full operating range in under one minute. The response rates of high head units will be curtailed to prevent damage from water-hammer in long penstocks, although even with such a limitation, units can provide very large power excursions if required.

A survey of nuclear plant performance published in 1986 [5.4] included information on plant to be commissioned up to 1990. This indicated that, in terms of range and load capability, nuclear plants compared favourably with fossil fired steam plant. Of the reactors sampled, 60 percent could operate down to 20 percent of full power or less. Almost half of the loading rates quoted were in the range of 2–5 percent per minute (the others higher) over substantial load changes of 50 percent or more. The paper noted that nuclear power plants include turbine-bypass arrangements for venting excess steam to the atmosphere, to dump condensers or to the main condenser. These systems provide flexibility during scheduled or unscheduled load changes, as well as permitting fast start-ups and shut downs. Governor droop settings were in the range from 2.5–5 percent, but to inhibit primary control action, a dead band was incorporated in the speed governor loop. Nuclear plant may also suffer 'poisoning out' (the accumulation of radioactive decay products in the nuclear core), which can delay restart for many hours.

The dynamic performance of generating plant must be specified with regard to thermal stresses, which will, for instance, set a minimum time between shut down and the following start up.

5.2.3 Measures for Demand Adjustment in the System Plan

In contrast to the measures for generation adjustment discussed in the previous section, those for demand adjustment by switching or voltage change will be relatively cheap to provide, but have more impact on consumers.

Table 5.1 Characteristics of some pumped storage schemes.* Environmental factors may place constraints on the operation of hydro plants, for instance, the maximum changes in water level permitted in one day

Station and operating cycle/size	Operating mode change	Generation change		Demand change	
		Power (MW)	Time (secs)	Power (MW)	Time (secs)
Muddy Run (USA) Weekly/1280 MW	Shut down one pump	–	–	−150	60
Hydraulically coupled to run of river hydro	Load one generator	100	300		
	Load whole station	800	900		
Dinorwig (UK) daily/1800 MW	Shutdown to full load (whole station)	1320	100		
8500 MWh	Full load to no load				10
	Full load generation to full pump				2200 (station) 1000 (unit)
	Full pump to full generation		500 (norm) 100 (emergency)		
Foyers (UK)	Standstill to full load generation	300	105		
	Standstill to full pump				350–780
	Full pump to full generation		180		
Northfield (USA) Weekly/1000 MW	Standstill to full load generation	250 per unit	360		
10465 MWh	Standstill to full load pump			≈250 per unit	600
	Full pump to full generation		960		

* Some data sources used: International Symposium and Workshop on the dynamic benefits of pumped storage (US Dep. of Energy and EPRI 1984); Operational aspects of Dinorwig pumped storage station on the CEGB system (D. A. Kidd, IEE lecture 1984); Foyers pumped storage project (D. J. Miller et al., *Proc IEE* **122**, 11, 1975).

Demand disconnection by under-frequency relays is one of the most effective and widely used methods of restoring the generation demand balance following the sudden loss of internal generation or import from neighbours. It will be necessary to include financial provision for the relays and installation in the system plan. Its main limitations are lack of sensitivity in protecting small parts of large interconnections, and difficulties may be experienced in equating the demand disconnected to the generation lost. It is also not possible to vary the location of disconnectable demand from that determined by the siting and settings of the under-frequency relays. As Concordia has said, it is a coarse tool for use in extreme situations. The action should be rapid and decisive, distributed across the system so as to avoid transmission overloads, not be dependent on communications links and, as far as possible, remain effective in the event of system splitting.

As well as advantages, interconnections to neighbours bring risks in that the utility is no longer entirely 'master of its own fate'. For instance, loss of generation within a section or of transmission connections into a section of a large interconnected system might result in low voltages or the overload of remaining circuits. The system frequency would not then necessarily indicate any abnormal situation. The most effective immediate action might well be to reduce demand; in fact, some utilities have provided schemes to disconnect demand on detection of low voltages or of overloads on transmission circuits. A number of utilities have provided facilities for the reduction of demand from control centres by voltage reduction or by disconnection. Demand disconnection by frequency-trend relays has been used in conjunction with under-frequency relays by some utilities. A typical criterion is that both a given fall and rate of fall of frequency should be exceeded, the value of the rate of change being used to determine the level of demand shedding. It has been argued that such a combination, compared to the more usual under-frequency criterion, will result in a smaller frequency drop and shorter periods of under-frequency running for a given power deficiency, and is also less likely to produce unnecessary tripping of demand for a slow asympototic frequency fall to a value below the under-frequency threshold value. However, compared to the use of frequency level only, the effects of random frequency changes will be magnified, and it is more difficult to calculate the demand to be shed for a given rate of change of frequency; this would be dependent on factors other than the generation-demand imbalance, such as system inertia. The calculation will usually take longer, since the frequencies at two different times have to be compared. Furthermore, frequency tends to have a more uniform value over larger areas of a system than does the rate of change of frequency. As far as is known, rate of change of frequency has not been used widely to initiate demand disconnection. However, if the configuration of the system is such that a frequency gradient is likely to exist immediately after a power unbalance, as in a 'long thin' system, it may be appropriate to use rate of change of frequency as a signal to initiate shedding.

Under-Frequency Relays

Guidelines for the application and setting of under-frequency relays have included the following [5.5, 5.9]:

- "As a practical matter perhaps 80 percent (of demand) should be included in the shedding schedule, some with appreciable time delay."
- The first step threshold should be sufficiently far from normal frequency to avoid tripping on severe but non-emergency frequency swings, and the last step threshold sufficiently high to prevent system frequency having to fall below the sustainable lower level. (This will be the level at which, for instance, the generation output decreases because the performance of station auxiliaries is affected.)
- The number and sizes of stages must be selected with several factors in mind: the sudden reduction of output which the generation will accept and still continue to operate viably; the avoidance of overvoltages on the system, the avoidance of over-frequencies; the avoidance of damage to generation which would be caused by operating too long at off-nominal frequency; the accuracy with which the relays can be set; the lowest transient frequency from which, after allowing some overswing, the system will recover; and the highest frequency at which the first stage can be set without unnecessarily shedding demand or interfering with the settings of other plant, such as pumped storage or gas turbines.
- It is commonly accepted that all possible measures should be taken to prevent system frequency falling to below some 95 percent of nominal value, that is falling below 57 Hz on a 60 Hz system or 47.5 Hz on a 50 Hz system. Within the 5 percent band, however, the range of frequency threshold values and amounts of demand shed vary quite widely. Some of the differences will be caused by the need to allow thresholds for other actions, for example switching gas turbine and pumped storage plant.
- Time delays can be added to provide a further way to discriminate between stages of disconnection.

Most under-frequency relays have two settings–the system frequency at which the relay will operate, and the time delay between this and the trip signal/s to breaker trip relays. The minimum time between detection of low frequency and breaker operation may be some 150 msec. The adjustable time delay gives another mechanism to discriminate between stages of demand disconnection.

Two models which can be used to estimate the power imbalance/frequency characteristic needed to determine settings are outlined below. The starting point will be a method to estimate the relationship between the generation-demand imbalance and the effect of this on the system frequency, that is the system frequency regulation.

Governor and System Droop, System Stiffness and Frequency Regulation – a Simple Model

'Droop' and 'system stiffness' are terms used to quantify the relationship between power change and frequency change. Governor droop is the frequency (or speed) change which will open the governor of a turbine from the no load to the full load position. (Thermal plants typically have a droop in the range 3–5 percent, with a response time of some seconds, and hydro units a droop of 5 percent). Related characteristics are:

- the incremental droop, which is the rate of change of the steady state frequency (or speed) with respect to output at a given steady state operating point;
- dead band, which is the change in steady frequency or speed within which there is no change in the position of the governor values;
- maximum inaccuracy or non-linearity, which is the maximum deviation in terms of rated output from the output given by the average droop.

When applied to a system, the frequency sensitivity of the demand must be included, and the equation defining the operation will be

$$G_a = G_o + r_g(f_a - f_o) \tag{5.5}$$

$$L_a = L_o + r_l(f_a - f_o) \tag{5.6}$$

where

G_a, G_o are the actual and base generations on the system in p.u. (base generation is the generation at nominal frequency),

L_a, L_o are the actual and base loads on the system in p.u. (base load is the load at nominal frequency),

f_a, f_o are the actual and base (nominal) frequencies,

r_g is the generation frequency characteristic (p.u. generation change per hertz), and

r_l is the load frequency characteristic (p.u. load change per hertz).

The overall system characteristic will be

$$G_a - L_a = G_o - L_o + (r_g - r_l)(f_a - f_o) \tag{5.7}$$

or

$$\begin{aligned} M_a &= M_o + (r_g - r_l)(f_a - f_o) \\ &= M_o + r_g(f_a - f_o) \end{aligned} \tag{5.8}$$

where

M_a and M_o are the actual and nominal generation-demand imbalances, and

r_g is the natural regulation characteristic of the system.

The term $(r_g - r_l)$ is sometimes called the 'stiffness' of the system, denoted K.

In practice, the coefficient r_g will be several times bigger than the coefficient r_l. If the sensitivity of demand to frequency is neglected and, as above, all the system dynamics, the simplest frequency–imbalance relationships will be obtained as

$$G_a = G_o + r_g(f_a - f_o)$$

or, in incremental terms,

$$\Delta G = r_g \Delta f \quad \text{and} \quad \Delta f = \frac{\Delta G}{r_g} \tag{5.9}$$

Frequency regulation – more exact models Many other models to estimate the frequency regulation characteristic of a power system can be found in the literature. The more comprehensive of these will include the dynamics of the energy conversion plant and control systems, for instance in the case of a conventional thermal station the boiler, turbine speed governor and demand. Laplace transform methods are often used, the end product being the frequency-time trace for the system over several minutes (5.6), e.g.

$$f_t = \frac{1}{S}\frac{\sum G_i - L_o(1 + r_l(f - f_o))}{2\sum H_i f_a \sum G_{in}} \tag{5.10}$$

where

$S =$ Laplace operator,

$H_i =$ inertia of plant i,

$G_{in} =$ nameplate capacity of plant i, and

$f_t =$ calculated value of frequency at time t.

Models of this type seem capable of estimating the shape of system transient frequency responses quite reasonably. Figure 5.2 gives a comparison of measured and model results for tests in the UK some time ago, in which the system was split into two roughly equal parts, with differing transfers between the two parts. Frequencies were measured over three time periods: over a few seconds (immediately before to immediately after the switching); over some 20 seconds; and

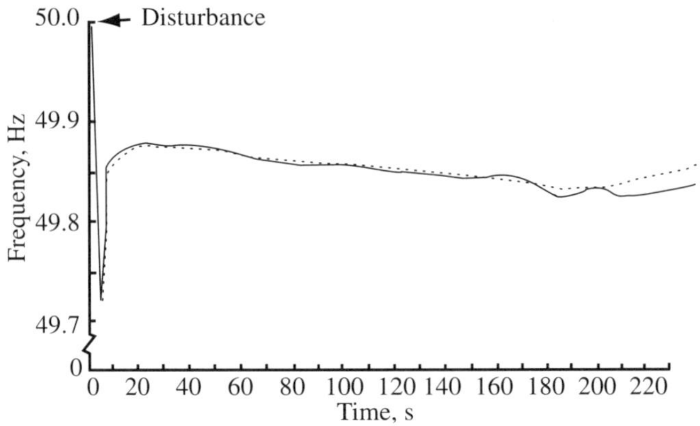

Figure 5.2 Comparison of frequency changes by test and from model. —: test result; ----: model result. (Reproduced by permission of IEE from [5.6]).

over several minutes. The responses in the first series would be mainly determined by the system inertia; in the second series governor action would come into play; and in the third series, the boiler-fuel system would influence the result. (The excellent agreement shown was not obtained in all comparisons; the value taken for the system inertia was shown to be a critical parameter.)

5.3 MEASURES IN THE OPERATIONAL TIMESCALE TO MINIMIZE THE RISK AND IMPACT OF A DISTURBANCE

Three classes of measures which are used in the operational timescale are described below: load disconnection and other automatic switching systems; memoranda and procedures adopted throughout the utility; and *ad hoc* measures implemented at short notice, often based on equipment overloading.

5.3.1 Under-frequency Load Disconnection

The settings used on under-frequency relays may be reviewed annually to ensure that the evolving system needs continue to be met optimally as far as the relay scheme permits. Models capable of simulating large frequency disturbances can be used if they include under-frequency load disconnection. Typically, the simulation will be run with a range of setting parameters, and the solution chosen which best fits the criteria outlined in Section 5.2.3.

In one scheme studied for use on a system of some 25 GW, nine stages were proposed with steps of 5 percent, 7.5 percent and 10 percent demand disconnection [5.8]. Other system and study parameters were system inertia (8 MW seconds per MW), load frequency characteristic (2 percent per Hz), and no spinning reserve. Figure 5.3(a) shows the estimated minimum system frequency over a range of generation deficits. Figure 5.3(b) gives the margins between the proposed load disconnections and the maximum limit, to avoid excessive generation surplus (above 10 percent), and the minimum limit, to ensure frequency recovery to above 48.0 Hz. Figure 5.3(c) shows the generation reductions required over the range of generation deficits. Thus, for a deficit of 25 percent, the minimum frequency would be some 48.55 Hz, the load shed 27.5 percent and the generation reduction required some 4 percent. It has to be said that this was not a 'green field' situation. An existing 40 percent disconnection scheme employing four stages, each of 10 percent, was to be extended. The composite measures are shown in Table 5.2.

In general, frequency deviations on smaller systems will tend to be bigger than on larger systems. Thus, load shedding schedules may have to accommodate bigger frequency swings, and more drastic actions may be needed in the shape of fewer and larger disconnection steps, with minimization of time delays and a higher percentage of load subject to shedding.

A description of a comprehensive study on a small system is given by Concordia *et al.* [5.9]. The 500 MW system was supplied by two oil fired power stations with a total capacity of 615 MW. The study comprised three successively more detailed phases. In the first, a lumped model of a single equivalent generator and single equivalent load was used. A dynamic model including the network, and similar to that used in automatic generation control, was used in the second phase, and a full transient stability model in the third. The criteria used to assess the alternatives were the sample standard deviation of the minimum frequency, the sample standard deviation of the steady state frequency post-shedding, and the maximum frequency excursion. The preferred solution for the 500 MW system was as shown in Table 5.3.

Table 5.2 A practical load shedding schedule (large system)

Stage	% initial load disconnected	Frequency setting (Hz)	Total of initial load disconnected (%)
1	5.0	48.8	5
2	5.0	48.75	10
3	10.0	48.70	20
4	7.5	48.6	27.5
5	7.5	48.4	35
6	7.5	48.1	42.5
7	7.5	47.7	50
8	5.0	47.3	55
9	5.0	47.0	60

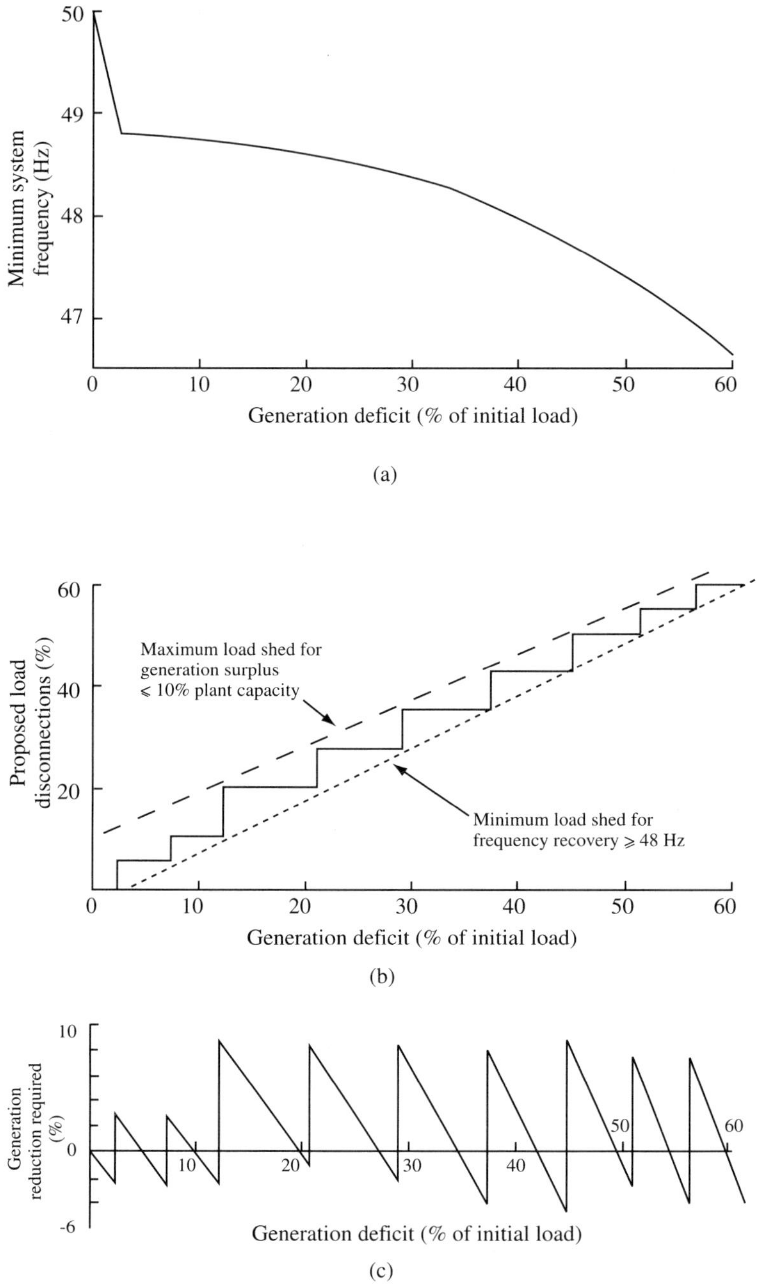

Figure 5.3 Example performance of load shedding scheme. (a) Minimum system frequencies; (b) load disconnection; (c) generation reductions required. Reproduced by permission of the National Grid Company plc

Table 5.3 An example of under-frequency relay settings (small system). Load disconnection schedules for future system demands up to some 880 MW were also studied

Frequency threshold, percent nominal	98	97.6	96.8	95.6	94
Hz	49.0	48.8	48.4	47.8	47.0
Load shed percent	15	20	25	10	10
Delay, seconds	0.20	0.20	0.35	0.35	0.35
Cumulative percent shed	15	35	60	70	80

5.3.2 Other Frequency Control Mechanisms

Under-frequency load disconnection and other mechanisms, manual and automatic, to control frequency deviations should be an integrated package covering a frequency range from, say, nominal plus 10 percent to nominal minus 6 percent.

Safety and legal considerations will place a limit on allowable overspeeds, and hence determine the times in which throttle valves must close. Station operators may also have standing instructions on actions to be taken in the event of a high frequency alarm.

Demand disconnection will be avoided if at all possible, and hence measures to increase generation will be implemented at a higher frequency level than any disconnection. These will include connection of pumped storage in the generation mode, and disconnection of pumped storage in the pumping mode. Controlled reduction of demand will follow by instructing first voltage reductions across the system, and then load reductions, often called load management. For industry this will mean making arrangements with the larger consumers to disconnect agreed amounts of demand, and for domestic consumers, disconnection by time switch or radio/mains propagated signal of water and space storage heating systems. Advantage will also be taken of contracts with neighbouring utilities which allow adjustment of transfers to relieve stressed conditions.

5.3.3 Memoranda and Procedures

Although the operating ethos of a utility should, in my view, leave some room for individual initiatives, an operator should be supported by unambiguous rules and guidance. Intentions in line with these should generally ensure their acceptability by concerned parties throughout the utility, and will be predictable by colleagues. An example of those which were used in a large utility can be found in summarized form in Reference [5.10], from which the following information with relevance to control in emergencies is obtained.

Interfaces between operational groups; duties and responsibilities within each group The operational groups included are Grid Control, Generation, Transmission, Distribution and Principal Consumers. Amongst the topics covered are system voltage control, operation of switchgear, protection requirements, auto-reclose, neutral earthing, control of generation, commissioning and decommissioning, interconnection of supply points, adverse conditions, interruption of supplies and abnormal occurrences.

Operational planning and programming The interfaces, with timetables for key events, between Operations on the one hand, and Generation, Transmission and Distribution on the other, are defined. The key events will include demand and energy estimates, energy studies, generation margin planning, network outage programming, protection co-ordination.

Operational standards of security of supply These are discussed in Chapter 4.

Procedures for instructed reduction of load including warnings The issue and cancellation of warnings from the grid control centre to other control and switching centres; procedures for instructed reduction in load.

Emergency action in the event of exceptionally severe breakdown of system This can take the form of action sheets to be implemented by local staff in the event of very serious breakdown of supply, or of system separation occurring. In some cases, action will be required without further reference to control centres. An indication of the content of such action sheets appropriate to

Table 5.4 Summary of power station action sheet in the event of an exceptionally serious disturbance

Actions in the event of low frequency
1. If the frequency falls to 49.5 Hz – synchronize and load available gas turbines
2. If the frequency falls to 47.6 Hz – shed demand manually without reference to control centres and continue to prevent frequency falling below 47 Hz
3. At this frequency, all pre-arranged demand blocks should have been disconnected
4. If generation has to be disconnected this should be done so as to avoid disruption of the main transmission system and also so as to maintain the generation operational, supplying its own auxiliaries
5. If generation is available for restoration, request the control centre for permission to start restoration and the manner – either synchronize with the main system or supply local demand
6. When the frequency has risen to 49.5 Hz ask the appropriate control engineers for permission to start restoring demand
Actions in the event of communication difficulties
1. Try all alternative means to contact control centre/s
2. If there is a complete failure of communications with control, contact the senior local engineer and with his agreement take action to restore demand (in reverse order to disconnection) and generation, and synchronise with the system
3. Continue trying to make contact and when successful, report all action taken.

a large thermal system with auxiliary gas turbines is given in Table 5.4 (taken from MPSP Vol. L, Chapter 5).

Day to day operational responsibilities of control centres at different control levels The responsibilities of the control centres and the interfaces between them. Actions required in emergency may be specified. Actions and transfer of responsibilities to outstation staff in the event of loss of communication with outstations are defined.

Procedures for data checking and state estimation The warning 'rubbish in, rubbish out' applies strongly in the area of computational aids, not least because a result appearing in a computer printout may be given undue credence just because it is computer derived. It is good practice to check all data against criteria of syntax, reasonableness, consistency and completeness; basically:

- *syntax* – is the alphanumeric data in the format required?
- *reasonableness* – are numerical values reasonable, for instance has a decimal point slipped?
- *consistency* – is the data consistent within itself?
- *completeness* – is the data complete?

The general procedures to confirm these points may be called 'data checking', and often application programs will be preceded by data check programs for this purpose. Failure to satisfy the check will often cause the applications to stop. 'State estimation' is the term usually reserved for a more formal numerical check, typically to identify 'rogue' values of raw data (telemetered or submitted by the operator), and to estimate missing data. A 'least squares algorithm' is often used in which the optimal estimate $\hat{x}$ of a vector x is that value of x for which the scalar sum of the least squares differences between the measured and estimated values are a minimum.

Procedures for generation scheduling The policy, staff responsibilities, procedures, and timescales necessary to issue and implement optimum generation schedules are defined. The schedules should be viable having regard to system security, voltage control, trading and agreed external transfers.

Procedures for manual generation dispatch These will fulfil the same role for mainly manual generation dispatch as the previous one for generation scheduling.

Procedures for automatic generation and control On-line computational aids can require a considerable amount of manual support which should preferably be defined for the operator in the form of procedures. Some of those for a.g.c. and for real-time security transmission are summarized below. Frequently, a distinction is made in these between the data which is constant or changes infrequently (for example, plant ratings), and that which can change between studies ('run data').

(1) *Automatic generation control and economic dispatch (AGC and EDC), excluding network model.* The information required will be specified:

- Initiation and termination of the a.g.c. function,
- Control of operating mode (tie line bias, tie line time bias, constant frequency, constant net interchange),
- Generator data (capacities, response rates, upper and lower limits overload capabilities),
- Control modes of each generator
- Alarm settings for area control error
- Set points and participation of individual generators, if not part of EDC
- Parameters for calculation – initiation and frequency, a.g.c. models to be used if alternatives are provided
- Economic data required for EDC including that for determining interchange with neighbours
- Manual overrides for missing or incorrect telemetered data (tie line flows, generator outputs, system frequency)
- Outputs required

(2) *Automatic generation control and economic dispatch, including network model.* It will be necessary to include a network model if the a.g.c. or economic dispatch functions include a check on the security of the network. The model and the amount of data, off line and on-line, needed will obviously depend upon the exactness of the system simulation sought, as discussed in the next section.

On-Line Security Assessment – the D.C. Model

D.C. load flows, or cruder approximations, such as the comparison of group transfers with group import/export capabilities, were used in early security assessment programs, not least to reduce computing times, and hence allow more frequent assessments to be made. A minor extension to this model will enable three-phase fault levels to be estimated.

The information needed for the d.c. model will be the active power transfers at each node and the reactances of the connections between each of these nodes. In their simplest form, the nodal transfers can be specified as such. In a more complex form, the connection point of each generator and load (a connection point to other voltage networks) will be specified and a front end program will determine the net nodal transfers from this data. In the latter case, the information required will probably be separated into static data (changes infrequently) and run data (may change between runs):

- Static data
  - Generator data (active power capacities, response rates, upper and lower limits, overload capacities, subtransient reactances (for fault levels), connection point);
  - Demand and transfer data (upper limits, connection point);
  - Network data (substation configurations including circuit breaker and operational isolator nomenclature, circuit reactances).
- Run data
  - Generator outputs at each node; } or alternatively, transfer at each node
  - Demands at each node; }
  - Switching state of all circuit breakers and operational isolators;
  - Over-ride data set in by the operator to replace erroneous or missing data;
  - Alarm limits for circuit power flows;
  - Format and content of the output (a small part of an output from a security assessment run is shown in Figure 5.4. The output should include any override data. The existence of a 'system split' should also be indicated.

On-Line Security Assessment – the A.C. Model

The potential of the a.c. model is much greater than that of the d.c. model. In addition to better estimates of current loading conditions, it provides a starting point for voltage, transient and dynamic stability studies. The data needed for the basic load flow will be as outlined above for the d.c. model, but with the inclusion of reactive terms (e.g. demand $= P + jQ$, circuit impedance$= R + jX$ generator limits in terms of terminal voltage and active/reactive power outputs).

5.4 SPECIAL PROTECTION SCHEMES

It is sometimes necessary to supplement the capability of the power system by protective and control measures on the primary plant, variously called 'automatic switching' in the UK, and 'Special Protection Schemes' (SPS) in the USA. These will be designed to detect and alleviate conditions which would otherwise

ROUTINE OUTAGE ASSESSMENT 1101 PAGE 1 OF 5

OVERRIDES AND PREVIOUS DATA USED NO- OF NODES +260
DATA ERRORS 19 FLOWS DISAGREE 0

OUTAGES CAUSING SYSTEM SPLIT		METR FLOW
*CELLARHD	- WILNGTN4	-140
*WILNGTN	- WILNGTN4 SGI	+284
*RATCLIFF	- DRAKELO4	-10
*CELLARHD	- WILNGTN4	-140
*WILNGTN	- WILNGTN4 SGI	+284
*FERYBDGC	- M FRYS T	+152
*BRADFRDW	- M FRYS T	-184
*M FRYSTN	- M FRYS T	+35
*FERYBDGC	- M FRYSTN	+154
*BRINSWTH	- TEMPLBOR1	+94
*BRINSWTH	- TEMPLBOR2	+31

ROUTINE SHORT CIRCUIT ASSESSMENT 1111 PAGE1 OF 4

O/R & PREVIOUS DATA USED MFC 754 DATA ERRORS 16

SUBSTATION		SWITCH	INFEED	SC LEVEL	SC RATING
W BURTON	NODE1	X405	3631	29736	35000
SC LEVEL	33367+	X205	1789	31578	35000
SC RATING	1000	X105	5792	27575	35000
		X305	12848	20519	35000
		X505	1558	31809	35000
		X705	1558	31809	35000
		X490	1562	31805	35000
		X390	1570	31797	35000
		X210	0	33367	35000
		X110	0	33367	35000
		X290	1570	31797	35000
		X190	1566	31801	35000
COTTAM	NODE1	X605	1672	30633	35000
SC LEVEL	32305+	X105	1876	30429	35000
SC RATING	1000	X805	15255	17050	35000
		X405	6111	26194	35000
		X305	4353	27952	35000
		X290	1524	30781	35000
		X190	1515	30790	35000

ROUTINE OUTGAGEASSESSMENT 1101 PAGE 2 OF 5

OVERRIDES & PREVIOUS DATA USED NO- OF NODES +260
DATA ERRORS 19 FLOWS DISAGREE 0

CIRCUIT OUTAGE	LOAD FLOW	LINES OVERLOADED	FLOW BEFORE	FLOW AFTER	LINE RATING
BRAMLEY -		DIDCOT -BRAMLEY 1	+750	+1260	900
DIDCOT 2	-750				
DIDCOT -		BRAMLEY -DIDCOT 2	-750	-1260	900
BRAMLEY 1	+750				
DIDCOT -		W WEYBDG -IVER	-277	-688	550
BRAMLEY 1	+750	MELKSHM4 -BRAMLEY 2	+553	+990	970
BRAMLEY -					
DIDCOT 2	-750				
CILFYND -		ABERTHAW -TREMORFA	+333	+623	504
WHITSN T	+295	WHITSON -USKMTH T3	-289	-579	504
MELKSHM4		TREMORFA -USKMTH T3	+268	+558	504
CILFYNYD	-297				
MELKSHM4 -					

Figure 5.4 Example of output from a security assessment run

cause unusual stress on the power system, and (to distinguish these from normal protection schemes) which perform a function other than or beyond the tripping of elements directly required to clear a fault. As discussed in Chapter 3, such measures may be adopted at the planning stage to reduce capital or land requirements, during construction to overcome construction delays, or in operation to meet unforeseen changes in the system.

The main components of system protection schemes and some of the design problems will be considered first, followed by outline descriptions and applications of schemes in different parts of the world.

5.4.1 The Elements of a Special Protection Scheme

A flow diagram to illustrate the main features of a Special Protection Scheme (SPS) is shown in Figure 5.5. It should be noted that SPS are often not required to operate at the same speed as fault protection relays but that their potential to damage the system operating state may be greater. More operator input may be needed to 'tune' the operation of the SPS to the current system state, and more validation of system and operation data inputs may be provided. Referring then to the numbered boxes of Figure 5.5:

- Box 1 – this is an algorithm to check the completeness and stability of the system data. It would, for instance, be necessary to eliminate transients in measurements.
- Boxes 2 and 3 – in conjunction with box 1, box 2 is an algorithm to replace missing or suspect data from history files, and box 3 is an interface to allow the operator to replace such data deficiencies. The output from boxes 1, 2 and 3 should be a complete and consistent data set for use in the SPS.
- Box 4 – this is the heart of the SPS. The pre-determined operational parameters are computed and compared with limits, often provided by the operator from off-line studies. Both the critical parameters and their limits may depend upon the current configuration of the system.
- Box 5 – the actions to be taken on detection of critical conditions are input by the operator or selected from a stored list, in accordance with the current system conditions.
- Box 6 – if the SPS allows automatic action on the primary system in the event of critical conditions being detected, the final decision on whether such action should be taken will often be left to the operator, i.e. should the SPS be 'armed' or not.

In general SPS will involve one or more of the following actions on the system:

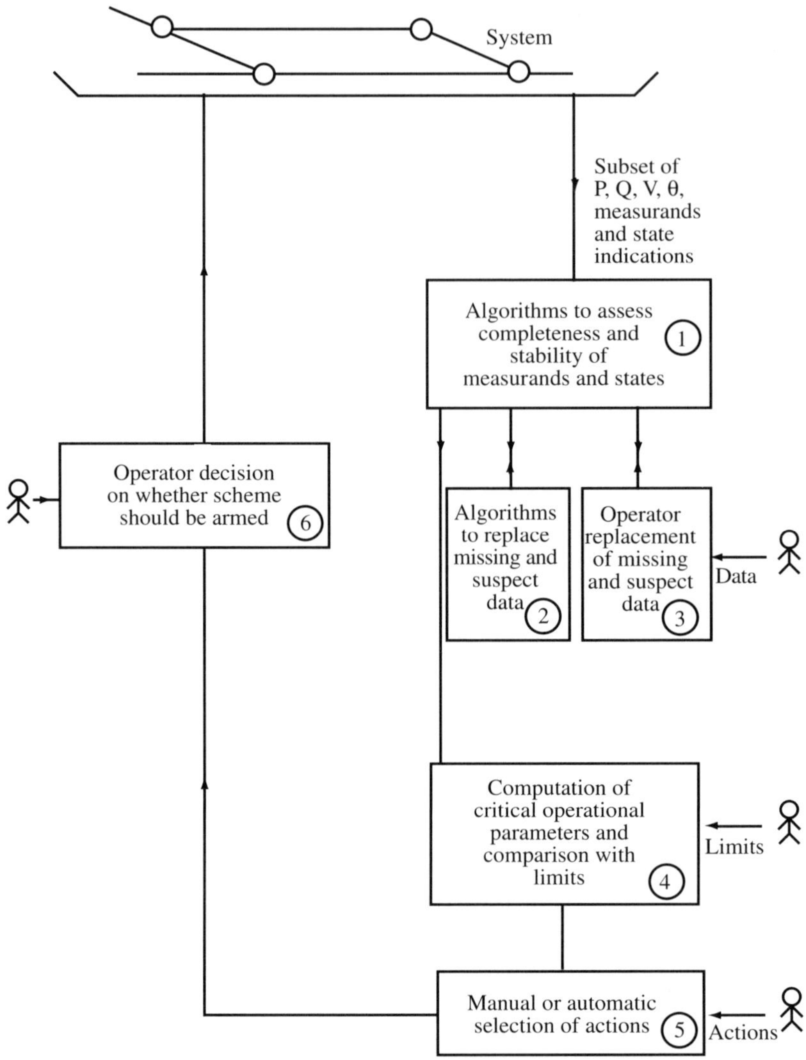

Figure 5.5 Flowchart of special protection scheme

(1) *Generation adjustment* – this is the adjustment of the amount or distribution of generation so as to balance demand and generation as necessary for the whole system, or to meet constraints on transmission flows. 'Generation rejection', a subset of generation adjustment, is the deliberate tripping of preselected generation manually or automatically.

(2) *Demand rejection* – the reduction, often tripping, of demand to balance demand and generation over the whole system and/or to meet constraints on transmission flows.

(3) *Transmission switching* (often automatic) – the reconfiguration of the transmission network to eliminate excessive flows or voltage problems following generation or demand adjustments, or network changes.

(4) *Reactive power adjustment* – the adjustment of reactive power infeeds (capacitance or inductive) to satisfy voltage constraints.

Referring to boxes 4 and 5, the types of computation and logic frequently found will be:

(1) Computation of power transfer out of exporting areas; comparison with pre-determined limits for the current configuration, and if these are exceeded, reduction of generation in the area/s through advice to station operator/s and/or a signal to station a.g.c.'s. or even to trip generators. The pre-determined limits may be based on circuit thermal limits or system voltage or stability limits. Often, this would be the $(n-1)$ or even $(n-2)$ limit that is the transmission capacity remaining after the worst credible contingency.

(2) The inverse of (1) for an importing area, i.e. the power import would be compared with the pre-determined import limit for the current configuration, and action taken as necessary to increase generation or to reduce demand in the area/s.

5.4.2 The Performance of SPS

It does not overstate their importance to say that SPSs are an essential component of modern power systems. Planners need to know their probable performance in terms of reliability and probabilities of correct and incorrect operations. Operators will need to know these statistics and also be aware of the impact of their malfunctioning, of whatever type, on the performance of the power system. This concern has led to surveys of experience with SPS, a recent one of which was carried out by the IEEE and CIGRE by means of a questionnaire addressed to designers and operators of SPS (believed to be in 1992) (5.11). Responses were received from 49 utilities in 17 countries for a total of 111 schemes – USA, Canada and Japan each about 20 percent, Europe 16 percent, Australia 9 percent, and others 14 percent. Some of the salient features which emerged from this survey are summarized below.

Installation and life durations of SPS The need for SPS seems to be increasing. About 64 percent of all SPSs reported in this survey have been installed since 1980. As of the early 1990s, 95 percent of all those installed were still in service.

Types of SPS installed Table 5.5 shows the number of installations by percentage of the commonest types of SPS.

Table 5.5 Percentage of commonest types of SPS

SPS type	Percentage
Generator rejection	21.6
Load rejection	10.8
Underfrequency load shedding	8.2
System splitting	6.3
Turbine valve control	6.3
Load and generator rejection	4.5
Stabilisers	4.5
HVDC controls	3.6
Out of step protection	2.7
Other[1]	31.5
	100.00

[1] 'Other' includes dynamic braking, reactive compensation, generation deloading, combination of schemes, etc.

Design and operational features of SPS Some design and operational features of SPS have been summarized in Figure 5.6(a,b) (see [5.11]). In general, these give comparisons of problems, for instance dominant causes of SPS failures in Figure 5.6(a), rather than absolute values, and the authors of the paper add that very often the absolute values indicate practically zero rates of problems and failure. Some interesting points emerging from these statistics are:

- *Causes of SPS failures*: the very low figure attributed to software failure probably reflects the substantial efforts that are put into software verification during the testing and commissioning phases.
- *Primary effects of SPS failures and effects of unnecessary SPS operation*: These figures show the proportionate effects on the power system of incorrect operation of SPS. The high contributions from 'generator instability' and from 'generator trips' may again be because many SPS are installed to handle problems with the operation of generation. The same effect may also be evident in the high contribution from 'loss of load'.
- *Estimated frequency of SPS operations*: These estimates demonstrate the problem found with many protection and allied systems – how to guarantee that a scheme will operate correctly if its operating time 'in anger' will be measured in milliseconds per year, or less. The popularity of the methods available are compared under 'Verification of reliability of SPS'. It seems to the author that of these, only 'monitoring' addresses the problem in a definitive manner.

Winter and Le Reverend [5.12] proposed definitions and numerical indices to compare the performance of control aids as follows:

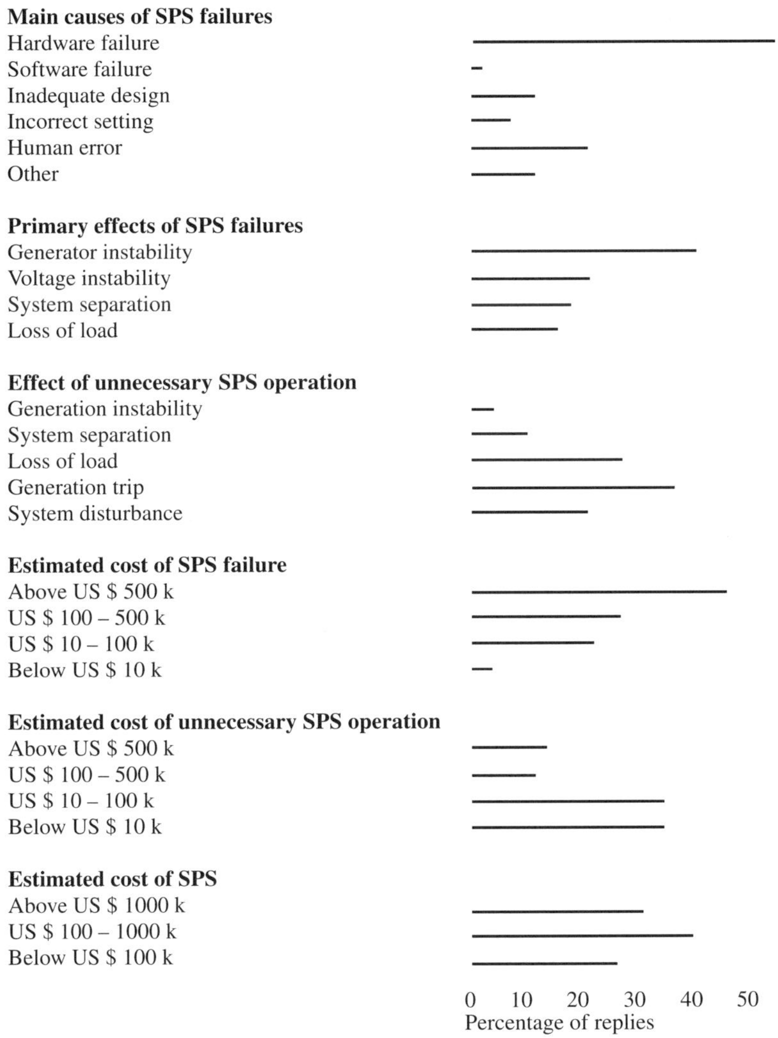

(a)

Figure 5.6 Operational and design statistics of special protection schemes (Reproduced by permission of Cigre from [5.11])

- successful operation–an operation of the scheme that achieves or betters the performance objective of the supply systems (number = n_1);
- failure – a scheme operation that (i) fails to prevent or minimize the effect of a disturbance for a contingency of severity equal to or less than specified, or (ii) should not have occurred, but results in or contributes to a disturbance (number = n_2);

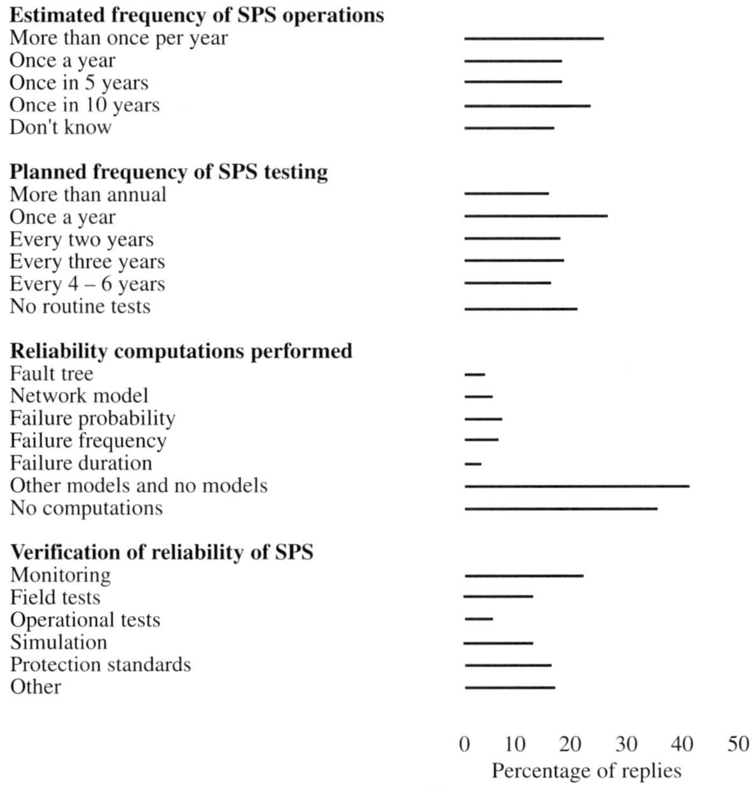

(b)

Figure 5.6 *(continued)*

- unsuccessful operation – a scheme operation that fails to prevent or to minimize the effect of a disturbance of greater severity than that specified in the scheme design (number $= n_3$);
- unnecessary operation – a scheme operation that should not have occurred (e.g. human error) but that did not result in or contribute to a disturbance (number $= n_4$);
- effectiveness index = measure of the extent to which the scheme achieves its purpose $= \dfrac{n_1}{n_1 + n_2 + n_3}$;
- dependability index = measure of the extent to which the scheme achieves its design performance $= \dfrac{n_1}{n_1 + n_2}$;
- unnecessary operation rate $= n_4/$(number of years of operation).

Values of the indices for the schemes included in the CIGRE/IEEE survey are as follows:

n_1 (successful operations) = 1093

n_2 (failures) = 36

n_3 (unsuccessful operations) = 20

n_4 (unnecessary operations) = 306

From these, the 1992 performance indices were calculated for generation rejection and load rejection schemes (Table 5.6).

Table 5.6 Performance indices of special protection schemes

effectiveness index for generation rejection schemes	= 95 percent
effectiveness index for load rejection schemes	= 78.6 percent
dependability index for generation rejection schemes	= 97.1 percent
dependability index for load rejection schemes	= 85.8 percent
unnecessary operations per year for generation rejection schemes	= 0.54
unnecessary operations per year for load rejection schemes	= 0.09

5.4.3 Prevention of Overload and Instability

Typically, generation is reduced when the power flows from the points of connection exceed the firm transmission capacity. This elementary concept is discussed at some length, since it is the basic logic used in numerous SPS. Referring to Figure 5.7(a) (for one substation) and Figure 5.7(b) (for a group of substations), $\sum G$ is the aggregate generation, $\sum L$ the aggregate demand, $\sum R$ the aggregate transfer from the remainder of the network (excluding the circuits under consideration), and T_n, T_{n-1}, T_{n-2}, etc. the aggregate transmission capacity of these circuits, with all circuits available, with the most critical (in

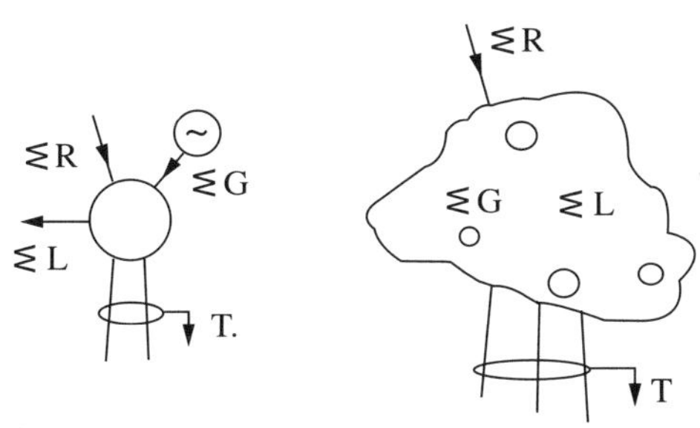

(a) (b)

Figure 5.7 Substation and group parameters for security criteria

terms of transfer capacity) circuit not available. With the two most critical circuits not available, the various conditions which may obtain will be one of the following (the net transfer will be $\sum G - \sum L + \sum R$):

(a) Security criterion – all circuits required to meet maximum transfer (n criterion).

$$\text{If net transfer} > T_n, \text{ reduce } \sum G \text{ and/or } \sum R \qquad (5.11)$$

(b) Security criterion – all but one (the most critical) circuits required to meet maximum transfer (($n-1$) criterion)

$$\text{If net transfer} > T_{n-1}, \text{ reduce } \sum G \text{ and/or } \sum R \qquad (5.12)$$

(c) Security criterion – all but two (the most critical) circuits required to meet maximum transfer (($n-2$) criterion)

$$\text{If net transfer} > T_{n-2}, \text{ reduce } \sum G \text{ and/or } \sum R \qquad (5.13)$$

In these criteria, the transmission capacities will be the minimum of those determined from thermal or stability considerations using the appropriate security criterion. This is not a trivial task. It can involve the following steps:

(1) For the adopted security criterion, choose the base generation, demand and transfer condition to be used (e.g. the peak or the minimum demand condition).

(2) List the system configurations to be analysed.

(3) For each of these, list the conditions to be analysed. If an ($n-2$) criterion is being used, this could, for instance, involve $\{1 + n + n(n-1)/2\}$ conditions (that is, the 'all circuits available' case, the n 'single outage cases' and the $n(n-1)/2$ 'double outage cases'). There is no guarantee that satisfying all double outage cases means that the single outage cases are satisfactory, as demonstrated in Figure 5.8. If both circuits *e–f* and *g–h* trip as in Figure 5.8 centre, the flow in the other connections will increase to $T + 2P$, satisfactory if a double circuit criterion is being followed. If however only one, *e–f* trips and the impedance of *e–f* and *g–h* is relatively small compared to that of the other connections, the flow in *g–h* will increase to approximately $2P$, possibly overloading that circuit.

(4) Reverting to the main computational sequence for each of these, determine the maximum transfer at which there are no overloads and/or stability is just maintained.

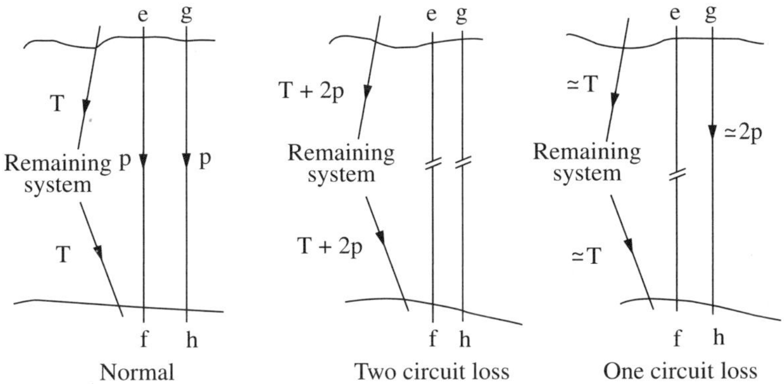

Figure 5.8 Poor redistribution of flows following a circuit outage

(5) The largest of these in each set of cases will be the T_n, T_{n-1}, T_{n-2} for comparison with the net transfers in the conditions (a), (b), and (c) above.

The commonest actions will be to adjust $\sum G$ or $\sum R$, possibly by tripping generation when the potential overload or instability is detected.

The setting parameters of such schemes will typically relate to the parameters $\sum G$, $\sum R$, T, and any necessary actions, for example can be presented in tabular form as follows ($\bar{T}$ is the maximum value of T set by thermal loading and/or stability considerations):

$\sum R$	$\sum G$	$\sum L$	Corrective action
Up to $\bar{T}$	Any value	Any value	None
Above $\bar{T}$	Any value	Any value	Reduce $\sum G$

5.4.4 System Applications of SPS

Several SPSs ranging from the simplest single substation to complex area schemes are described in the following pages. These include examples reported from the mid-1980s to the present. They have been selected based on the availability of information, and to give a round view of the application and usefulness of SPS in the planning and operation of power systems. Their inclusion is not in any way a reflection on the planning or operational practices of the utilities involved.

Improvement of Transformer Utilization when Fault Level Constraints Exist

Improved utilization can be obtained by increasing the number of transformers operating in parallel at transforming substations, but the rupturing capacity of the switchgear at the lower voltage will place a low limit on this number – quite often only two. A compromise between utilization and fault level duty can be achieved by automatic switching of the transformers, generally using the lower voltage circuit breakers as illustrated in Figure 5.9. A transformer utilization (total load/total transformer capability) of 50 percent is achieved if there is no connection between the two busbars (Figure 5.9(a)). This can be increased to 75 percent if the busbars are paralleled, but the h.v. switchgear fault rating (taken as that which would be supplied over two transformers) would be exceeded (Figure 5.9(b)). The utilization can be increased to 67 percent at three transformer substations, but still keeping within the switchgear fault rating, if two transformers are normally on load with automatic switching to switch the third transformer into service if one of those on load should trip (Figure 5.9(c)). If switchgear rated to interrupt the fault infeed from three transformers is used in conjunction with automatic switching, a utilization of 75 percent can be achieved (Figure 5.9(d)).

The United Kingdom – North Wales [5.13]

North Wales is a small area of some 9000 km^2 on the western side of England. Located close to the industrial North West and Midlands, its amenity value is

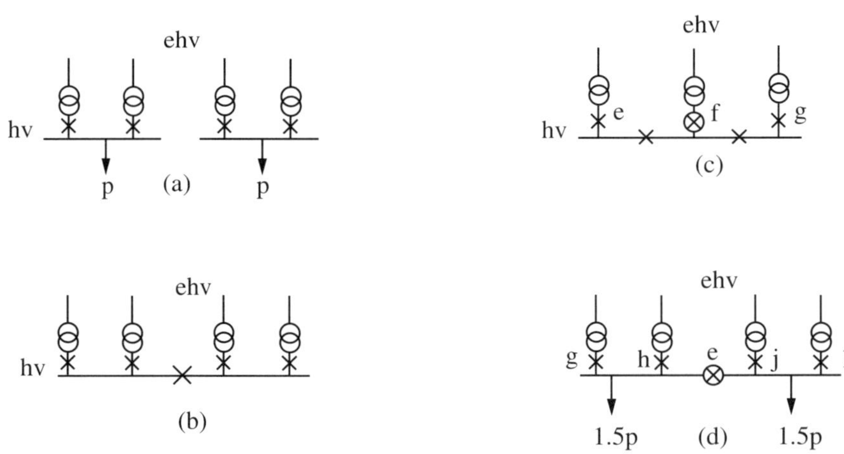

Figure 5.9 Fault levels, transformer utilization and automatic switching at ehv/hv substation. (a) Transformer capability = P; Max load per busbar = P; Max substation load = 2P; hv fault level = two transformer infeeds; transformer utilization = 50 percent; (b) Max substation load = 3P; fault level on hv switchgear (four transformer infeeds) above switchgear rating; transformer utiliza-tion = 75 percent; (c) Max substation load = 2P (close breaker 'f' if breakers 'e' or 'g' open); hv fault level = two transformer infeeds; transformer utilization = 67 percent; (d) Max substation load = 3P (close 'e' if 'g', 'h', 'j' or 'k' open); hv fault level = three transformer infeeds; transformer utilization = 75 percent

high. Geographical features and location led to the construction of two hydro stations (≈2200 MW), and two nuclear stations (≈1220 MW) in the area in the 1960s and 1970s, adding to the existing small hydro and diesel capacity (some 50 MW). The outlets to the rest of the system consisted of two 400 kV double circuit lines, each some 80 km long, much over mountain and moorland. These lines could be subject to severe weather conditions.

The then configuration of the local system is shown in simplified form in Figure 5.10. The demand at Wylfa was much smaller than the capacity of the station. To improve security to the 132 and 33 kV system, these lines were normally operated in parallel with the 400 kV circuits.

Potential problems found or anticipated with this system and the solutions adopted or proposed were as follows:

(1) Depending on the generation at Wylfa, tripping of both circuits of the Wylfa–Pentir line could lead to overloading, and possible instability, of the lower voltage Wylfa–Pentir system. The solution adopted was to intertrip 132 kV and 33 kV circuits within this system (in practice at three points, denoted '*a*' in Figure 5.10) to interrupt this lower voltage path. This would leave the local demand isolated on the Wylfa generation, reducing the load rejection imposed on the station. If the generation was insufficient to meet the demand, the isolation points '*a*' would be moved closer to Wylfa.

(2) Tripping of the Pentir–Deeside and Pentir–Trawsfynydd 400 kV circuits would leave generation at Wylfa and Dinorwig, plus any small generation and minus local demand to flow to the main 400 kV system over the 132 kV and 33 kV networks between Pentir, Trawsfynydd and Deeside. Depending on the amount of generation, pole slipping would occur, and to prevent this, the lower voltage network would be opened at points '*b*'. Without further action, the Wylfa and Dinorwig machines would accelerate rapidly, reactor gas circulator motors at Wylfa would stall and trip, and the Wylfa generation would be lost. To prevent this, the Dinorwig-via Pentir–Wylfa connection had to be opened, leaving Wylfa supplying its own local load.

    Other scenarios had to be analysed (e.g. Wylfa generating and Dinorwig pumping), but the central theme was to minimize disruption at Wylfa.

(3) Similar problems to those of Wylfa and Dinorwig would be encountered for the Trawsfynydd and Ffestiniog stations, with the loss of the Trawsfynydd–Pentir 400 kV circuits. A similar scheme was implemented.

(4) A fault resulting in the loss of the both Pentir–Deeside circuits whilst either the Trawsfynydd–Deeside or Trawsfynydd–Legacy was switched out would leave only the Pentir–Trawsfynydd–Deeside or Legacy circuits connecting the North Wales stations to the remainder of the system. Depending on the

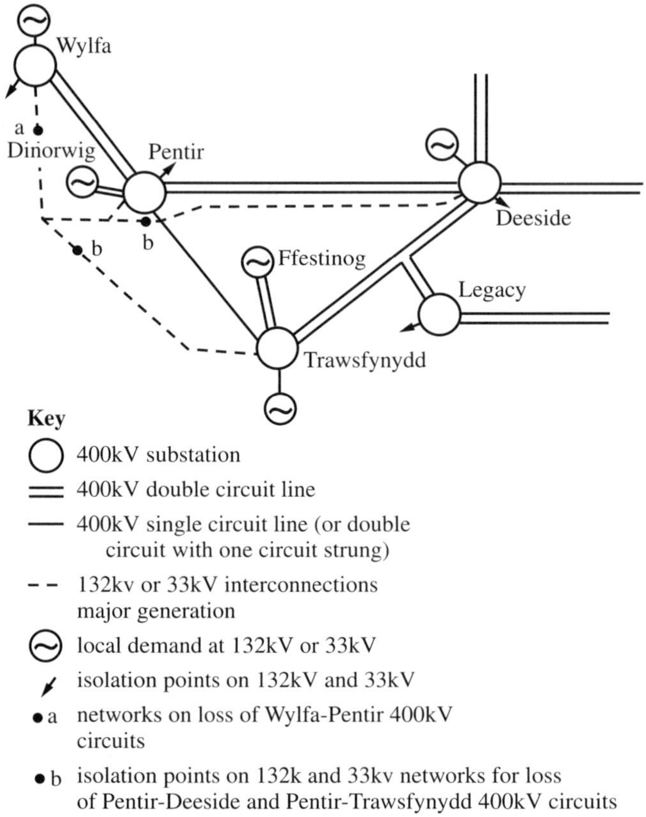

Figure 5.10 North Wales system in simplified form – example of automatic switching

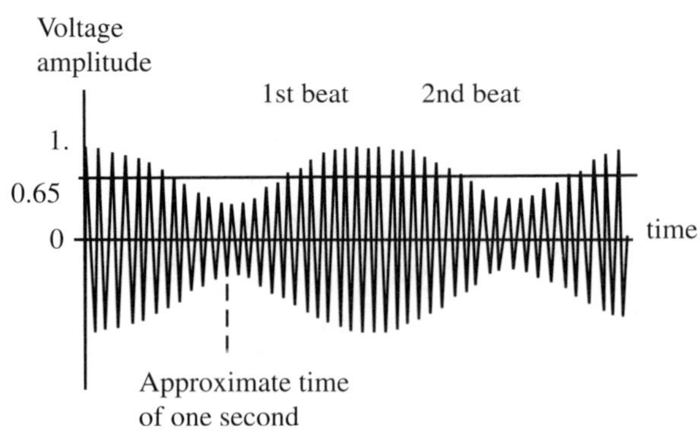

Figure 5.11 Voltage beats indicating potential loss of synchronism in the DRS scheme (see Reference 5.14)

generation/pumping conditions, instability and overloading of this remaining connection could result. The situation would be alleviated by automatically tripping machines at Dinorwig. The scheme involved monitoring the generating/pumping power at Dinorwig and the status of the Pentir–Deeside circuit. When a potentially critical configuration/loading condition was detected, an intertrip signal would be sent to Dinorwig, normally arranged to trip two machines to ensure that the remaining transfer into or out of the station would not exceed 1250 MW. The selection of the machines to trip and the arming of the scheme would be done by the operator. The maximum time from fault inception to opening the station circuit breakers was assessed at 180 msec.

(5) Other components of this SPS included pole slipping protection and underfrequency protection.

This example illustrates the potential complexity of some SPSs, and the attention to detail needed to ensure their satisfactory operation.

France – Co-ordinated Defence Plan [5.14]

Electricité de France (EDF) has been a leader in the application of automatic schemes to section a large system into a number of isolated parts in the event of a severe disturbance. The defence plan (known as the 'DRS' plan), in use for about 35 years, comprised load shedding and system sectioning. Four stages of load shedding were used, each of about 15 percent, at the medium voltage level at frequencies of 49, 48.5, 48 and 47.5 Hz. The operating time was about 0.2 sec. Pumped storage in the pumping mode was tripped at about 49.5 Hz. Thermal units were also isolated on to their auxiliaries if frequencies too low for reliable operation were reached. System sectioning was initiated on the detection of voltage beats, a symptom of loss of synchronism caused by generators operating at different speeds. Sectioning points were located at the boundaries of groups of generators which analysis indicated would tend to swing together. Equipment to detect and count the beats (DRS relays) was located at these points, normally set to operate between one and four beats (Figure 5.11).

This system operated satisfactorily on numerous occasions, but also had problems – each of the DRS relays operated autonomously, and hence the circuit trippings were staggered (detrimental to the overall operation). In some cases, the islanding of regions out of synchronism was not sufficient to prevent the incident spreading; it could in fact worsen the situation. The detection and counting of beats was not a very selective way of determining the isolation points.

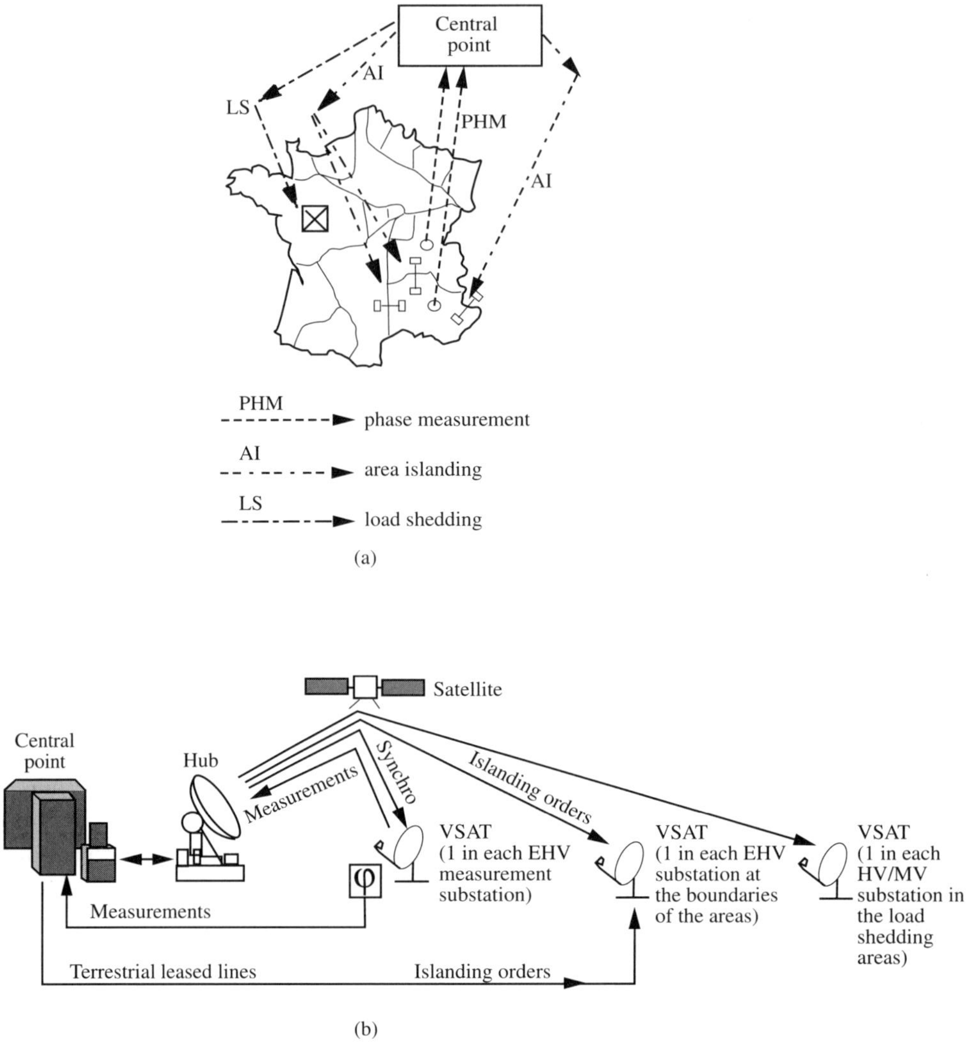

Figure 5.12 A special protection scheme to increase transmission export capability (see Reference 5.14, reproduced by permission of Cigre))

The replacement system also aims to preserve the integrity of the system by system splitting and load shedding. The logic is based on a comparison of voltage phase angles measured in elementary areas of the system, and telemetered to a decision making centre (Central Point (CP)) in Figure 5.12. There, the phase measurements are compared about 20 times per second and orders telemetered to trip circuits to island areas out of synchronism and/or to shed load, this according to power-balance in the areas and a phase angle criterion.

Very high integrity is specified for the system – a mean time between false operations greater than 1000 years and a probability of correct operation of 0.999. The response time of the system (loss of synchronism to islanding/load shedding) should be as short as possible, and 1.3 seconds was anticipated. Clearly, communications could be critical and a satellite based system was selected to transmit system phase measurements between the outstations and the CP.

Canada – Load and Generation Rejection at a Major Power Station

This scheme was installed in the 1980s to minimize the impact of delays in completing transmission on the operation of a large nuclear power station. Four 800 MW nuclear units were to be added to the Bruce generating complex of Ontario Hydro, bringing the total capacity up to 6400 MW. The planned addition of a two-circuit 500 kV line was, however, to be delayed for several years, and the non-availability of this line would, with the normal security criteria, result in the station output being limited by 3000 MW, at a cost of $1 billion. A scheme to reject four generating units would allow the station output to be increased and the cost penalty reduced to $175 million. However, as the system could only withstand a generation rejection of some 1500 MW safely, it would also be necessary to reject 1500 MW of demand simultaneously with the generation [5.15].

The operational requirements to be met by the 'Bruce LGR' (Bruce Load and Generation Rejection) scheme included:

(1) In accordance with the security criteria of the North East Power Co-ordinating Council, in simplified terms, the system should be operated so that under normal conditions it would withstand a double-circuit fault, and under 'emergency' conditions (i.e. demand would otherwise be interrupted) it should withstand the loss of any single element. Both ampacity and transient stability limits should be considered.

(2) Only the minimum number of generators and minimum amount of demand to maintain viable operation should be tripped.

(3) The demand-generation balance should be restored within ten minutes.

(4) Power-flow limits should be available to shift staff at all times.

(5) All demand should be restored within 30 minutes, assuming the rejected generators had tripped successfully to house load and were immediately available for generation, and that not more than one 500 kV circuit was permanently faulted.

(6) Following a contingency, the system should be rapidly returned to a state able to withstand a further single circuit loss (the post-contingency fault criterion).

The ampacity viability was determined on-line, comparing present flows and flows following a credible contingency to the ampacity circuit ratings. In regard to transient stability, the system at the time consisted of some 35 circuits and six busbars, with two or more elements out of service for 45 percent of the time. It was felt that the number of possible configurations would be so large as to make off-line calculation and storage of all possible cases infeasible, and hence a hybrid off-line/on-line scheme was implemented. Some of the salient features of this scheme were:

(1) Displays showing the operation of the scheme:

- the Bruce units and generation which had been rejected,
- the demand rejected,
- the system frequency and area control error,
- the voltages before and after operation of the scheme,
- out of limits values identified in the monitoring programs with poke point access to the alarm messages.

(2) Generation resources available for system restoration.

(3) Recommendations for restoration, updated as transmission was restored, together with a prognosis on outcome. Manual rotational load shedding would be used if it was anticipated that requirements would not be met.

(4) A programme for load restoration would be implemented as the area control error was projected to reach zero. This recommended the demand to be picked up at minute intervals, based on the projected trajectory of the area control error.

(5) Pre-defined limits were stored in DACS (the Ontario Hydro SCADA and energy management system) for the power system with all elements in service, and for the system with any one or two elements removed. Special limits were stored for more than one element out of service. Comparison with stored results was made in terms of the system connectivity, rather than breaker and isolator states, thereby considerably reducing the number of conditions to be stored. If no equivalent could be found, for a post-contingency case, this was reduced to a single branch whose impedance was compared to that of the same branch in the normal (non-outage) case.

The transient stability limit was then taken as $(Z_{normal})/(Z_{outage}) \times$ limit normal$\times\kappa$, where κ was less than one to ensure a conservative limit.

The degradation in system reliability which would occur with the introduction of this scheme was estimated and considered acceptable in view of the large operational savings that would be obtained.

Canada – a Generation Rejection Scheme to Increase Transmission Export Capability

This scheme was implemented in the mid-1980s on a medium sized utility – the New Brunswick Electric Power Commission [5.16]. The scheme was somewhat unusual in that generation rejection was used to release transmission capacity for through power flows. The outline structure of the relevant parts of the interconnection of that time are shown in Figure 5.13. The Maritime interconnected systems were connected via d.c. back-to-back connections rated at 700 MW total to the Hydro Quebec system, and via a single circuit 345 kV circuit to the NERC interconnection (New England, Ontario, north east USA). This interconnection

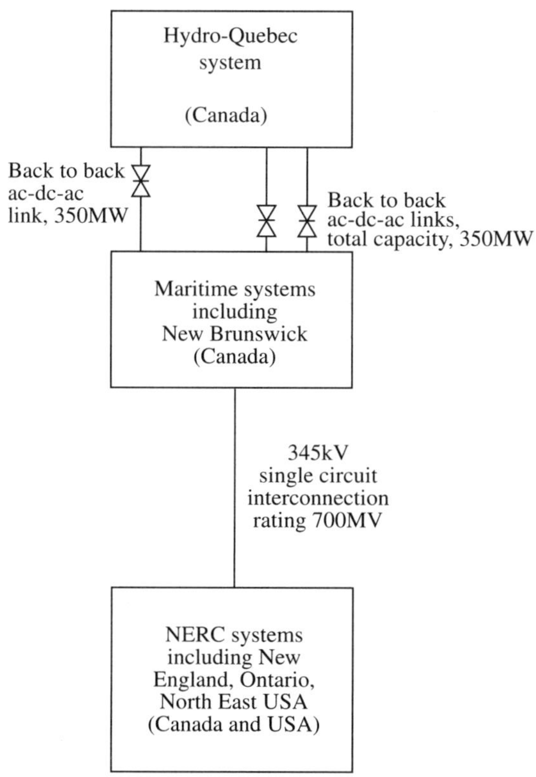

Figure 5.13 Special protection scheme in Canada

was rated at 700 MW, and tripping of the line when carrying this export would result in a generation surplus of 25–50 percent of the Maritimes demand. Studies and operational experience had indicated that a separation from New England with a generation surplus greater than 10 percent of the demand would cause overspeed, instability and system collapse. The alternative to limiting the export flows to some 300 MW at peak Maritime demand or 150 MW at minimum demand was a generation rejection scheme, which would allow exports up to the 700 MW limit, irrespective of the demand in the Maritimes. There was also a need to trip Maritimes generation to prevent overload and/or instability, and tripping of the interconnection as a result of major demand loss in the Maritimes or system disturbances there or in New England.

On detection of a sudden reduction in export flow over the interconnection and a frequency rise (to above 60.3 Hz) in the Maritimes, the d.c. import from Hydro Quebec would be reduced and selected generation in New Brunswick tripped. The figure of 60.3 Hz was considered to be a level below which governor action could be relied on and above which generation rejection would be necessary. Detection of coincident high active and reactive power flows implied that the system was in an abnormal state, and was used to initiate reduction of d.c. import or tripping of New Brunswick generation. The generation rejection schemes were armed manually by the system operator, the amount and units being set hourly in a specified priority order.

Japan – an On-Line Transient Stability Control System

This dual computer system was designed to perform transient stability studies on-line at about five minute intervals, using the as-telemetered state of the system [5.17]. If potential instability on the occurrence of a fault was detected, generation would be rejected. It was to be implemented on a 500 kV system in the vicinity of Tokyo (Figure 5.14). The functions within each of the boxes of the outline flowchart in the figure are:

(1) *System model* – the power system configuration and circuit parameters are assembled from the telemetered state indications and the stored database.

(2) *State estimation* – derives the complete and consistent load flow.

(3) *Network reduction* – reduces the lower voltage system to an equivalent generator-circuit-load.

(4) *Case screening* – uses a simplified calculation method to separate the cases into ‘probably stable’ and ‘doubtful’ categories.

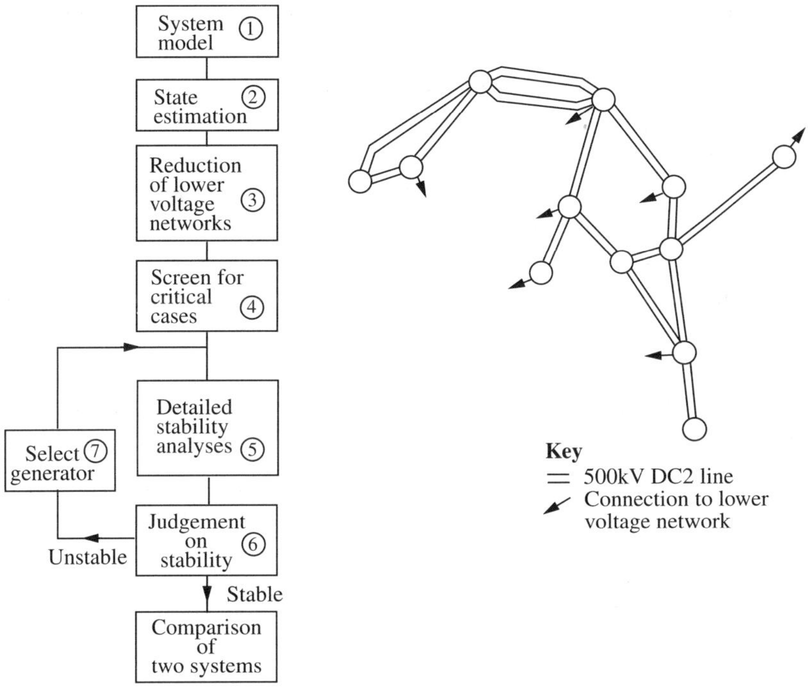

Figure 5.14 Online transient stability system in Japan

(5) *Detailed stability calculation* – a detailed Parks model is used to assess the transient stability of the system. AVR and saturation effects are included.

(6) *Judgement on stability* – this package assesses whether the transient oscillations will decay after a few seconds.

(7) *Generator selection* – selects the best generator to trip to ensure stability in the event of a fault. The choice may be a compromise between the one most effective towards assisting restoration of stability or the one capable of fastest restoration. Steps 5, 6 and 7 are iterated until a 'system stable' judgement is reached.

(8) *Fail safe comparison* – the results of the two computer systems are compared. Although hardware redundancy is provided in most computing systems, it has been relatively unusual to find as in this case different software in the two systems.

The computing system was sized to deal with power systems of up to 100 generators, analysing up to 100 contingencies on lines, buses and transformers in about five minutes.

Russia – Special Protection Schemes on the Unified Power Systems

It is judged from the literature that, whereas special protection schemes have been installed by many utilities on an *ad hoc* basis, their use has been more systematic in the territories of the former USSR, where they have been used as a planned alternative to plant capacity [5.18]. They are referred to collectively as the 'UPS (Unified Power System) emergency control system'. It has been developed on a four level hierarchy with as its main purpose localization of the emergency and prevention of its spread to neighbouring parts of the system, as follows:

- first level – these comprise the local devices which operate directly during emergencies (protective relays);
- second level – this is the central complex of an emergency control region; adjustments are determined to the first level devices in the pre-fault conditions;
- third level – co-ordination of the second level complexes and if necessary the settings of the first level devices are adjusted to cope with inter-regional emergencies;
- fourth level – co-ordination of the third level complexes throughout the whole UPS of Russia, with, if necessary, adjustment of the first level devices to handle inter-area faults.

The automatic schemes provided include out-of-step protection, under-frequency load shedding, tripping to house-load, and generation start up and loading. Voropai *et al.* [5.18] state that high reliability is obtained, faults on the main grid and supergrid resulting in outages of 5–6 system minutes in 1997, with no system collapses for many years. Another statistic suggests that 0.014 percent of total generation was lost from all blackouts, which is equivalent to some 30+ system minutes.

5.5 REDUCTION IN THE SPREAD OF DISTURBANCES

This is a more complex problem than reducing the risk of disturbance, and has perhaps engendered more recommendations, and remedial work. Reports from many disturbances indicate that the features essential to reasonable containment of a disturbance are:

(1) Rapid clearance of faults so that:

    (a) the operating conditions of power stations are not so disturbed by abnormal frequencies and voltages as to result in loss of generation;

(b) sequential tripping of transmission circuits, leading directly to loss of demand or to islanding, is avoided.

(2) Rapid achievement of sustainable system conditions, implying that

(a) generation and transmission conditions conducive to further loss of transmission through overload or instability are avoided;

(b) there is equality between demand and generation, both on the whole system and within any islands, which have been formed as a result of loss of transmission;

(c) adequate information to system and station operators is provided, in spite of any loss of normal power supplies for instrumentation, telemetry and communications.

5.5.1 Rapid Clearance of Faults

Some principal effects of slow fault clearance on station performance can be identified as: tripping of generators on overcurrent, under- or over-voltage; underfrequency; negative phase sequence protection (the latter with unbalanced faults) and possibly overspeed; tripping of station auxiliary drives on for instance under-voltage; and various maloperations of station auxiliaries. Some of these measures are taken for protection of the generating plant itself from damage, and in these cases, the main additional safeguard for system purposes is to ensure that the generating units should be capable of running stably, supplying only their own auxiliaries. Testing and training of station staff to achieve this is practised by some utilities (see Chapter 8). Independent emergency power supplies for essential station auxiliaries provide the other main safeguard, and these of course can also provide 'black-start' capability.

Apart from the effect on generation, prolonged fault clearance times can cause disruption of the network through tripping of healthy circuits on zones 2 and 3 of distance protection, and on backup overcurrent and earth fault protection. Clearance of faults in this way may well lack discrimination or even guaranteed clearance if high power flows in normal operation have to be allowed for, or if there are significant infeeds between the point of fault and the positions of the backup relays. One question is whether forms of protection providing more certain clearance of the 'stuck breaker' fault should be widely provided.

5.5.2 Sustainable Conditions Following the Initial Fault Clearance

The manual and automatic actions necessary to achieve a viable operating state must be put in hand as soon as the fault is cleared. Care must be taken that these actions do not themselves precipitate problems, for instance the sudden changes in output power to which the generation is subjected should be kept within its capabilities. Static and dynamic ratings of plant must be observed.

5.5.3 Restoration of Normal Conditions

Once the system state has been stabilized, manual or automatic actions must be taken to restore the system to as near normal, in terms of voltage and frequency conditions and amount of demand supplied, as the available plant capacities permit. This subject is considered at some length in Chapter 7, and only the main points will be mentioned here.

- If at all possible, the cause of the initial disturbance should be determined early in the restoration process; this to ensure that the conditions which contributed to the disturbance are not repeated.
- Supplies at normal voltage and frequency either from neighbours or from still healthy sections within the disturbed area are of great value in providing focal points from which to build up the system.
- Care should be taken to avoid overloading the remaining transmission and generation as the system is restored. 'Make haste slowly' is applicable.
- Staff with the technical responsibility for restoring the system should not at the same time have to deal with the media and external bodies (e.g. government).
- Plans for restoration of the system, or for major parts of it, from a dead state should be prepared. These should be discussed with operational staff and practices held.
- It may be necessary in severe disturbances to bring in staff from elsewhere in the utility, for instance operational planners and planners, to support the control staff. Such staff should be made aware of these possible duties, and be given information on what would be involved.
- If the damage to the system is such that repair will take days, as might follow flooding, hurricanes or system-wide gales, support might be sought from or offered by neighbouring utilities. Outline plans to accommodate, feed and deploy such staff should exist.

5.6 MEASURES TO MINIMIZE THE IMPACT OF PREDICTABLE DISTURBANCES

The author uses the term 'predictable disturbance' (not generally found in the literature) to describe those upsets to normal operation which are anticipated, and which will involve changes to established procedures. The warning period for predictable disturbances can vary from a few minutes to weeks even months. The duration of the disturbance again will vary from minutes to months. Table 5.7 gives examples from opposite ends of the spectrum of events.

The measures which can be taken to minimize the effect of predicted natural events will be discussed first, followed by a brief comment on plant breakdown, and then a review of the technical and organization steps available in the event of labour unrest.

5.6.1 Natural Phenomena

The potential end effects of earthquakes and other natural phenomena on the power system will be system faults, and disconnections with in the more severe cases physical damage to plant. However, warning of the event may give enough time for precautions to be taken, the general objective being to reduce possible dependence on supplies judged most at risk. As examples, if heavy lightning storms were anticipated, power transfers via overhead lines from more distant generation would be replaced by local generation (probably at increased cost, otherwise the local generation would have been running in the first case). The output from stations subject to flooding would be decreased. With more warning; it might be possible to adjust operating conditions so as to minimize the risk of faults in the anticipated ambient conditions. For instance, a long period of below zero weather could leave overhead line insulators coated with a deposit of ice and dirt. On thawing, a conducting film would be formed, leading to flashovers and faults. In anticipation of this, the operating voltage of the ehv network could be reduced by 5–10 percent in advance of the predicted time of the thaw, thereby hopefully reducing the incidence of faults. This strategy would require studies to establish the technical viability of operating the system at the lower voltage, including simulation of the method of bringing the system to that voltage.*

5.6.2 Incipient Breakdown of Plant

Generation and transmission plant is fitted with monitoring devices to warn of impending problems or breakdown [5.18]. These include winding and oil temperature thermometers and oil level gauges on transformers, oil temperature and level and pressure gauges on cables, oil level and temperature gauges on bushings. Thermal plant will be fully instrumented with temperature and mass flow instrumentation for operational as well as condition monitoring purposes. Rotating plant will often have vibration sensors attached. The outputs from these

* The author does not know whether this has actually been done. Voltages have been lowered as a method of reducing demand during periods of plant shortage.

Table 5.7 Warning periods, durations and spread of various types of disturbance

Event	Range of warning period	Duration of event	Spread
Lightning	Minutes to hours	Tens of minutes (in any location)	Localized but moving
Storm	Minutes to hours	Tens of minutes to hours	Square kms
Tornado/cyclone/ twister/whirlwind[1]	Hours (path may be uncertain)	Usually less than one hour over any area	The 'footprint' of a tornado is not likely to be more than a few 100 meters wide and its length some tens of kilometres
Hurricane	Up to a few days	Hours	Many km^2 (diameter say 500 km). Wind speeds can be in excess of 300 km per hour
Earthquake	Normally very short	Minutes	Many km^2
Floods	Usually hours but very occasionally minutes	Up to days	Depends on location, from a single valley in mountains to many km^2 in flood plains
Incipient plant breakdown	Hours to weeks	Plant repair time	Localized
Labour problems within the supply industry	Often weeks	Generally days or weeks, but can extend to months	System-wide
Labour problems outside the supply industry	Often weeks	Generally days or weeks, but can extend to months	The spread is likely to be system-wide, but the impact will be in particular areas

[1] These are alternative terms used in different parts of the world.

devices will be transmitted to a plant room on the site, and if the site is not manned, the more important measurands will be re-transmitted to a convenient manned site. Depending on the importance of the monitored plant, including its location on the system, data for major items of plant may be re-transmitted to the system control centre.

In general, the system operator will take whatever steps are necessary to disconnect the suspect plant items from the system and isolate it for inspection. The timescales in which this should be done may be stated in the system operational memoranda, for instance it may be required to disconnect wound type voltage transformers within a few minutes of the operation of the alarm.

5.6.3 Labour Problems*

The pervasive need for electricity throughout modern life, plus the fact that it can only be stored at the point of consumption in kWh quantities at most rather than the GWh quantities required by consumers, means that the industrial 'muscle' held by the modest number of workers within the industry is considerable. The industries supplying the raw materials, mainly fuel, will also have some share in this muscle but because their products can be stored, the impact will be less.

The measures which can be taken to contain possible industrial action are discussed below. In summary, these are to diversify suppliers, stockpile, modify operating procedures, and in the extreme, to ration supplies to consumers. The applicability of each of these depends upon which group of workers are taking the action, whether inside or outside the industry, the attitude of workers in related industries and on the type of action. An 'overtime ban' and 'work to rule' often precedes a full withdrawal of labour. These actions could mean that there would be no flexibility in operating shift rotas, no temporary transfers of duty, no call-outs from home, and no overtime. During such a period, the workers' income will probably fall (no overtime), as will the companies' fuel and other stocks. Both sides will have the opportunity to test the resolve of the other. Four short-term effects will be certain whether limited or full industrial action is taken: the workers' income will decrease; the resources and stockpiles of the companies affected will fall; the operating costs of the companies will rise; and maintenance plans will be disrupted.

Problems in Industries Supplying Raw Materials

The problem faced by supply companies in these circumstances is to maximize their endurance, that is, the time for which they can continue to supply

* Acknowledgements are made to Messrs Ledger and Sallis, from whose book *Crisis Management in the Power Industry, an Inside Story*, some information for this section was obtained [19]. The book was published in 1995, and reprinted in 1995 by Routledge.

consumers in spite of interruptions to the flow of raw materials, fuel in particular. The industry will have the options of increasing imports from neighbours, finding alternative sources of fuel, changing type of fuel, adapting operating regimes to maximize the electricity produced from the available fuel, or to ration electricity supplies. As an example, an 'efficiency based' merit order can be used, that is, one in which generation is biased towards stations with high efficiencies (electrical energy out/fuel energy in).

In practice, the scope for increasing electricity imports will often be limited. Typically, normal imports will only be a few percentage points of the maximum demand, effectively placing an upper limit on achievable imports. The use of alternative sources of fuel is likely to be more promising; this may only involve an increase in normal transport patterns, for instance, of coal from Australia to the UK in the 1980s. It has the advantage that some stations will be accustomed to burning the imported fuel.

It may be possible to convert some stations to an alternative fuel. In preparation for an anticipated strike by coal miners in 1984/85, the UK's Central Electricity Generating Board (CEGB) increased its oil and gas generation capability from just under 8 GW in March 1984 to nearly 18 GW in 40 weeks. This was done by modifications to lighting up oil burners, fitting oil pipes and pumps at some large coal stations, improving oil handling at others, and returning to service or deferring closures at a number of smaller stations. The oil burn was increased from 62 000 tonnes per week at the beginning of the strike to a maximum of 550 000 tonnes per week some 45 weeks into the strike. The decrease in oil burn when the strike finished was equally rapid, some 350 000 tonnes in almost one month.

Modify Operating Procedures

Subject to the overall fuel situation, generation can be biased towards stations with high energy conversion efficiencies (electrical energy out/fuel energy in), and the operating margins of individual generators adjusted to maximize their efficiency. As an aside, this could have a substantial impact on the policy for holding system reserves.

Turning to the consumption of electricity, this can be reduced by rationing supplies or by lowering voltage or frequency levels. The latter will yield small economies, but without much impact on consumers; the former can give much greater reductions in energy consumption, but with much more effect on consumers. In either case it may, as was so in the UK, be necessary to obtain governmental dispensation to reduce voltage and/or frequency below statutory limits, and to disconnect consumers deliberately.

Disconnection and other forms of reduction of consumption can be implemented in a variety of ways:

- banning some less essential uses of electricity (advertising displays, flood-lighting at sports events, etc.);
- reduced 'comfort' levels in offices and shops, etc.;
- banning use of electricity totally to industrial consumers on selected days;
- piecemeal disconnection of residential consumers;
- this can be replaced by 'rota disconnection' (which can be paraphrased as 'equalising the misery') if a prolonged supply shortage is anticipated; in an application experienced by the author at home, each consumer was allocated to a supply code. Local papers published the days and the four-hour periods within these days on which the different supply codes might be disconnected, together with an expected risk of disconnection.

Technical as well as organizational problems have to be overcome when implementing such schemes. For instance, the necessary switching would probably be done on the distribution network both to minimize the perceived effect by consumers at large, and to prevent abrupt changes in the demand. However, it might be necessary to increase staffing in the distribution network to achieve this.

Apart from withdrawal of labour, more direct action in the shape of secondary picketing might be used by the striking work force. Typically, the objective of this would be to prevent delivery of commodities essential to operation of stations; these could include lighting up oil, hydrogen and water treatment chemicals. The quantities involved would not be negligible in the order of tonnes per week, and in some cases, delivery by helicopter might be used.

Disruption of heavy transport can also hazard fuel supplies, although the magnitude of the problem will largely depend upon the usual arrangements. If rail transport is normally used, its partial or total withdrawal can be met by road transport. If, on the other hand, the normal transport is by road, it is unlikely that rail could completely replace it because there would be locations not accessible by rail links. The potential of road transport workers to disrupt everyday life was shown by the fuel tanker drivers in the UK in autumn 2000.

Problems Within the Industry

Subject to any sympathetic action, such as happened on a small scale during the 1984/85 miners' strike in the UK, it should be possible for the supply industry to adopt strategy and tactics which will maximize its endurance in the event of strikes external to the industry. This may not be the case for labour problems within the industry, when the short-term objectives and actions of the industry at large, and of the workers involved, can be directly opposed. It is also possible that it would be more difficult to take specific and short-term preparatory action.

Thermal generation will probably be more susceptible to industrial action than any other power source, the effects being a cumulative shortage of plant capacity, and inflexible operation of the still available plant. As always the imperative will be to balance the demand and generation. The main control mechanism will be adjustment of demand to use the generation available, albeit with poor frequency control or, if the utility is part of an interconnected system, poor control of transfers with neighbours.

Industrial relations within the British supply industry have historically been good. Ledger and Sallis [5.19] record only five occasions in the last 75 years when there have been actions in the UK – in 1926 during the General Strike, in 1949 when workers at a small number of stations in Greater London went on strike for a week, in 1970 when a ban on overtime and working to rule was implemented at many area board depots and power stations for one week, in 1973 when the Electrical Power Engineers Association (the union to which a large percentage of professional engineers in the industry belonged) placed a ban on out-of-hours work, and in 1977 when an unofficial ban on overtime and work to rule was implemented for one day, and then a few weeks later for some 18 days by industrial staff in some power stations.

The 1973 action resulted in the shutdown of 5000 MW of plant and load shedding by voltage reduction, late return of plant to service, and to a higher coal burn because of reduced output from nuclear stations. The availability of plant was carefully managed by the staff taking action so that, although considerable difficulties were caused to the CEGB, there were no disconnections, the impact on consumers being limited to voltage reductions. In the 1977 action, 4500 MW of usable output was lost on the first day of the second period. As a result, there were 5 percent disconnections of supply in each Area Board lasting between 11 and 25 minutes. The rota disconnection system was implemented, and on the worst day of the action, disconnections up to three hours were applied throughout the country. Load reductions at peak were sometimes over 20 percent. The action was unofficial, and members of the Electrical Power Engineers Association continued to operate some stations.

Support from Neighbours

It has already been noted that transfers between neighbours are often relatively small in terms of maximum demands in most systems, and often well below the capacity of the transmission. It seems, therefore, that the support available from neighbouring systems during industrial actions is more likely to be determined by financial and contractual considerations, and the attitude of operational staffs in the utilities, than by technical issues. The situation could be further complicated in the case of support between non-contiguous neighbours, with energy transfers occurring through third parties.

5.7 AN APPROACH TO MANAGING RESOURCES

It will be apparent from the earlier descriptions that the crucial factors in maximizing endurance will be comprehensive monitoring of fuel and other essential stocks, an endurance model to estimate the period for which electricity supply can be maintained for various assumptions on the input of resources and patterns of supply, and mechanisms to implement agreed (and perhaps unusual) operational procedures. Suggested preparatory tasks are summarized in Table 5.8 and those appropriate during the action in Figure 5.15. The preparatory tasks have been divided into physical measures and organizational measures. The former are to put the system into the best possible shape to withstand a potentially long period of resource deprivation. The detail of the latter will depend upon the form of the industrial action, in particular, whether it is internal or external to the industry and, if internal, which group/s of workers are in dispute.

The endurance model is a vital component of the procedures. The objectives and detail of such models will vary between utilities, and often be commercially confidential, but one may infer some characteristics from the circumstances in which they will be used and from the history of past industrial actions. A comprehensive model will have several objectives:

(1) To assess the impact of the different policy options which might be adopted by the workers in dispute, including when to act and for how long, the type of action (overtime ban or withdrawal of labour, etc.), and the system functions to be involved (generation, transmission, distribution, control, etc.).

(2) To determine targets for stock levels and stocking.

(3) To outline the salient features of day-to-day operation, including projections of endurance.

(4) To provide a summary of operation during the action.

(5) As part of this model or elsewhere an estimate of the ongoing operational costs.

The core of the model will be a loading simulation algorithm. This will have to be very flexible in view of the abnormal operating situations which may occur, and it is judged that this will be achieved more easily using a period-by-period simulation than a convolution method. Not least, convolution would involve

Table 5.8 Some suggested preparatory tasks

Physical Measures	Organization Measures
(a) Increase stocks of materials expected to be in short supply	(a) Agree or confirm endurance targets
(b) Improve availability of all plant as far as possible, including termination of maintenance programmes	(b) Update the various models to suit the expected changes in operating conditions and constraints
	(c) Clarify the situation with major consumers and agree procedures for reducing energy and power demands
	(d) Clarify the situation with neighbours with respect to expected trading and support patterns
	(e) Clarify the situation with suppliers of services such as fuel transort and material supplies
	(f) Advise staff on possible patterns of work

constructing load duration and generation output–fuel consumption histograms, not easy tasks in the circumstances of unusual operating conditions.

Logically, it would seem that the main difference between simulation for normal and for emergency conditions will be that emergency simulation will need to be more flexible, for instance

- unusual generation and demand patterns, and hence the possibility of unusual transmission constraints, unusual demand profiles in respect to both magnitude and shape;
- different fuel sources and fuel characteristics;
- different fuel routes;
- possibly unusual plant availabilities.

The repercussions of these changes can generally be foreseen when the loading simulation program is specified.

Additionally, and perhaps not usually included in loading simulation programs, the stock and deliveries of other essential commodities such as water treatment chemicals, hydrogen, carbon dioxide, lubricating oils, etc. must be monitored.

Summing up, simulation for emergencies will be more constrained – easier for manual treatment, but more difficult for mathematical optimization. The ques-

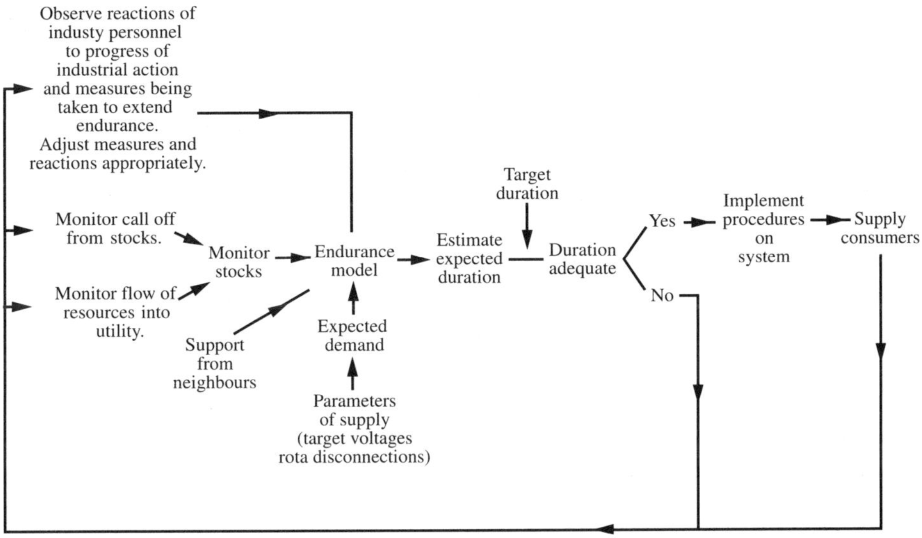

Figure 5.15 Suggested procedures for managing resources during a period of industrial unrest.

tion is, how much mathematical optimization (e.g. maximum endurance) should be attempted against straightforward simulation?

5.8 THE CONTROL CENTRE

System control facilities have been discussed in Chapter 4. The comments below on those particularly needed for control in emergencies have been extended to include emergencies affecting the control structure itself.

5.8.1 SCADA

The main impact on the SCADA system is likely to be the frequency at which SCADA information must be acquired and processed to be fully valuable to operators. (It may be noted that there will not be a lot of impact on the amount of data to be collected.) Representative figures from several surveys are quoted in Table 5.9.

It will be appreciated that the extreme condition would only occur during an emergency on the system, and that sizing the SCADA system for such events will in fact dictate its processing power. The correspondence between the level of emergency and SCADA facilities will not be so definite in regard to staffing and number of operator positions in the control room. These will be determined by the need to handle normal switching duties expeditiously, as well as switching for emergencies.

Turning to displays it is judged that emergency display needs will be largely met by those provided for normal operation. A few exceptions will be a system split display to indicate if the system is operating in two or more disconnected sections, and an 'extended frequency' display. This will show system frequencies within a range of nominal ±6 percent, say, as against a normal display of nominal ±2 percent. A refinement of this will be to show frequencies on a semi-geographical system display. Undoubtedly, the main feature which it can be argued is provided to assist in handling abnormal situations is the mimic diagram, often free-standing and constructed of small plastic tiles so that it can be modified in line with changes in the power system. These diagrams are costly (equivalent to several VDU displays), and will play a big part in dictating the size and layout of the control room. It is also quite common to drive the on-

Table 5.9 Some SCADA response targets

Specification A	
System activities specified	normal, high peak
Response targets	
Status update, normal	1–5 sec
Status update, high peak	1–5 sec
Measurand update, normal	5–10 sec
Measurand update, high peak	5–10 sec
Specification B	
System activities specified	steady, high, peak, overload
Response targets	
Display response, steady	1 (0.5) secs
Display response, high	1.5 (0.75) secs
Display response, peak	2.0 (1.0) secs
Display response, overload	2.0 (1.0) secs
(Note 1(0.5) = expected response (standard deviation))	
Specification C	
System activities specified	normal, emergency
Analogue sampling rate, normal	6 sec
Analogue sampling rate, emergency	10 sec
Status update	these are included in the analogue cycle on occurrence
Some responses in practice	
Measurands – 1.5 to 20 sec (most in range 2 to 10 sec)	
Status – 2 to 24 sec (most in range 2 to 5 sec)	
Alarms – 2 to 50 sec (most in range 2 to 15 sec)	
Typical cycle times – 5 to 10 sec.	

line information on mimic diagrams independently of VDU displays, and from different sources of data on the system.

The above deals with normal emergencies, that is those which occur as a result of the usual operational hazards. Those which develop from man-made circumstances, for instance industrial action, will often engender displays aimed at monitoring critical aspects of the particular class of emergency, and will usually include stock holdings and rates of replacements of essential materials. The costs of developing these displays will be allocated to the emergency.

Telecommand and a.g.c. control will be provided as normal control facilities, and only minor additions should be needed to handle emergencies. These could include a demand shed instruction, whereby a single telegraph instruction at the control centre would enable demand to be disconnected at numerous selected substations simultaneously. The maximum generation changes allowed in a.g.c. installations are quite small, and the bigger changes needed during emergencies will be obtained by telephone discussion between the control centre and station operators.

5.8.2 Main, Standby and Backup SCADA/EMS Systems

It is common practice to duplicate or more SCADA and EMS systems within one control centre building. Some utilities go further than this and provide backup centres. These would only be activated in earnest if as a result of some major incident both main and standby systems were unavailable. Often the facilities will be rudimentary, comprising indications of frequency and some strategic voltages and line flows, with heavy dependence on telephones for communication with outstations. Their location needs careful consideration, 'near but not too near' the main centres. Their costs, including the associated maintenance and the communications and data links, should be allocated to emergency control.

5.8.3 Communications

Utilities need very secure (physical rather than content) communication links for data and speech. Some will be system-wide and some only local, between neighbouring substations, for instance. Basically, the security of communications can be improved by reducing dependence on any one carrier. For instance, the utility can itself provide channels over its own equipment (power line carrier, pilots with cables, etc.), or it can hire channels from external carriers such as public communications networks or industries which have invested in widespread communication networks, such as railways. It is suggested that only system-wide communication and data networks, probably low speed and of limited capacity provided for use when the main operational systems are not available, should be charged as an emergency control facility.

It is quite possible that operating procedures would have to be modified to accommodate limited facilities. For instance, generation changes might be restricted to fewer stations, the others being held at fixed outputs, or station output targets might be revised less frequently, with stations ramping between these targets without further instructions. Switching might be minimized by deferring outages for maintenance and new construction. Nonetheless, the costs of emergency communications systems could be considerable.

REFERENCES

5.1. *Frequency Control Capability of Generating Plant*, *IEE Digest* 1995/028.

5.2. Rowen, W. I., 1995. 'Dynamic response characteristics of heavy duty gas turbines and combined cycle systems in frequency regulating duty'. *IEE Digest* 1995/028.

5.3. Grein, H. L. and Jaquet, M., 1984. 'Operation flexibility of various designs of pumped storage plant'. *Int. Symposium and Workshop on Dynamic Benefits of Energy Storage Plant Operation*, US Dept. of Energy/EPRI, Boston.

5.4. Carvalho, F. L., 1986. 'Nuclear power plant performance in power system control: a review of international practice'. *Cigre Paper 39-14*.

5.5. Concordia, C., 1968. 'Design of electric power systems for maximum service reliability'. *Cigre Paper 32.08*.

5.6. Ashmole, P. H., Battlebury, D. R. and Bowdler, R. K., 1974. 'Power system model for large frequency disturbances', *Proc. IEE*, **121**(7).

5.7. Ward, R., 1987. 'System performance requirements'. Opening remarks at IEE Discussion Meeting on Emergency Load Shedding Requirements in the UK and Associated Low Frequency Relaying Techniques.

5.8. Symons, O. C., 1985. 'Automatic disconnection of load at low frequency'. *CEGB Report TPRD-ST/85/001/R. CEGB, UK.*

5.9. Concordia, C., Fink, L. H. and Poullikas, G., 1995. 'Load shedding on an isolated system'. *IEEE Trans. Power Systems*, **10**(3).

5.10. *Modern Power Station Practice*, Vol. L, Chapter 4.

5.11. Anderson, P. M. and LeReverend, B. K., 1996. 'Industry experience with special protection schemes (IEEE/Cigre Report)'. *IEEE Trans. Power Systems*, Vol. II(3).

5.12. Winter, W. H. and LeReverend, B. K., 1989. 'Operational performance of bulk electricity system control aids'. *Cigre Electra*, **123**.

5.13. Harker, K., 1984. 'The North West supergrid special protection schemes'. *IEE Electronics and Power*.

5.14. Trotignon, M., Connon, C., Maury, F. *et al.*, 1992. 'Defence plan against major disturbances on the French EHV system: present realisation and prospects of evolution'. *Cigre paper 39-306*.

5.15. Winter, W. H. and Cowbourne, D. R., 1983. 'The Bruce load and generation rejection scheme'. *Cigre-IFAC Symp. on Control Applications for Power System Security*, Paper 207-03, September.

5.16. Patterson, W. A., Jensen, B. M., Picot, T. J. and Brown, G. W., 1985. 'Generation rejection scheme increases transmission capability for power exports'. *Cigre Study Committee 39 Meeting*, Paper EM 85.05, Toronto.
5.17. Chubu Electric Power Company, *On-line Transient Stability and Control.* Brochure.
5.18. Voropai, N. E. *et al.*, 1998. 'Reliability in the restructured Russian utility industry'. *IEEE Power Engineering Review.*
5.19. Ledger and Sallis, 1995. 'Crisis Management in the Power Industry, an Inside Story'. Routledge.

FURTHER READING

Jabeeli, N., Van Slyck, L. S., Ewart, D. N. *et al.*, 1991. 'Understanding automatic generation control'. *IEEE PES Winter Power Meeting*, Paper 91 WM 229-5-PWRS.

Anderson, P. M. and LeReverend, B. K., 1994. 'Industry experience with special protection schemes'. IEEE/Cigre Working Group 39.05. *Electra.*

Logeay, Y., Jeanbart, C. and Musart, M., 1988. 'EDF simulator for control centre operators'. Cigre paper 39-12.

Kundur, P. *et al.*, 1998. 'Power system disturbance monitoring: utility experience'. *IEEE Trans. Power Systems*, Vol. 3(1).

6
The Natural Environment – Some Disturbances Reviewed

6.1 INTRODUCTION

Nature imposes an environment on power systems which man can influence in the long term – usually it seems for the worse – but hardly at all in the short term. In spite of their relatively short duration, the abnormal, extreme weather and other environmental conditions will determine many of the plant and system design criteria. Hence, it is important to have an appreciation of the 'bullets' which Nature may fire. The first part of this chapter reviews these, mainly qualitatively. The second part describes some of the disturbances which have actually occurred. The author has found that the reviews of large scale disturbances attracted more interest than most other topics when discussing emergency control – a case of 'there but for the grace of God go I'. However, although there is a basic pattern common to the evolution of many disturbances, as illustrated in Figure 3.3, within this there are very many alternatives, to the extent that it is difficult to select 'typical' cases. Thus, rather than describe a few incidents in detail, a rather larger number have been summarized, concentrating mainly on significant features in the initiation and spread of the disturbance, and its restoration and the lessons learnt. The criteria used to select the incidents are technical, organizational and operational features, complexity and the amount of information available. Only published incidents have been included.

A common format which includes the sources of the information has been adopted in the descriptions. The disturbances are listed by geographical area: Europe and the Middle East, Scandinavia, the Far East and Australasia, and America.

6.2 USEFUL SOURCES OF INFORMATION

Extreme environmental conditions are reported in the press and, with particular attention to electricity supply, in the technical press and journals. There have in the past been some five main sources of published information on disturbances, ranging from government or similar commissioned reports to single line entries in annual reports. These are summarized below.

6.2.1 Government and Similarly Sponsored Inquiries

Occasionally, the impact of a disturbance has been so large that government has been prompted to set up an inquiry into causes and proposed actions to prevent a recurrence, for instance the blackout on the eastern seaboard of North America in 1965, and the failure of supply to Auckland, New Zealand in 1998.

Inquiries at this level are wide-ranging, backed by all the technical resources of the utilities involved, and quite often bring in external consultants. The main report is usually complemented by technical reports, and will be available to purchase from government/utility sources. As an aside, verbatim reports of operators' conversations can make fascinating reading!

6.2.2 Utility Inquiries

Utilities often set up inquiries into major incidents, particularly if loss of supply is involved. The inquiry will have ready access to the utilities' technical and other resources, with the work done by a multidisciplinary team under a senior manager. Summaries of reports will be prepared for the press. In the course of time, technical reports often appear in the technical press.

6.2.3 Annual Reports

Annual reports of utilities sometimes include summarized information on fault performance. Very large incidents may be given a separate page!

6.2.4 International and National Surveys

Surveys made by international societies such as Cigre and published in their journals or at conferences are one of the best sources of information. The surveys are often made by utility staff serving on Study Committees or Working Groups. As such, they will have ready access to information. The published surveys tend to be short on detail, but point the reader towards more information. The societies sometimes commission individuals or groups to collate several papers into a 'brochure'.

By comparison, there are relatively few national surveys, probably because the number of disturbances in individual utilities and/or the study group structure do not permit this. North America is one of the exceptions, for instance the reports produced by EPRI.

6.2.5 The Internet

Many organizations have provided web sites. A few of those currently relevant in the power systems area are:

North American Electric Reliability Council	http://www.nerc.com/
Electricity Association	http://www.Electricity.org.uk
EdeF	http://www.edf.fr/
EPRI	http://www.epri.com/
FERC	http://www.fedworld.gov
IEC	http://www.hike.te.chiba-u.ac.jp/ikeda/ICE/home.html
IEE	http://www.iee.org.uk
IEEE	http://www.ieee.org
Virtual Library for Power Engineering	http://www.ece.iit.edu/~power/power.html/
US Department of Energy	http://www.doc.gov/
Open University	http://eeru-www.open.ac.uk/

Numerous manufacturers and supply companies have their own sites (see, for instance, *Energy Guide to the Internet*–Utility Data Institute/McGraw Hill, Schuman III, R. W. and Schapp, J. F., editors, 1995).

6.3 EXTREME ENVIRONMENTAL CONDITIONS

Although the perception of UK weather is that it is moderate, over the years some extreme conditions have occurred, to quote over some 60 years up to 1990:

- Severe gales–1927, 1953 (with the North Sea storm surge, 300 died in the UK, 1800 in the Netherlands), 1962, 1965 (two cooling towers collapsed at one

UK power station), 1968, 1969, 1976, 1987 (the 'Great Storm'), 1990 (47 die, gusts over 170 km per hour).

- Very heavy rain or snow – 1940 (freezing rain), 1943, 1947, 1955, 1968, 1975.
- Prolonged cold or snow – 1947, 1963, 1975, 1979.
- Tornado – 1950 (track some 200 km long across south Midlands, four dead).

It seems that the world is being increasingly subjected to large scale disasters. To quote isolated cases, February 2000 has seen severe flooding in Mozambique; 1999 a succession of tornadoes in Oklahoma (one with the fastest winds ever recorded on earth to that date); 1998 the flooding in Honduras and Nicaragua, wildfires in Florida, a tsunami in New Guinea and ice storm in North-east USA and Canada. The director of the Federal Emergency Management Agency (FEMA) in the USA is reported as saying [6.1] that

> *If, as many experts believe, we are entering a period of more frequent and more severe weather events, this year may be just a precursor, a hint of what is to come. ...Increased evaporation from the ocean caused by higher global temperatures is likely to increase the number and severity of floods, severe winter storms and mud slides. The shifting of rain events may bring widespread drought and an increased incidence of wildfires. The odds are that tornadoes and hurricanes will be more intense.*

6.3.1 Hurricanes [6.2][6.3]

Hurricanes are rotating tropical storms which usually originate between latitudes of 7 and 15 degrees north or south of the equator. They develop when rising air currents over warm oceanic waters create areas of intense low pressure. As the air spirals upwards, it cools and the water vapour it carries condenses rapidly, forming dense cloud and torrential downpours of rain. The latent heat released in this process further feeds the development of the hurricane. The centre or eye of the storm will be some 25–40 km across, with quite low internal wind speeds of about 24 km/hour. Typically, it progresses at about 16 km/hour initially, in a westerly veering to the north west/north east direction in the northern hemisphere, and a westerly veering to the south west/south east in the southern hemisphere. The strongest winds rotate around the eye, and can reach speeds up to 350 km/hour. Triggered by the low air pressure and spiralling winds, vast amounts of water are sucked from the sea and form huge waves and storm surges that can reach some eight metres in height, and cause severe flooding if they hit land. Hurricanes occur more frequently in the summer and autumn when the sea is at its warmest.

Hurricanes which originate over land can cause enormous damage and significant loss of life. Hurricane Gilbert (1998), with a severity factor of 5 (the highest possible), tracked from the Caribbean over Mexico and Texas, caused damage estimated at over $10 billion and killed over 350 people; this was small compared to Hurricane George, which devastated Honduras and Nicaragua in 1998. Hurricane Mitch, also in 1998, changed the geography of Honduras. Days of rain saturated the ground and hillsides collapsed. River courses changed and water levels were metres above normal. Some 6000 people died. Hurricane Hugo hit Charleston in West Virginia, USA in September 1989, affecting nearly 40 percent of the customers of three local supply companies, damaging 600 transmission structures, 1600 poles and nearly 28 000 distribution transformers. Hurricane Gloria blew through the service area of Long Island Lighting Company in 1985, interrupting power to 80 percent of the utility's customers for up to 11 days. Hurricane Andrew cut across southern Florida in August 1992 causing damage estimated at some $25 billion. Such severe hurricanes seem to be occurring more frequently.

As a general comment, it may not be possible to provide economically plant such as overhead lines guaranteed (more or less) to withstand the effects of such storms. Hence, much effort is placed on recovery measures.

6.3.2 Tornadoes

A tornado (sometimes called a twister, whirlwind, waterspout, landspout or dust devil) is a spinning funnel of air formed within a stormcloud which builds up so much energy that it bursts out of the cloud mass and its tip may touch the ground. The worst tornadoes can have wind speeds up to 800 km/hr, lateral speeds of over 100 km/hr, and some reach diameters over 1.5 km. Severe tornadoes occur in the American mid-west, and 'storm chasers' use cars with radar tracking equipment both to study their development and to track their path. Some 30 tornadoes occur annually in Great Britain, occasionally causing structural damage. Tornadoes form or move over water, and occur most frequently over the Gulf of Mexico and adjacent land areas, and the west coast of Africa (small ones have been experienced in the UK). Dust devils or whirlwinds occur when tornadoes form over hot dry land. The spinning funnels of flying dust, sand and debris obtain their energy from the heat of the ground, and usually occur over the arid regions of the USA, Australia, India and the Middle East, although they are found worldwide. In some areas, these are frequent events and can last up to several hours: some 700 per year, for instance, in the Gulf of Mexico and up into the USA. Although the energy released in a tornado is far less than that in a hurricane, the energy density at the point of contact with the earth is several times higher than that in a hurricane, this accounting for their high, but localized, damage potential. Typical damage caused by a severe tornado would be a swathe of destruction some kilometres long and tens of metres wide, where the tip of the vortex has touched down to earth. People have been injured and killed by the direct effect of the wind, and also by flying objects which it picks up.

As far as is known, the impact of tornadoes on electricity supply installations has been limited. Although damage has mainly been to overhead distribution systems, there have been instances of 500 kV and 230 kV steel towers being blown down. Tornadoes are graded by the damage they cause on a scale of F0 (minimum) to F5 (maximum). Some estimated wind speeds are F1 up to 180 km/hr, F3 300 km/hr and F5 510 km/hr.

6.3.3 Gales

Gales may be less dramatic than hurricanes and tornadoes, but are longer lasting and potentially more widespread. Effects which have been observed in some gales have been:

(1) duration up to several hours; perhaps days if sequential gales move across a system;

(2) wind speeds gusting up towards 200 km/hr;

(3) systems at all voltages are susceptible to damage. That at ehv is mostly caused by flashovers to tower steelwork, sometimes conductors clashing. The majority at lower voltages is caused by falling trees and flying debris resulting in broken conductors and snapped poles;

(4) the distribution of faults is not uniform, some circuits suffering multiple faults;

(5) automatic reclosure of circuits will be a major help in maintaining the integrity of a system when it is exposed to gales causing frequent faults over many minutes.

6.3.4 Hail, Snow and Icestorms

Although hailstones can, very rarely, weigh towards 1 kg, have killed and injured people and annually cause millions of dollars worth of damage to property, there have been few reports of problems caused to the supply industry. This may be because, although frequent, hailstorms are localized phenomena of short duration. In contrast, snowstorms can have widespread and prolonged impacts, ranging from damage to overhead lines (mainly distribution) to interruption of staff movements and fuel supplies by road and rail. It seems likely that utilities in countries such as the UK, where severe and prolonged snow conditions are infrequent, will be disrupted more than those which experience such conditions every year. Some utilities use helicopters to ferry personnel and material.

Ice storms can coat structures and overhead conductors with ice causing problems with weight loading, possibly aggravated by winds. One of the most severe occurred in Canada/North east USA in January 1998 [6.1]. It was caused by a stream of warm air, unusual so early in the year, flowing under very cold air at higher levels. As the warm air rose, it turned to ice which coated all external surfaces, including overhead lines, as it fell to earth. Ice up to 10 cm thick formed on lines, and its weight and the wind forces snapped wooden poles, crumpled steel towers and brought transmission and distribution lines to the ground. One USA utility lost over 85 percent of its transmission and distribution infrastructure in the affected area; in one region over 95 percent of the customer base lost power. There were 70 transmission lines out, and 200 transmission structures damaged. Losses exceeded \$125 million. The impact on the community was severe. Schools shut down, postal services were halted and businesses closed, some permanently. In the absence of power, petrol stations without backup generators could not deliver fuel, houses were dark and cold, food spoiled in inoperative freezers, and homes depending on wells using electric pumps lost water supplies. Some people could not obtain money when cash machines ceased to operate. Many restaurants were unable to assist because they themselves were without food.

This utility deployed more than 4000 workers into the field. Transmission and distribution repairs were made simultaneously so that the repair crews could be fed and sheltered. Emergency generators with a capacity of some 17 MW were deployed. Power restoration took 23 days, with residual repairs continuing for several weeks.

6.3.5 Earthquakes and Tsunamis

The worst earthquakes can be short, sharp and vicious. It seems, however, that whilst the general damage caused by an earthquake may be widespread, that inflicted on a particular utility can be quite limited in geographical area, and its impact will depend critically upon where the utilities facilities and the earthquake zone coincide. The earth's surface rests on a small number of tectonic plates. These are about 120 km thick and float on molten viscous rock. Earthquakes occur where the plates abut, forming well defined earthquake prone zones principally and approximately east to west across southern Europe, Turkey and Asia Minor; across northern China, northern India and eastward/southwards down through Indonesia and the South China seas into the Pacific, Japan and east across the Northern Pacific to North America and southwards towards

Indonesia and New Zealand; and the whole of the west coast of the Americas (the San Andreas fault). Earthquakes usually occur in two stages – a first, main shock followed hours or days later by a secondary shock. The secondary shock is usually of low intensity, but may nevertheless aggravate the damage of the primary shock and cause new damage. Their strength is measured on the Richter scale of one to nine; this is logarithmic in powers of ten, for instance:

Richter scale	Structural damage
Less than 3	May not be felt but will be detected by instruments
4 (mild)	May cause cracks in walls
Under 6	Possibly slight damage to well designed buildings, but major damage to poorly constructed buildings
6–6.9	Can be destructive over largish area
7–7.9 (strong)	Numerous buildings destroyed. Can cause serious damage over large areas
8 and over (great earthquake)	Can cause serious damage over hundreds of kms.

The damage and death toll in an earthquake can be strongly dependent on the integrity of buildings in the area affected. Fortunately, power system civil works will be constructed to a high standard which will go some way to explain why these seem to survive well in these conditions.

Earthquakes which occur under the sea can produce devastating tidal waves, known as tsunamis, capable of causing immense damage if they hit land. Tsunamis can travel many thousands of kilometres across the ocean before subsiding, at speeds of over 150 km per hour, with wavefronts 30 metres or more high. They have been responsible for some 50 000 deaths in the last century, but fortunately the really destructive ones occur quite infrequently. A centre to warn of the approach of tsunamis has been set up in Hawaii. Precautions are also taken on the west coast of the USA and a warning system established. Loss of life may be averted if people in low-lying areas adjacent to the sea are moved inland to higher ground.

6.3.6 Vegetation Brushfires

The most serious environmental risks from vegetation are judged to be flashovers to trees and fires e.g. [6.4]. The former are controlled by monitoring tree growth and pruning programmes, keeping branches sufficiently far from overhead line conductors to eliminate risks of flashover, even when windy. Coastal areas may be prone to current leakage and flashover of insulator strings caused by wind blown salt contamination.

Smoke from brushfires has caused flashovers when blown across overhead lines, and could presumably contaminate insulator strings.

6.3.7 Thunderstorms, Lightning and Overvoltages [6.5][6.6]

High voltage systems should operate satisfactorily both under normal operational conditions, but also be able to withstand without permanent damage the transient overvoltages sometimes experienced. These will be principally caused by lightning and switching operations, the former predominating on systems of 100 kV and under, and the latter on systems from 300 kV upwards. (In addition abnormal voltage conditions may sometimes occur as a consequence of non-linear electromagnetic phenomena.) Insulation strength and characteristics of the system components must be selected to reduce the frequency of supply interruptions and component failures from overvoltages to levels low enough to be operationally and economically acceptable.

Lightning is a transient discharge of electrical charges developed within the atmosphere. Its commonest source is the thunder cloud formed when packages of warm, moist air rise through cool air. The rising air expands as the air pressure falls with increasing height. It cools adiabatically and condensation occurs at the dewpoint temperature. The water droplets freeze as they are carried higher on the ascending air currents. Electrical charge separation occurs and the cloud in general becomes positively charged in its upper regions, negatively charged elsewhere. The electrical fields within the cloud increase as the charges build up until the air insulation strength is exceeded, when breakdowns occur – between clouds, within clouds, or between cloud and earth. These latter discharges may affect electrical plant.

In the temperate zones, lightning discharges or flashes are usually negative. Negative charges are first lowered to the ground in a downward leader strike, followed by a return strike up the leader channel. As many as 40 strokes may be found in one flash, although the average number is three. The maximum current occurs in the return stroke ranging between 1 kA and up to 200 kA.

The lightning surge protection needed on a power system will obviously depend upon the lightning activity in its area of supply [6.5]. One measure is the thunderstorm day number (T), that is the average number of days per year on which thunder is heard in a given locality. Values in the UK range from under 3 to up to 21, average 9. A more significant parameter for the power system engineer is the average number of flashes to ground (per square kilometre per year (Ng)). The effective 'collection area' of a structure will also be important, that is the 'footprint' of the structure (usually an overhead line) within which the strike will terminate on the structure. For purposes of comparison and standization, the waveforms to be produced by impulse generators have been standardized by the IEC.

Overvoltage transients can be caused on an overhead line by induction when a stroke discharges to earth close to the line, or by direct contact when the stroke discharges to any part of the line. A strike to the tower of a line with earth wires will usually be innocuous because the voltage rise caused by the lightning current flowing through the tower to earth impedance will usually be below the withstand strength of the line insulation, although back flashover (tower to phase conductor) can occur. A strike to a phase conductor will usually result in a voltage rise sufficient to cause flashover with, hopefully, successful auto reclose of the faulted circuit.

Estimates of the proportions of strokes causing flashover are given in the reference quoted above. From these it appears that tower footing resistance is, as would be expected, very important in determining performance against strikes to towers (e.g. a doubling can increase the percentage of strokes causing flashover several times; this percentage decreases rapidly as the system voltage increases). For strikes to phase conductors, the proportion of strokes causing flashover is small (e.g. 0.9 percent at 400 kV), and decreases slowly as the system voltage rises.

System Generated Overvaltages

Although somewhat out of place in the general plan of this book, it seems appropriate to consider system induced overvoltages here. These can reach considerable amplitudes, for instance on interruption of inductive or capacitative currents, demand rejection and circuit switching, particularly when trapped charge is present. These transients are increasingly important as the system voltage rises. Their salient features are [e.g. 6.6]:

- they may be oscillatory or aperiodic, most often damped oscillations superimposed on the power frequency wave;
- the oscillation frequencies are typically in the range 0.1 to 10 kHz;
- the amplitudes range from 1 p.u. to some 3 p.u., even above 4 p.u.

The overvoltages may be truly transient as with inductance or capacitance switching, or temporary, lasting many cycles even minutes. Some causes of these are

(1) Load rejection on a simple series generation complex/line/demand complex. The Ferranti effect can cause a substantial and steady state voltage rise defined by circuit parameters, by leakage over insulation and by transformer saturation. Voltages above 1.3 p.u. can result for long lines.

(2) If an earth fault occurs on one phase, the maximum voltages on the healthy phases of a three phase system will be 1.4 p.u. on a solidly earthed system and $1.2 \times \sqrt{3}$ p.u. on a Peterson coil earthed system.

(3) Steady state resonance in which inductive components of a series circuit (as in (1)) resonate at some harmonic with the shunt capacitance of the line. Measurand voltages of the order of 1.4 p.u. have been observed. Resonance may also occur for harmonics present in transformer inrush currents.

(4) Reflection of a voltage wave at the remote end of an open circuited line.

(5) The presence of a trapped charge left on a line when it is opened. Leakage from the line, over insulators or through permanently connected equipment such as a voltage transformer will discharge this but any remaining charge will add to the impressed voltage when the line is reclosed.

The Control of Overvoltages

Fundamentally, overvoltages are controlled either by equipping the system with devices whose resistance drops from a very high to a very low value at some threshold value, or by operating the system so as to avoid the phenomena which result in internally generated overvoltages, or by installing devices to control post switching voltage rises [6.6].

The simplest device is the co-ordinating gap consisting normally of a rod/rod geometry air gap, often fitted across the bushings of transformers. If a transient overvoltage appears on the transformer terminal of amplitude sufficient to cause the gap to flash over, the voltage collapses to zero and the transformer winding is protected. Unless the flashover occurs at the zero of the supply voltage, a power arc will develop causing power frequency fault current to flow and tripping of the affected circuits. The flashover voltage of the gap (the impulse strength) is strongly polarity dependent. It may need to be determined by test with the actual geometry on the gap arrangement, including adjacent surfaces.

The surge arrester is a more advanced approach to the limitation of surge voltages in that the arrester 'seals', after operation, there is no follow through power current and hence no circuit breaker operation for fault clearance. This is achieved through the use of material often mainly zinc oxide, with a very non-linear current voltage relationship $I = V^n$ where n may be as high as 40. Judged from the literature, the choice of surge arresters will require the following information:

(1) The rated voltage, that is the maximum power frequency phase-earth voltage (the type of neutral earthing will affect this).

(2) The magnitude and waveshape of the most severe discharge current likely to occur.

(3) The withstand strength of the protected insulation (normally designed to a standard level plus safety factor).

(4) The service conditions, including the type of equipment protected.

(5) The position of the arrester in relation to the protected equipment.

Alternatively a statistical approach may be attempted in which the flashover voltage distribution function of the insulating element is convoluted with the overvoltage probability density function to obtain the risk of insulation failure. The mathematical concept is the same as that used in estimating the probability of insufficient generation capacity.

Co-ordinating gaps and surge arrestors will be applicable for the control of atmospheric or system induced overvoltages. The latter can also be contained by other mechanisms some of which are outlined below (and except for the first one involve additional equipment)

- avoidance of switching actions which can lead to overvoltages, for instance energising a transformer feeder from the line end with the transformer unloaded. Overvoltages over three times nominal have been recorded in tests. In general, the circuit should be energized from the end which has the lower source side impedance, and yields the lowest receiving end overvoltages. This technique has been called 'best end' switching;
- resistor switching is successfully used to reduce transients on energising lines. The initial connection is via a resistor having a value somewhat less than the line surge impedance. The resistor is short circuited by the breaker main contacts after a few milliseconds;
- shunt (line to earth) connected equipment to drain away trapped charges;
- point on wave switching, in which the closing or opening of the three phases of a circuit breaker are timed to minimize switching overvoltages.

Overvoltages and the Operator

In practice, the responsibilities of the operator to avoid overvoltages will be limited. Computational aids provided in the control room are unlikely to include any for estimating overvoltages; it is more likely that precautions to avoid overvoltage conditions will be included in operational memoranda or standing switching instructions. An area in which there may be some overlap however is in the precautions to be taken if thunderstorm/lightning conditions are anticipated. These may include reducing transmission power flows by increasing the output of local generation, cancelling circuit outages for maintenance or construction, where possible reinstating plant already off load, and prohibiting staff access to outdoor plant.

6.3.8 Floods

Floods appear in various ways, from torrents of water pouring through narrow valleys draining high grounds (flash floods) to the tens of thousands of square kilometres of flooded plains found in, for instance, the Ganges delta in Bangladesh or in China. These floods can be caused by overspill from rivers and canals, perhaps compounded by the bursting of banks or levees, or by waterlogging of land which prevents natural drainage through the solid. The source of the water may be high rainfall, as in flooding of some tropical islands and, increasingly, in temperate climates as in Europe in Autumn 2000, or the release of accumulated precipitation as snow and ice melts. Occasionally, the sea has contributed to the havoc by high tides and/or on shore winds which have caused water surges to back up on to land, as occurred on the eastern coast of England in the 1950s.

Floods do not seem to have excited much comment with regard to electricity supply in the first world countries. There have been cases of substations, power stations and control centres being flooded, but it is thought recovery was rapid – days rather than weeks. It seems likely that the very nature of the primary installations, in most cases much of it above ground level, means that only the secondary equipment for protection, control and communications will be prone to damage. This can be dried quite rapidly with air blowers. The author has seen substations built with low surrounding walls to prevent the ingress of liquids, and presumably equipment could be built on rafts raised above the anticipated flood levels.

The 'El Nino' phenomenon has been much discussed in the 1990s as a possible cause for a perceived worsening of weather worldwide. It is the appearance from time to time of warm sea surface water in the central and eastern Pacific. 'Time to time' has been put variously at 4–7 years, 5–7 years, 5–6 years, etc. One of its main manifestations is the appearance of warm surface water off the coasts of Ecuador and northern Peru, sometimes northern Chile. It lasts about 12–18 months. Its severity varies: for instance, in a very strong event there will be extreme rainfall in Peru with flooding and destruction; in a moderate event there will be above normal rainfall and coastal flooding. Other suspected patterns are for India and Indonesia to be warm and also dry on the eastern fringes, the east and west seaboards of northern USA and of Canada to be warm, southern USA to be wet and cold and Brazil to be wet and warm.

6.3.9 Geomagnetic Storms [6.7][6.8]

Sunspot activity that leads to Solar Magnetic Disturbances (SMDs) on earth has historically increased and decreased in an eleven year cycle. The basic mechanism is that some of the charged particles emitted from the sunspots will cause perturbations in the earth's magnetic field. These cause differences in electrical potential up to 5–10 volts per mile of very low frequency (almost d.c.) along the earth's surface in a generally east-west direction. Quasi-d.c. currents will flow in the earth, and some will enter the power system through its neutral earthing points. The magnitude of these currents will depend upon location of the system and the earthing points, their orientation, circuit resistance and ground resitivity. The effects have been observed as far south as a latitude of some 55° in the UK, becoming stronger as the North Pole is approached. Scandinavia and significant parts of Canada and Russia lie above this latitude. There will be similar phenomena in the southern hemisphere, although there the only land masses at latitudes of 55° south and more are only partially inhabited.

The impact on the power system can be quite serious. If the star points of the ehv windings of ehv/lower voltage transformers are earthed, the quasi-d.c. currents flowing in the transformers may saturate their cores causing waveform distortion and reducing the phase voltages. (In one study a drop of 20 percent was estimated to occur on a 500 kV line during a severe SMD.) The harmonic currents can overload shunt capacitors, which may then be tripped by protective relays leading to a further voltage drop. Immediate and long-term damage has been caused to transformers as a result of overheating of core and windings. Effects have also been reported on SVCs.

Short-term, operational measures to alleviate the impact of SMDs have included reducing power flows, particularly on transformers believed to be at critical locations, increasing reactive power margins in anticipation of voltage problems, and using higher relay settings on shunt connected plant such as capacitors and SVCs. There would also seem to be the possibility of adopting configurations which would increase the impedance of the network to earth currents. Some equipment and system changes have included installing more series capacitors on lines, and fitting capacitors in ehv transformer earths, both measures aimed at reducing the flow of quasi-d.c. currents.

6.3.10 Disaster Control

Specific measures can be taken against some of the natural disasters discussed above, particularly where these are known from experience to strike in defined areas. In others, the effects are quite capricious, but in either case, the impacts may be so large that the only feasible actions will be to expedite recovery rather than to prevent the immediate damage. This often takes the form of organising mutual assistance between utilities, for instance national register and pools of spare equipment.

The United States has a strong programme in disaster control [6.9]. FEMA launched a national initiative (Project Impact) aimed at building up disaster-resistant communities in 1997. The insurance industry is sponsoring a Showcase Communities programme to demonstrate what communities can do to reduce their vulnerability to disaster and EPRI established its Disaster Planning and Mitigation Technologies Target. This covers 175 disaster related technologies. Future disaster related products will include a post-storm damage assessment system which will use remote sensing data from aerospace programs and computer imaging technology to provide pictures of damage in near real time. This should give views of transmission and distribution equipment related to the geographical surroundings.

Much of the work in disaster recovery relates to physical aspects, but EPRI is also pioneering Disaster Recovery Business Alliances. These are alliances of businesses in a community which work with chambers of commerce, local, state and federal government agencies, and with volunteer organizations to help the recovery of business markets after a disaster. Each will include the local power company.

6.4 NOTEWORTHY DISTURBANCES

Some actual disturbances will be described in the remainder of this chapter. These have been selected on a combination of availability of information, technical, managerial and operational interest, and societal impact. ‘Peter’s Principle’ (if something can go wrong it will) has also played a part. Much of the information used came from Cigre surveys of disturbances, was obtained in response to questionnaires circulated under the aegis of one of the Cigre study committees and was published in Cigre papers or brochures. The format of the questionnaire is first outlined in the following.

6.4.1 The Questionnaire

The questionnaire has been summarized in Table 6.1. Questions 1, 2 and 3 requested a description of the system in which the disturbance had occurred. Questions 4 to 8 dealt with the disturbance, including statistics on its magnitude, its evolution and restoration using figures attached to the questionnaire to illustrate these, factors judged by the utility to be important in the event, recommendations and actions emerging from the event and finally significant factors. The questionnaire includes figures (6.1, 6.1a and 6.2) on which to show the sequence of events and 6.3 the sequence of restoration.

Table 6.1 Outline of questionnaire for the international survey on major disturbances

1. Country and Company.
2. Main parameters of the system
   maximum demand to date
   maximum generation capacity
   transmission (voltages, circuit lengths, numbers of substations, interconnections to neighbours).
3. System control
   levels of control
   levels at which decisions (dispatch, switching, etc.) are made
4. Description of disturbance
   date and day of week, special day (e.g. public holiday)
   time
   severity (system minutes)
   general system conditions (demand, generation, import/export, system frequency, etc.)
   weather at the time.
5. Any significant abnormalities in own or neighbours' systems at the time.
6. Quantitative data on the disturbance (time duration, demand disconnected, generation disconnected, frequency variations, times to reparallel system if split(s) occurred, time to restore various levels of demand.
7. Description of main sequence of events in the development of the disturbance including reference to Figures 6.1, 6.1a and 6.2. Thus, in Figures 6.1 and 6.1a, lines between boxes show the sequence.
8. Description of main sequence of events in the restoration, including reference to Figure 6.3.
9. Significant factors in the spread of the disturbance and in restoration.
10. Recommendations and actions emerging from the disturbance and covering facilities, policies, procedures.
11. Significant factors helping in the analysis of the disturbance, including instrumentation, methods of analysis used.

6.4.2 An Example (a Complex Fault on a Simple System)

It is assumed that three groups of substations *A*, *B* and *C*, are interconnected as in Figure 6.1. A fault occurs in *B* (event a) as a result of which two circuits *A* to *B* trip through maloperation of protective gear (event b) and the remaining circuit trips on overcurrent (Figure 6.2 event c). The remaining circuit between *B* and *C* then trips on instability, Figure 6.4, event e. Following the separation of sections *A* and *B*, there is a significant imbalance between demand and generation in *A*, contained by demand disconnection in *A* (Figure 6.3, event d). With a circuit out of commission and the high extra power flows between *B* and *C*, instability develops, the remaining circuit *B* to *C* trips (Figure 6.4, event e). Again, there is a significant imbalance between demand and generation in *C*, resulting in a need to reject generation. Too much is rejected and there is significant fall in frequency,

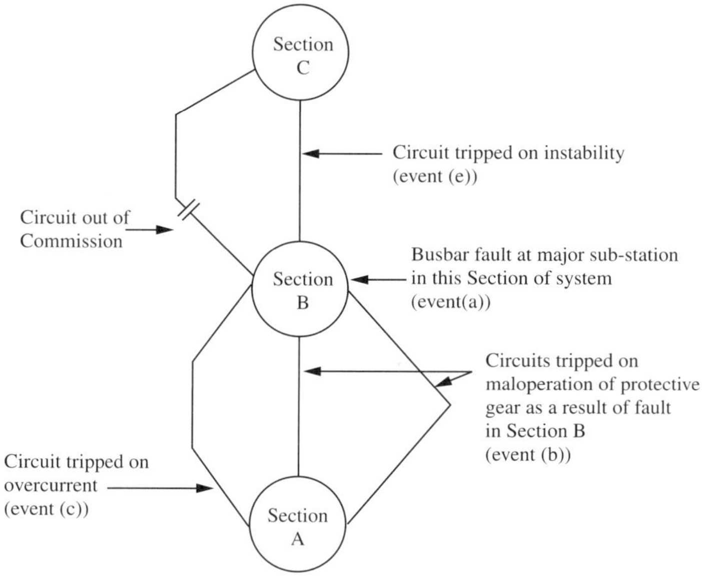

Figure 6.1 System used as example in the questionnaire

reduction in the output of station auxiliaries and as a result stations, and finally island *C* collapses (Figure 6.4, event f). Frequency in the other island *B* is stabilized by demand disconnection or generation rejection and conditions in that island then stabilize (Figure 6.5, event g).

6.4.3 Tabular Information on Disturbances

Tables 6.2–6.5 list the salient points of some disturbances which occurred in Europe and the Middle East (13), Scandinavia (2), the Far East and Australasia (5) and North America (6).

6.4.4 Descriptions of Disturbances

Nineteen disturbances are described below. These are allocated by geographical area and date as follows:

UK	3 (1981, 1986, 1987)
France	2 (1999)
Scandinavia	1 (1997)
Malaysia	1 (1996)
New Zealand	1 (1998)
Australia	2 (1977, 1994)
USA and Canada	9 (1986 to 1998)

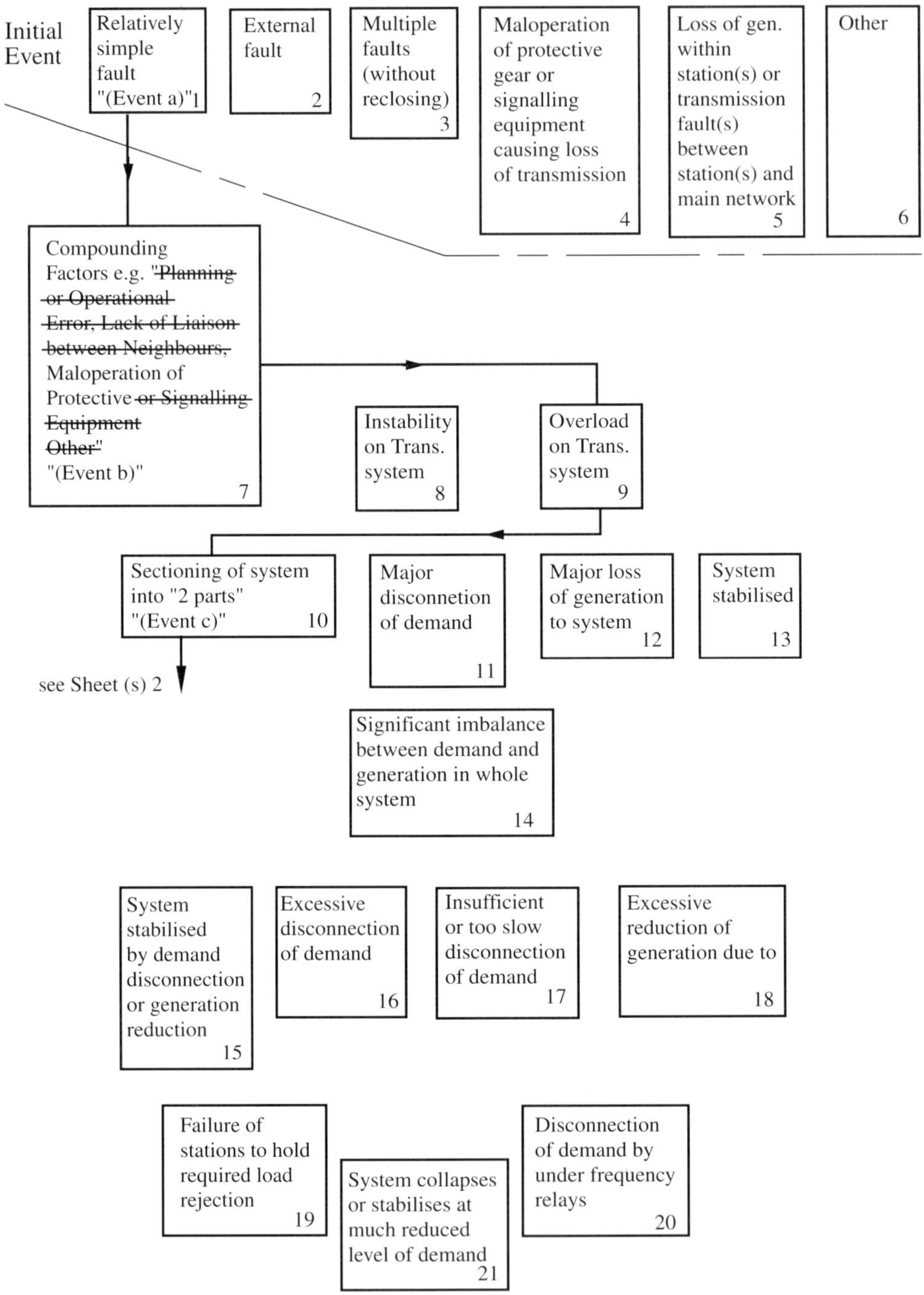

Figure 6.2 Evolution of disturbance in the example (Reproduced by permission of Cigre)

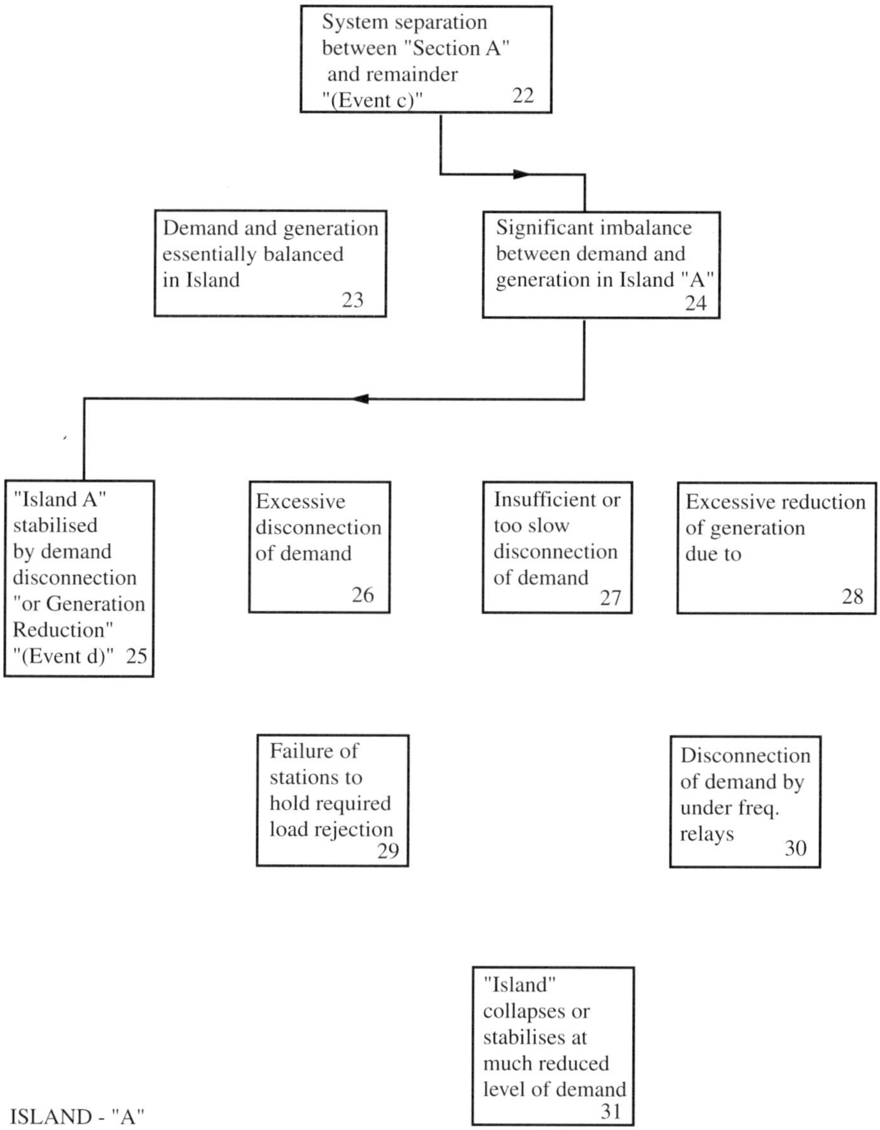

Figure 6.3 Evolution of disturbance in the example (Reproduced by permission of Cigre)

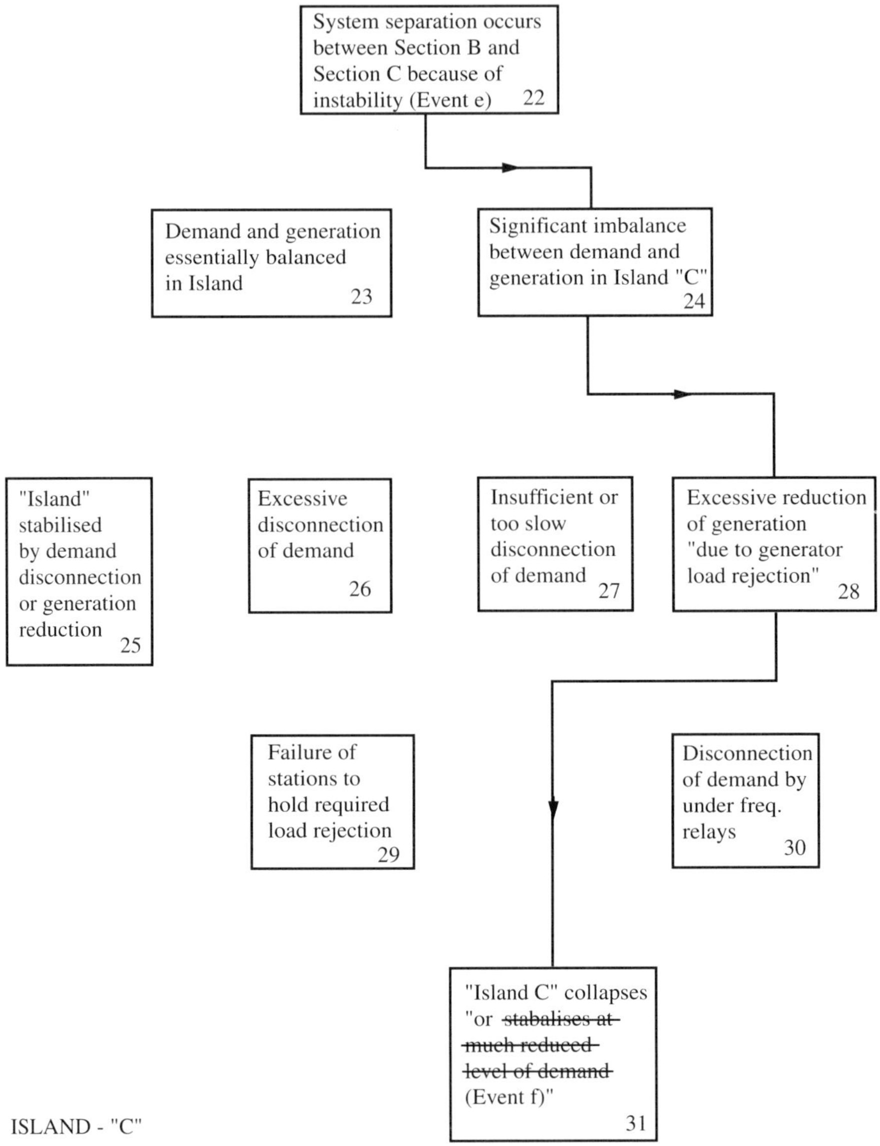

Figure 6.4 Evolution of disturbance in the example (Reproduced by permission of Cigre)

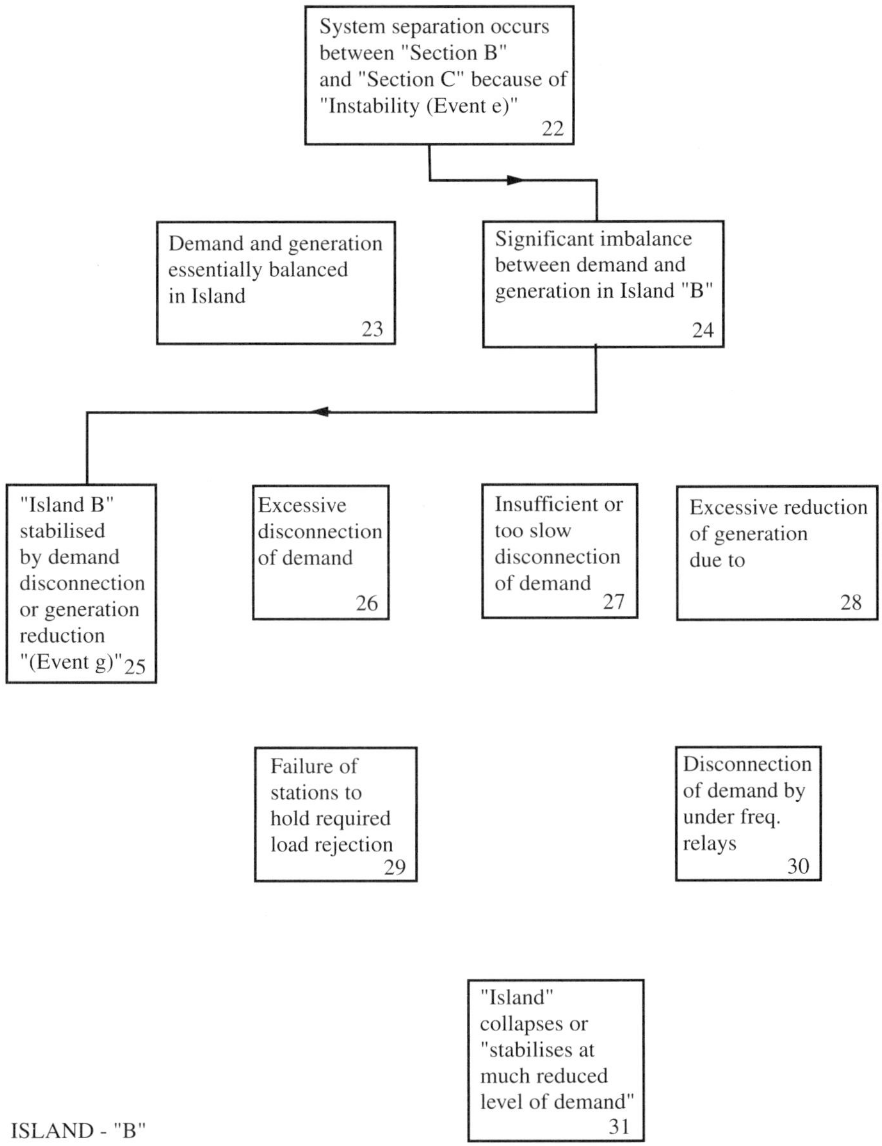

Figure 6.5 Evolution of disurbance in the example (Reproduced by permission of Cigre)

Table 6.2 Summary of some disturbances – Europe and the Middle East

Location	France	France	United Kingdom	United Kingdom	United Kingdom	Greece	Eire	Portugal	Spain	Belgium	Belgium	Netherlands	Israel
Year/ Season	1978 Winter	1987 Winter	1981 Summer	1986 Spring	1987 Autumn	1983 Summer	1992 ?	1985 Summer	1993 Summer	1985 Autumn	1990 Winter	1984 Summer	1995 Summer
Day/ time	08:26		Wednesday 9:08	Thursday 16:10	Friday 02:37	18:58		Sunday 13:52	16:30	Monday 11:02		Thursday 1:06	13:15
Weather	Cold	Low temperatures in west	Hot, sultry	Thunderstorms and torrential rain	Very high winds		Severe lightning storm	Fine and warm	Violent summer storm	Good	Severe hurricane	Good	Hot
System conditions	Low voltages, no reserve, high power flow	Reasonable	Normal, but some circuits out for maintenance	Normal	Normal	Normal	Believed reasonable	Normal, but substantial import	Two 400 kV circuits and one 400/220 kV transformer were under maintenance	Normal	Main system normal		Stressed
Max. frequency deviation			+1.4%, −5.4%						−1%				
Fault severity (system mins)	140		5.9			≈150		19.2		0.8	30	39.5	8.6
Initial cause	400 and 225 kV circuits tripped on overload, followed by loss of generation and external ties. Instability resulted and system sectioning scheme operated	Five generators tripped	Flashover to trees	Circuit outages caused by lightning faults	Many circuit trips caused by high winds	Control error	Generation loss in north of system	Fire under line	Weather conditions	Trip of large unit	Weather	Fault on bus coupler	Fire under double circuit line

Contributory factors	Continuing and rapid increase in demand	Low voltages resulted in further loss of generation	High power flows (oil burning stations shut down for economy)		Incorrect weather forecast on direction of storm	Loss of synchronism (system weakened because of outages)	Protective gear maloperation		Operator error worsened situation		Insufficient disconnection of demand by l.f. relays	Low voltages lead to the tripping of other plant
Loss of demand %	75	14	25 (of affected area)	0	≈16	100	Total	15	3	78	Total	70
Loss of generation %		≈21	0	0		100	Total	Nearly 6.7	16		Total	62
Duration of spread of disturbance (minutes)	Less than 1 minute			≈6	≈110	≈35 secs	8 secs				3.8 secs	14
Time to reparallel (mins)	30 mins but further trips occurred at 09:08		75	–			3 mins		60 mins		20 mins	
Time to external ties restored (mins)												
Time to demand restored	5 hours (90%)		90 mins (90%)		≈500 mins (90%)	300(?) (90%)	50% 43 mins 100% 128 mins		50% 10 mins 100% 60 mins		50% 237 mins 100% 572 mins	70% ≈180 mins 90% ≈250 mins
Sequence of spread (1) (2)	6, 9, 14, 10, 8, 12, (26), (26), (26), (26)		1, 1, 9, 12, 21, (23), (26, 27)			6, 1, 14, 11, 14, 8, 12, 16, 19, 23	1, 7, 9, 12, 24 (26, 29, 33) (25)		1, 14, 7, 10, 17			
Cost						0.03% annual production $6 million to economy						

Notes (1) The numbers refer to the box numbers in Figures 6.2–6.5.
(2) The bracketed numbers indicate the events in the network sections if the system splits.

Table 6.3 Summary of some disturbances – Scandanavia

Location	Mainly Sweden	Sweden
Year/Season	1979, Winter	1983, Winter
Day/time	20:42	Tuesday, 12:57
Weather		Normal
System conditions	Normal	High power flows
Maximum frequency deviation	+1.5 to 0 Hz	Collapse in one island +3.9 in other
Fault severity (system minutes)	10	72
Initial cause	Protective gear maloperation	Faulty isolator caused busbar fault, opening a major interconnection and lower voltage connection
Contributory factors	Possibly insecure operating state	Other parallel lines overloaded and the voltage fell rapidly
Loss of demand (%)	13 (of southern) 100 (of northern)	63
Loss of generation (%)	100 (of northern)	64
Duration of spread of disturbance (minutes)	13	1
Time to reparallel (minutes)		50
Time to external ties restored (minutes)		38
Time to demand restored (minutes)	30 mins for total (north) 2 hours for network restoration	≈5 hours
Sequence of spread	4, 7, 8, 12 (24, 26, 27) (24, 26, 30, 33)	6, 1, 7, 9, 10, 8, 12 (26, 32, 33) (26, 27)
Cost		200–300 million SWKr

6.5 INCIDENTS

6.5.1 UK – August 1981 [6.4]

This and other incidents in the UK illustrate the range of hazards with which the system operator may be faced, even in a temperate climate. The 400 kV system was large and strongly interconnected. Operation of the system was characterized by transfers of some thousands of megawatts from the centre to the south, and hundreds of megawatts from the southeast to the southwest. Conditions at the time of the incident conformed with those foreseen during the operational planning work.

Table 6.4 Summary of some disturbances – Far East and Australasia

Location	Japan	Thailand	Thailand	New Zealand	Australia
Year/Season	1987, Summer	1985, Winter	1992, Winter	1998, Summer	1977, Summer
Day/time	1:19	Saturday 15:25		From late January to early March	
Weather	Unusually hot	Normal		Unusually hot	Bushfires occurring
System conditions	Normal until 19 mins before failure	Normal		The supply capacity decreased over about 4 weeks as cables progressively failed	
Max frequency deviation		51.96, 48.28			48.7, 50.35, 49.16
Fault severity (system minutes)		10.8	5		
Initial cause	Impedance protection operated because of low system voltage and high currents	Switchgear maloperation	Undergrowth fire caused breaker fault		Fault on 300 kV line
Contributory factors	High demands with rapid rate of increase. *P-V* characteristic of demand	Protective gear problems	Problems on system protection scheme		High resistance fault led to load shedding. Maloperation of boiler protection
Loss of demand (%)	21	28		Approximately 150 MVA	Approximately 7
Loss of generation (%)		47			
Duration of spread of disturbance (minutes)	≈ 10	8			Overall time to substantial recovery of frequency approximately 100 secs
Time to reparallel	≈ 4				
Time to external ties restored (minutes)					
Time to demand restored (minutes)	≈ 75	50% 13 mins 100% 36 mins			
Sequence	6, 10, 8, 13	6, 1, 7, 10, 13, 14, 16, 22, 23			1, 7, 16, 17, 5, 17, 23
Cost (energy lost cost)	INA				

Table 6.5 Summary of some disturbances – North America

Location	USA	USA	Canada	Canada	Canada	USA
Year/Season	1997, Summer	1996, Summer	1989, Spring	1988, Spring		1985, Summer
Day/time	20:37	14:24	02:45	Monday, 20:08	Tuesday	Friday, 11:47
Weather	Violent thunderstorms, heat wave	Very hot				Very high temperatures
System conditions	Significant import but transmission conditions comfortable. Collapse in island	High imports	Believed normal; 2 lines out for voltage control	Believed normal	Some 300 MW unavailable from previous day	Believed normal but demands high due to weather
Fault severity	≈400					72
Initial cause		Flashover from 345 kV line to tree	Intense geomagnetic storm (10 year peak). 7 static compensators tripped	Four concurrent earth faults tripped 2 lines and 2 busbars at major substation	Flashovers at same substation which isolated major source of generation	Severe fire in undergrowth caused line trips
Contributory factors		Protective gear maloperations. Insufficient voltage support, possibly demand characteristics		A special protection scheme failed to operate correctly and circuits tripped on instability	Special protection scheme operated correctly to trip 3200 MW demand	The fire exposed the lines to an adverse environment for many minutes. Protective gear maloperations
Loss of demand (%)		4750 MW		Total	Cyclic load shedding of 800 MW	Total
Duration of spread of disturbance (mins)		≈40 secs		3 secs		Total, 11 mins
Loss of generation (%)					Approx. 25	
Time to reparellel (mins)						
Time to external ties restored (mins)						
Time to restore demand (mins)				85% in $8\frac{3}{4}$ hours	Cyclic load shedding was implemented for $4\frac{1}{2}$ hours	$3\frac{1}{2}$ hours
Sequence of spread	1, 1, 6, 1, 1, 6, 2, 1, 9, 11, 14, 13, 8, 31			3, 12, (26, 30, 3) (26, 1, 30, 8) (26, 3, 3, 33)	3, 14, 13, 17	3, 7, 4, 9, 10, 14, 12, 23
Cost	\$310 m (say 20% of net generation and transmission assets)		13.2 million Canadian dollars			

As a consequence of conductor sag in the hot humid weather, flashovers to trees occurred on three circuits between the south east and south west over a period of some 15 minutes. This split the south west and part of the south east networks from the remainder of the UK system. Generation and demand were reasonably balanced in the main part, but there was a significant generation deficit in the smaller section (south west/south east). Local shedding by under-frequency relays, tap changer action and manual action reduced the local demand (probably by some 30 percent). There was a high voltage problem following the load shedding which was controlled by switching demand back. Restoration was slightly hampered by a failure at first to appreciate that a system split existed (steps were later taken to identify on control room displays when a substation was operating in two or more unconnected sections, or when different parts of the system were operating at different frequencies).

The three critical circuits effectively paralleled the south west/part south east network with the remainder of the country. Operational planning advice had been that in the event of one of these circuits tripping, the bus coupler breaker at a substation operating split should be closed, thereby providing a fourth connection between the two parts of the system.

6.5.2 UK – 1986 [6.9]

Many disturbances, even major faults, are survived without consumers realizing that there have been problems with their supply of electricity. This incident can be described as a 'near miss', in that subsequent analysis showed that a major load area was close to voltage instability.

Lightning strikes tripped circuits into the load area, leaving remaining circuits heavily loaded. There was also torrential rain. Voltages fell rapidly over the next five minutes (some parts to under 90 percent nominal). Instructions were immediately given to put gas turbines on load and implement two voltage reductions of 3 percent. The tripped circuits were reclosed. The voltage decline was halted after about five minutes, and satisfactory flows and voltages achieved within 11 minutes. The interactions of automatic tap changers attempting to restore voltage and the various measures instituted to reduce demand were complex, and it was argued at the time that the demand reduction provided by the voltage response characteristic of the consumer demand, and the different response times of the supergrid and lower voltage tap changers helped to avoid a severe disruption of supply.

6.5.3 UK – October 1987 [6.10, 6.11]

Gale/hurricane force winds crossed the south-east of England during 16th October 1987 resulting in widespread damage to property, and almost total disruption of public services. The supply system was severely affected.

Prior to the disturbance, several 400 kV circuits were out of service for maintenance or construction. Some of the salient features of the disturbance were:

(1) With some generation being two shifted, approaching half of the demand in the affected area was being met by local generation.

(2) Some 6 percent of the total system demand was being imported, all via a connection into the affected area from a neighbour. When this was lost due to faults in that system, which was also affected by the high winds, under-frequency protection, quick start gas turbines and pumped storage generation held the situation.

(3) Over 200 circuit breaker operations occurred during the disturbance and early stages of restoration; nearly all auto-reclose operations were successful.

(4) However, a little later, during the most critical few minutes, generation was lost at several stations, some because of transmission faults and some because of voltage and frequency variations at the generator terminals. Transmission capacity into the major load area was severely depleted. Automatic and operator action to prevent permanent damage on remaining circuits led to their tripping, resulting in some 16 percent of the total system demand being lost.

(5) System restoration started immediately. With no external supplies available, several large stations had to implement black start procedures. Attempts to restore the network were thwarted for some one and a half hours by continuing high winds which caused reclosed circuits to fault and trip again and again.

(6) The system was restored with bulk supplies offered, although at reduced levels of security, some six hours after the first faults. Acceptable levels of security and voltage were achieved within nine hours.

(7) Further faults occurred due to salt, dirt and other debris blown by the storm force winds. Extensive cleaning was required over several days.

(8) The damage to local distribution systems was so extensive that distribution utilities were unable to restore supplies for many hours, in some cases.

This disturbance was a very severe test of the SCADA and communications facilities, as well as the power system. At one stage, the mimic diagram and overload lists were the main source of on-line information. Response to load shedding instructions was delayed, caused in part by the intense operational activity in the control rooms.

6.5.4 France – December 1999 [6.12]

Very severe storms crossed France for some 7 hours during the morning of Sunday, 26th December and again overnight on 27th/28th December. In the first disturbance 400 kV lines in the Cherboug peninsula tripped, resulting in low voltages in Normandy and a considerably weakened supply to Brittany. Travelling east, the vicinity of Paris was struck at about 7:00 am and 400 kV lines from the south and northwest were tripped as well as much 225 kV, 90 kV and 63 kV equipment. The 400 kV network to the east of Paris was also affected leading to the isolation of some generating units on to local demand. The tripping of further lines weakened the interconnection between the east and west of France. Heavy damage to the 225 kV and 90 kV networks in the northwest led to the outage of most of the Reims region, and a vulnerable supply to Alsace and the Vosges. The maximum number of 400 kV lines out of operation simultaneously was 38, and at midday 5000 MW of demand was unsupplied. The main challenges faced during the disturbance were seen as maintaining supplies from major stations, supporting the supply to Paris, and maintaining interconnection between the east and west of France, with the southeast, Switzerland and Italy, and the dynamic stability of a station on the northeast periphery of the network.

The second storm crossed France from the west coast in the Vendée region to the Rhone–Alps area in the east of the country. First, two busbar faults caused by salt deposits, occurred within 30 minutes. The second followed almost an hour later by the tripping of lines connecting the Bordeaux region to Brittany and the Basque country to the remainder of France. This resulted in the southwest region of France being connected to the general network by only one 400 kV line, also a station in the vicinity lost its auxiliary supplies. Some 90 minutes later several lines tripped, isolating the network around Bordeaux, which continued to be supplied by a unit at the station. Further trippings resulted in a second isolated, but energised, network around Toulouse. Separation of the southwest area created a generation–demand imbalance for the remainder of France, rapidly corrected by loading pumped storage and hydro plant and increasing thermal output. Four-and-a-half hours into the disturbance, and with the progression of the storm to the east, the hydroelectric stations in the Massif Central became disconnected from the main network for some $4\frac{3}{4}$ hours but continued to supply the regional load. With many 400 kV lines tripped, the network was extremely strained, requiring generation in the east and south to be reduced and generation in the Alps to be increased.

In total, in the two disturbances, at various stages some 8% of the 400 kV and 225 kV circuits and 184 ehv and hv substations were unavailable. The connection of the majority of power stations to the network was maintained, allowing the system generation–demand balance to be well controlled. Generation protection and control facilities operated satisfactorily and a major factor in the successful handling of the incidents was the work of the system operation and transmission staffs. Of the substations without supply after the first disturbance, 35% were back in service by the next day and 64% by the day after.

6.5.5 Scandinavia – 1997 [6.13]

No consumers were disconnected in this disturbance on the Nordel system, which at that time consisted of several hundred generators ranging in size from windmills of 0.2 MW to nuclear units of 1100 MW, interconnected via a 400 kV transmission system of over 10 000 km. There was, however, considerable disruption to the power system.

The weather was normal for the time of year, but the system was heavily loaded and the pattern of power exchange between countries was unusual. A single phase earth fault on a 400 kV busbar at a substation was caused by an icicle contacting the conductor. The busbar protection operated correctly, clearing the busbar in some 60 msec. One of the remaining lines tripped some four seconds later, reducing the system connection to a major nuclear station to two relatively long radial lines. Power oscillations grew and the units tripped in, it appears, some 8 secs. During this period, the voltage of the 400 kV grid in southern Sweden varied from nominal −20 percent to nominal +10 percent. The oscillations on the system then decreased, and a quasi-steady state frequency below nominal appeared to be reached in about one minute.

6.5.6 Malaysia – 1996 [6.14]

A partially stuck breaker and incorrect protective gear settings in this incident resulted in the initial loss of some 10 percent (922 MW) of the system generation. With a generation mix of 67 percent gas turbine/combined cycle, 22 percent thermal and 11 percent hydro, the frequency dropped to 49.1 Hz within three seconds. Gas turbines in a free governing operating mode picked up generation rapidly, but a number then tripped on machine protective systems – turbine temperature limit or flame out. This resulted in a substantial further loss of generation (some 2140 MW). 1580 MW of demand was shed by motor frequency relays, but this was insufficient to stabilize the frequency, and the system blacked out 16 seconds after the first fault.

6.5.7 New Zealand – late January–early March 1998 [6.15, 6.16]

In this disturbance, the Central Business District (CBD) of Auckland, the largest city in New Zealand, suffered a supply failure that left it with a minimal supply of electricity for three weeks, and restrictions for another month or so. This lengthy duration is an unusual feature of the event.

Auckland is a coastal city on New Zealand's northern island. Its demand in 1999 was about 750 MW. The normal summer peak of the CBD was 150 MVA. It was supplied from two substations, Quay and Mount Roskill, connected into the northern island 110 kV transmission network via two pairs of 110/22 kV transformer feeders (Figure 6.6). The cables of the pair feeding Quay Street were paper insulated, gas filled, installed in 1958. The other pair were corrugated aluminium sheathed cables, vintage mid-1970s. The normal supply capacity was approximately 280 MVA.

A substantial reinforcement of two 110 kV cables from Penrose to the Liverpool Street area began two years later than intended because of wayleave problems. It was suggested in one description of the incident that this delay and the unusually hot summer, attributed to the El Nino phenomenon, which increased the air conditioning load well above normal summer loadings, and increased ground temperatures might have contributed to the failure.

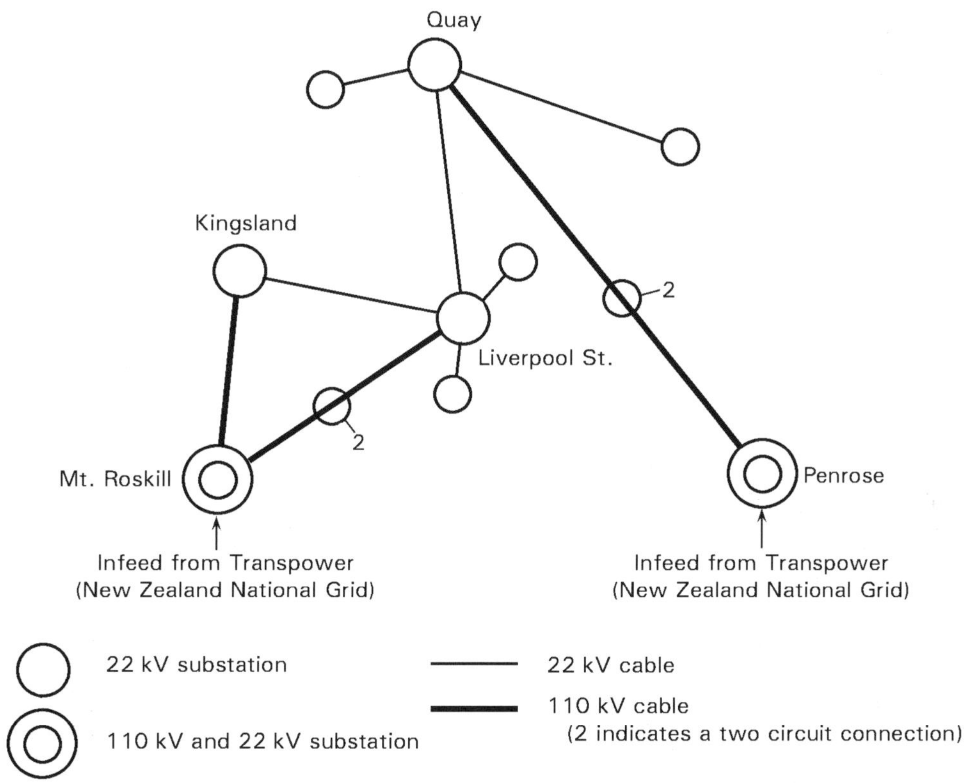

Figure 6.6 110 kV and 22 kV networks in Auckland, New Zealand (based on information in Reference [6.15].

The loss of supply was caused by the progressive failure of four of the 110 kV transformer feeders. A circuit into Quay Street faulted first, with no noticeable effect on the supply. Eighteen days later, the second cable of that pair failed; the utility requested the CBD customers to reduce power use for three weeks and to start up emergency diesels. However, one of the Liverpool Street cables faulted 10 days later, and there was an urgent call for more power reductions. The second cable failed the next day. Only about 20 MVA was then available from the Kingsland substation, and the utility announced that it could no longer supply the CBD.

Many firms moved staff out of the CBD into other accommodation. Small generators were used. Some companies had just enough power for lights and computers. Urgent measures taken to restore power included day and night work to repair the cables, including use of outside contractors, and construction of a temporary 110 kV double circuit line between Penrose and Quay Street, erected in three weeks. Emergency generators were installed totalling over 30 MW. A gas turbine driven ship was connected to the harbour network supplying 12 MW. Problems found with emergency generators that had been permanently installed in some buildings were insufficient fuel storage (only enough for a few hours generation), and fuel pumps not connected to emergency electricity supplies. Fuelling emergency diesels was a major logistical problem.

A total supply of 110 MW was available to the city some 17 days after the last cable failure.

The subsequent analysis of the incident suggested that the failures of the gas filled cables was not surprising, as these had a history of gas leaks. The oil filled cable failures were unexpected, and it was suggested that the shortcomings in installation (including the fact that the cables were bedded in sand with a higher thermal resistivity than was assumed when the cable ratings were determined) was a factor. In his first article, the author of the two articles from which this information has been taken suggests the following:

- a view that the privatization effected some years earlier had had no bearing on the failure;
- the real value of power to a professional office may be 100 times the normal price;
- laws and local regulations should ensure that services can be augmented in the same time frame as the developments overloading existing provisions;
- in the case of systems with winter peaks, summer loads accompanied by summer high temperatures, should also be checked;
- rationing by price may not be a workable option for regulating demand in the event of a sudden major failure;

- emergency diesels may have to run for several days;
- emergency spares should be checked regularly to confirm location and working order.

A ministerial enquiry set up by the New Zealand Government was completed five months after the incident, and concluded that the failure of the first gas filled cable was the result of either thermo-mechanical problems or gas pressure loss. The second gas filled cable and the first oil filled cable failure were caused by thermo-mechanical problems. One mechanism put forward was that thermal cycling caused the cable cores to 'ratchet' within the sheaths, causing distortion in the cable boxes and ultimate failure. The final oil filled cable failure was attributed to thermal runway (i.e. the insulation temperature had reached the level at which increasing dielectric losses became the main source of cable heating).

The problem of determining the safe rating of cables was being studied by the utility when the second paper was written. Options included sampling cable bedding materials along the route, X-raying joints, measuring conductor resistance and installing thermocouples at intervals along the cable routes.

6.5.8 Australia – 1977

Although dated, this incident has been included as an example of a high resistance fault. The system/plant design and environment make some systems more prone to these, as was the case in the South East interconnection of Australia at the time of this disturbance. (In one year, bushfires and contact with trees had been responsible for 60 percent of the 330 kV system faults. Tree and vegetation clearance had been restricted to reduce the environmental impact of the lines.) A 330 kV line fault was cleared at both ends by zone three impedance protection. However, the imposed resistive load of the fault, estimated at some 800–1000 MW on an interconnection with a generation capacity of about 11 000 MW, caused the frequency to fall to 48.7 Hz. Under-frequency relays operated, tripping about 280 MW of demand. The frequency rose to 50.35 Hz on clearance of the fault and about 70 seconds later, two 500 MW generators tripped because of high water level in the boiler drums. The frequency fell again to just over 49 Hz. It recovered over minutes, since the fast reserve standby had been committed in the first frequency fall.

As a result of this and similar incidents, the utility decided to install directional earth fault comparison (requiring fault detection from both ends of the faulted line to trip), relays to reduce fault clearance times, and also to reduce zone 3 tripping times from 3 to 1.5 seconds. Studies were also being made to improve boiler control systems, and to improve communication between system control staff and plant operators.

6.5.9 Australia – 1994 [6.14]

A combination of heavy pollution on insulators and high humidity caused numerous line faults in the vicinity of Perth, Western Australia, in 1994. At the time, some 670 MW was being imported into Perth from the Muja power station region. Synchronism between the two regions was lost when the interconnection fell to two circuits. The frequency in the Perth island dropped rapidly to about 40 Hz, stabilizing there for some 30–40 seconds. However, the load shedding resulted in overvoltages in excess of 10 percent, and generation was lost when a 300 kV line flashed over and tripped. The ensuing fall in frequency led to a blackout. The frequency in the Muja island rose to 55 Hz, and major generating units tripped. The resulting frequency fall was arrested successfully by load shedding.

This disturbance has a number of classical features – islanding with very different conditions in the two islands, low frequency and load shedding in one and high frequency in the other, but further problems developing in both.

6.5.10 USA – July 1986 [6.17]

This disturbance occurred in the western power system. Temperatures were high in the south west, leading to very high loads which were accompanied by high power exports from the Pacific northwest to California. A flashover to a tree occurred on a 345 kV line exporting power towards the Pacific coast. A paralleling circuit also tripped. The loss of two of the three circuits connecting a 2000 MW station towards the coastal load centres some 1300 km distant caused special stability controls at the station to operate correctly which should have ensured stability and prevented further outages. However, another line tripped and series capacitors on two lines were bypassed, resulting in a voltage depression to the west, accentuated by the distribution of generation. Some 24 seconds later, a 230 kV line tripped on zone 3 protection. This led to overloads, voltage collapse and an angular instability. The system broke up into five islands within seconds, with load and generation losses of nearly 9000 MW and 4000 MW, respectively.

Taylor and Erikson [6.20] suggest that insufficient voltage support led to angle instability, possibly compounded by irrigation and air conditioning motor loads.

6.5.11 USA – 1989

Although thousands of earthquakes are recorded daily by geologists, only a few hundreds of these are considered dangerous (over say 5.5 on the Richter scale). Earthquakes tend to occur in well defined areas, at the boundaries between tectonic plates or across the centre of these. An interesting point is that earthquakes are no respecter of location occurring through cities and urban areas as well as open countryside. An earthquake (severity 7.1) stretched across the Santa Cruz area of California, interrupting supply to about 1.4 million customers, including the city of San Francisco. Restoration commenced in about an hour, with over 93 percent (say 1.25 million plus) consumers on supply within two days.

Points which emerged were the value of the utilities private communications systems, and of emergency operations centres established in regional organizations.

6.5.12 USA – September 1989 [6.18]

Hurricane Hugo traced a path north west and north from Charleston in South Carolina, affecting some 1.14 million customers in three utilities (Duke Power, South Carolina Electric and Gas, Carolina Power and Light). Some 600 transmission structures and 16 000 wood poles were damaged, and some 27 600 distribution transformers had to be replaced. Some of the points which emerged from experiences in this hurricane were the criticality of logistics functions, from organizing lodging plans to preparing meals for emergency workers, the re-assignment of personnel to specific responsibilities, clarification of operating policies and procedures with visiting personnel (including those following their own rules and practices), nightly strategic planning meetings, and drills to provide practice to staff who may be included in handling an event.

6.5.13 USA – August 1996 [6.19]

This system in western USA consists of 500 kV, 345 kV and 230 kV lines extending over an area of some 4 million km^2 with a peak load in summer 1996 of some 118 GW. A 4 × 500 MW station at the eastern edge of the load area was connected by mixed transmission (three 345 kV lines in series with one 500 kV and several 230 kV lines) to the western load centres. One of the 345 kV lines tripped on flashover to a tree. An earth fault element on another of the 345 kV lines operated incorrectly, thereby losing two of the three outlets from the station. A system protection scheme had been installed to trip two units at the station in the event of the loss of two 345 kV lines, and this operated correctly. This should have stabilized the situation, but several near simultaneous switching events occurred – a 230 kV line some 500 km to the west tripped and series capacitors on two lines were bypassed. As a result, the voltages at the eastern end of the system fell and several relatively small hydro generators tripped. A key 230 kV interconnection to the north tripped on third zone impedance protection (high current and low voltage conditions). This caused oscillations as power redistributed to the west and four of the 230 kV lines constituting part of the east-west connection tripped. This, with the earlier trippings, interrupted a major source of power to the load areas. The major north-south intertie in the western part of the system opened about two seconds later. With further cascading the system split into five islands. The total loss of load was about 4.7 GW and of generation 3.9 GW.

6.5.14 Canada – January 1998 [6.1, 6.20]

Severe ice storms resulted in major disturbances to the Hydro Quebec system early in 1998. Freezing rain over a period of five days gave an accumulation of several centimetres of freezing rain and snow, and after a few hours H.V. lines started to fail due to ice accumulation on conductors and fallen tree trunks. Wooden poles snapped. Over the next three or more days, many towers on a 735 kV transmission loop and on underlying 315 kV and 230 kV circuits were down, and connections to neighbouring systems were out of service. One of the utilities affected, Niagara Mohawk lost over 85 percent of its transmission and distribution in the affected area. The losses totalled over £125 million. The impact on society was significant – schools and business closed, petrol could not be pumped, houses were cold and dark, money could not be obtained from cash dispensers, roads were closed due to fallen trees and wires. At the worst times, nearly 1.4 million Hydro-Quebec customers were without power, whilst Central Maine lost 275 000 customers.

Restoration work started within hours of the onset of the disturbance. Assistance was provided from far and wide. Nevertheless, some one million customers were still without supply five days later. Power restoration took 23 days, with residual repairs continuing into the early summer.

6.5.15 Canada and USA – January 1998 [6.21]

The same ice storm also caused disturbances in Ontario and New England. Amongst these, in chronological order were:

- freezing rain and heavy ice accretion led to tripping of three 115 kV and seven 230 kV circuits; 300–400 MW demand and 250 MW generation lost;
- tripping of a radial 230 kV line led to an island being formed with excess (130 percent) generation. Frequency rose by some 2 percent;
- a 115 kV circuit contacted a 25 kV circuit. Frequency and voltage fluctuations occurred. Some 215 MW of generation and 500 MW of demand were lost.

6.5.16 USA – January 1998 [6.21]

A loss of 900 MW affecting 290 000 customers occurred when a snow storm with strong winds moved through the Commonwealth Edison system.

6.5.17 USA – January 1998 [6.21]

Accumulations of between 30 and 60 cm of snow downed trees and power lines, tripping four transmission and numerous distribution lines. Supply to some 80 000 customers was interrupted.

6.5.18 USA – March 1998 [6.21]

As a consequence of the inadvertent opening of a 345 kV line during protection testing an exporting area with a demand of some 3000 MW was islanded. The frequency rose rapidly but at 60.3 Hz a generation rejection scheme operated, shutting down some 470 MW of generation and reducing the import over a d.c. circuit by about 40 MW. Governor action stabilised the frequency at 60.23 Hz and the island was resynchronized to the main system three-quarters of an hour later.

6.6 CONCLUSION

There has been no shortage of information on past disturbances. Presumably the future trend will be for this to decrease, although it seems likely that incidents including large losses of supply will still attract media interest and that summary reports, at least, will appear in the technical press. Much of the information will be qualitative but quite adequate for the reader to follow the evolution of incidents and to recognize the problems and lessons. Utilities may be willing to provide further information to research workers undertaking detailed studies.

The severity of the incidents often exceeds the credible contingency criteria. By and large credible contingencies relate to short duration disturbances whilst many of the most severe incidents evolve over hours or days, enabling restorative measures to be started whilst the disturbance is maturing. This time spread, the redundancy designed into systems and the fact that each part of a system is, when required, supported by the rest gives the resilience demanded in power system performance in many parts of the world.

REFERENCES

6.1. Lamarre, L., 1998. 'When disaster strikes', *EPRI Journal*, September/October.

6.2. Hensen, R., 2000. 'Billion dollar twister, Scientific American presents weather and what we can and can't do about it', *Scientific American Quarterly*, **11**(1), 2000.

6.3. Reed, J., 2000. 'Fleeing Floyd', *Scientific American Quarterly*, **11**(1), 2000.

6.4. Hawkes, N., 1981. 'How trees put the lights out in Britain', *The Observer*, 9 August.

6.5. Guile, A.E., Patermon, W., 1977. *Electrical Power Systems Vol. 2*, Pergamon.

6.6. *Modern Power Station Practice Vol. K*, ENV Transmission, British Electricity International, 1991.

6.7. Douglas, J., 1989. 'A storm from the sun', *EPRI Journal*, July/August.

6.8. Kappenman, J.G., 1988. 'Geomagnetic storm forecasting mitigates power system impacts', *IEEE Power Eng. Review*, November.

6.9. Dwek, M.G., 1988. Post-fault voltage recovery and automatic tap changer interaction, contribution to *Cigre Group 38* discussion, Cigre.

6.10. CEGB, 1987. '*Riding the Hurricane: how the CEGB's power system weathered the storm*', CEGB Brochure.

6.11. Simmonds, T., 1987. 'The six-hour battle', *Power News*, November.

6.12. Merlin, A., 2000. 'The storms in France and the grids', *Electra*, **188**, 11–15.

6.13. Hiskens, A., and Akke, M., 1999. 'Analysis of the Nordel power grid distribution grid disturbance of January 1 1997 using trajectory sensitivities', *IEEE Trans Power Systems* **14** (3).

6.14. Janssens, N., 1999. 'Analysis modelling needs of power systems under major frequency disturbances', *Cigre Electra* **185**.

6.15. Leyland, B., 1998. 'Auckland control business district power failure', *Power Engineering Journal.*

6.16. *Auckland lights out – from failure to recovery*. E.A. Technology.

6.17. Taylor, C.W., Erickson, D.C., 1987. IEEE Computer Applications in Power, January.

6.18. McGee, R., 1992. 'Preparing for disaster', *EPRI Journal*, September

6.19. Hoffman, S., 1996. 'Enhancing power grid reliability', *EPRI Journal*, November/December

6.20. Irwin, P., 1998. 'The freeze', *Electrical World*, February.

6.21. North American Reliability Council, 1998. '*System Disturbances*', July.

7 Restoration

7.1 INTRODUCTION

The objective of power system restoration is to bring the system to the point at which as much demand as possible within the capacity of the remaining generation and transmission plant is being supplied at normal frequency, voltage and security levels. This will be a moving feast; it will depend upon the initial plant margins, but is often taken as all demand being supplied. In the event, restoration will be a combination of operator decisions and automatic control actions.

There will be two levels of problems in restoration. In the less severe, the disturbance or loss of supply will be relatively localized, and with luck, the remaining healthy system will provide a stable source of frequency and power for start-up to the disturbed area*. Another variant will be the extent to which the disturbance was foreseen, ranging from no warning at all with a sudden fault, to hours or days with fuel shortages, or detected and incipient failure of plant. These will be considered in turn following a review of the conditions which may be encountered and the strategic/tactical decisions to be made.

7.2 THE RANGE OF DISTURBED SYSTEM CONDITIONS

The factors defining the severity of an actual or foreseen disturbance are considered below:

- Loss of generation – the most frequent form of disturbance; the magnitude of the loss in respect to the total initial amount of generation and system transfers, and to the capacity margins of generation and transmission, will be a measure of its severity. The worst form of this type of disturbance is likely

*This is not always the case. Some 30 percent of the running generation was lost during the exceptionally severe wind storm in England in October 1987. Although 70 percent of the generation remained available, black start conditions existed in the affected areas because of transmission problems, and it was necessary to start up stations in these areas from emergency diesels and then gas turbines.

to be a busbar or section breaker fault in the ehv substation of a major power station, or a multi-circuit fault on transmission lines between such a substation and the main body of the system. In the worst cases, the whole of the output of a power station will be lost.

- The restoration of generation may, therefore, range from making up a few tens of megawatts following the partial loss of output of a generator as a consequence of station auxiliary problems, to making up gigawatts following the loss of the total transmission connections at a large power station.

- Loss of transmission – this is also likely to be a fairly frequent disturbance, whose severity can range from the loss of one or two circuits following lightning strikes, to whole sections of a network. The latter may result from sequential trippings of circuits as protective gear operates because of disturbed current and voltage conditions spreading through the system, to the effects from system wide severe weather conditions. In this case, the depleted system may experience sequential tripping from overcurrent and depressed voltage conditions.
  The restoration strategies must be able to deal with network states from one or two circuits off load, to the whole network dead.

- Loss of demand – this is most likely to be a consequence of generation or transmission losses rather than an event in itself. The loss may vary from one substation to the whole system; one of the most frequent immediate causes of large demand losses will be the operation of under frequency relays.
  Strategies for restoring demand involve and guide all the other areas of restoration, guide because some types of demand will have a higher priority for restoration than others.

- Loss of reactive compensation – potentially serious in that voltages within or close to normal tolerances are critical to ensure viable operation. Low voltage conditions particularly if widespread will require rapid action – increased excitation on plant, demand disconnection, synchronizing fast response plant, switching in any available transmission and capacitor compensation, switching out inductive compensation, etc. If falling voltages are not halted, voltage instability and system collapse can result.

- Loss of external connections – the complete loss of external connections will, because of the configuration of most interconnections, be infrequent. Depending on the system conditions prior to the circuit loss, the effect will be covered by one or more of the disturbed conditions described above.

7.3 SOME GENERAL ISSUES IN RESTORATION

A number of questions must be reviewed before finalizing restoration strategies. Some pertinent issues will be:

- Should priority be given to the restoration of the system or to the restoration of the demand – it would be possible in the early stages of restoration to give priority either to switching in circuits which build up the network, or to those which supply demand. As will be indicated later, most utilities concentrate on establishing a 'backbone' network before restoring demand.
- Should the possibility of further faults be considered during the restoration period – this comes down either to switching in demand without ensuring its supply is secure, or alternatively, providing this security and generally delaying the restoration. Practice differs between utilities.
- How much attention should be paid to quality of supply during restoration – the main criteria here is that the frequency and voltage conditions should not be such as to damage or shorten the life of equipment, either consumers' or utilities'. If restoration involves reparallelling with another system, the frequencies of the two systems must be very close as must the voltage, magnitude and phase angle at the point/moment of connection.

7.4 RECOVERY FROM AN ABNORMAL OPERATING SITUATION, LOCAL ISLANDING OR LOCALIZED LOSS OF DEMAND

In contrast to recovery from large scale failures, published information on handling the more minor disturbances is limited. From some points of view, the range of actions and priorities will be greater than with large scale failures, since surprisingly, the abnormalities may be more diverse. A reasonable strategy for an isolated system would be:

(1) Determine the condition of the system, in particular those factors determining its short term viability – frequency and its trend, voltages outside limits and trends, severe overloads and trends.

(2) Restore frequency within operational limits.

(3) Implement essential urgent action (e.g. necessary within 15–30 minutes) on generation:
   (a) re-establish load on nuclear units to prevent reactor poisoning,
   (b) re-establish load on large thermal units to forestall problems from differential expansions.

(4) Take any immediate action to prevent further deterioration of the overall situation (this might include load shedding).

(5) Restore voltages within operational limits.

(6) Adjust generation and demand:
  (a) to make the system secure against credible contingencies,
  (b) to reduce overloads or unsafe power flows, e.g. outside transient stability limits, to continuous or long term values.

(7) Interspersed with these actions restore demand as generation and transmission capacities make possible, taking account of priorities.

Depending on the number of operators available, including any who might have been called in to assist in handling the disturbance, some of these tasks will be handled simultaneously with the senior operator co-ordinating the activities, in particular checking the on-going security of the system.

Assuming the system frequency is within limits, localized problems within a large system would usually be excessive power flows, abnormal voltages, security standards not met, or local disconnection of demand. Having checked the condition of the system, plant, network and consumer needs would dictate priorities, but typically the sequence of urgent actions would be:

(1) Implement essential urgent action on generation (as (3) above).

(2) Take steps to prevent further deterioration (as (4) above).

(3) Adjust generation and demand to make the transmission system secure (as (5) and (6) above).

(4) Interspersed with these actions, restore demand as generation and transmission capacities make possible.

Actions will be implemented simultaneously rather than serially.

7.4.1 Checking System Security during the Restoration Process

It will be advisable for the operators during the restoration process to check continuously that actions being taken will not precipitate a further disturbance, for instance

- circuit trippings caused by operation of overcurrent or impedance (third zone) protection;

- generator trippings or output reduction caused by instructing operation outside the plants' capability (too rapid output change, stator overcurrent or excitation limits exceeded, terminal voltages low, too large rotor angles);
- reactive compensation trippings caused by abnormal terminal voltages.

Although not stressed so far, the over-riding consideration in restoration will be to resume supply to consumers as quickly as possible. In general, there will neither be time nor facilities to make detailed assessments of network viability as changes are considered. Operators may have to rely on approximate and rule of thumb methods, for instance

- comparison of power transfers into groups of substations against earlier assessments of the group transfer capability;
- phase angle calculation [$\theta \approx \sin^{-1}(PX/E^2)$];
- voltage drop $\cong PR + QX$;
- factors giving incremental flows in circuits for nodal power changes or circuit outages.

7.5 THE 'BLACK START' SITUATION

A black start situation exists when supply fails in part or all of an interconnection. In the extreme case, all generation will have ceased although in practice, and depending on the control and protection mechanisms, isolated pockets of generation supplying surrounding demand may remain. Nevertheless, most utilities make contingency plans on the basis that they may have at some time to build their system up from an 'all plant dead' state. The small power sources – batteries, diesels, gas turbines – which will often be needed to achieve this (systems with access to hydro generation may be the exception) must be provided at the planning stage, and will be costly; some utilities may rely on supplies being available from neighbours.

It has been suggested that there are three functional areas to consider during the restoration process: the active power balance; the reactive power balance; and the status of the protection and control mechanisms. The first will be determined mainly by the evolving generation and demand, the second the network reactive sources and demand situation, and the third the viability of the protection and control facilities after the disturbance. Each of these will be discussed in turn, followed by their synthesis into paradigms for overall system restoration.

7.5.1 The Generation Demand Balance

Constraints and Initial Stages of Generation Restoration

The thermal/mechanical properties of turbines and generators will determine their acceptable loading cycles, and these will be dependent on the type of plant. For instance, with steam-driven type boilers, there will be maximum times available for hot-restart, and then within these maximum loading rates, minimum times to start and minimum loading levels on restart. Once through boilers will have similar constraints. Some typical restart times quoted in the literature for different types of plant are given in Table 7.1.

Constraints and Initial Stages of Restoration of Demand

There will be a number of relatively small but critical demands within the system which must be met if restoration is to proceed smoothly. These will include supplies for station auxiliaries, pumping plants for some cables, and auxiliary supplies for substations, control centres and strategic offices. Very often these will be obtained from auxiliary diesels at the various sites, but if these have not

Table 7.1 Times to full load for different types of plant

Plant type	Size range	Times to full load
Conventional steam, drum type boiler	up to 1000 + MW	Depends on state at shutdown, e.g. hot[1] –1 to 2 hours warm[2] –1 to 5 hours cold[3] –2–10 hours
Combustion turbine	up to, say, 200 MW	Say 15–30 minutes[4]
Combined cycle	up to, say, 500 MW	[5]
Nuclear		hot –from 10 to 200 hours cold –from 20 to 250 hours
Pumped storage	up to some 400 MW	[6]
Hydro		

[1]e.g. to full load, for shut down period less than 8 hours.
[2]e.g. to full load, for shut down period between 8 and 36 hours.
[3]e.g. to full load, for shut down period above 36 hours.
[4]the combustion turbine may have the same start-up profile whether hot or cold.
[5]for one particular configuration, gas turbines up to full load in some 24 minutes, steam turbines in some 50 minutes. The restart time may be significantly shorter than the hot start times.
[6]the 300 MW reversible pump-generate units in the NGC Dinorwig station have the following capabilities: stand still to full load generate = 1.5 mins; full load pump to full load generate = 8 mins. These exceptionally short times were stipulated to provide spinning spare on a large isolated thermal system. In general, the times for mode and output changes on pumped storage plant will be longer.

been provided, restoration of supply within an hour or less to prevent sulphur hexafluoride in switchgear and cables from liquifying, maintain air pressures, drive cooling fans, etc., will be needed. Some industrial demands will be time-critical, for instance electromechanical processes in which the electrolyte is maintained liquid by heating. Down times of only 30–45 minutes may be acceptable. Commercial time critical demands are likely to include public transport, traffic control signals and hospitals, but again, the responsible authorities will often provide auxiliary diesel plant.

A minimum demand must be provided for each generating unit as it is brought on load. The size of the load increments used will depend upon the response capability of the generation, and upon the need to maintain frequency within operational limits.

An allowance should be included for the system I^2R losses. These will be small, say 2–4 percent of the demand at peak for the whole system, and could be ratioed from that value (strictly, the ratio of squares of power flows, but a linear ratio could take better account of the depleted network condition).

7.5.2 The System Reactive Balance

The system reactive balance will be determined by the reactive capability of the on-line generation, the reactive component of the restored demand, the capacitance and inductance of the network, and of the shunt compensation plant, as these are restored and the series reactive losses in the network. The reactive power supplied by the shunt elements will vary with the square of the system voltage.

One of the commonest problems during the early stages of restoration is to prevent overvoltages. This requires minimizing the circuits switched in (e.g. use only one circuit of a double circuit line), operating generators at minimum voltage levels, minimizing shunt capacitance and maximizing shunt reactance, adjusting transformer taps and restoring loads with lagging power factors at an early stage.

As an example of the magnitudes of the reactive powers, the balance for the NGC system reported in the *1995 Seven Year Plan* was as shown in Table 7.2.

7.5.3 Status of the Control and Protection Facilities

It will be necessary to activate numerous organizational procedures once a blackstart situation has been recognized. These will have been promulgated through the organization as part of the procedures for dealing with a blackstart

Table 7.2 Reactive power requirements. (a) lines and cables, (b) transformers and quadrature boosters
(a)

		System	Voltage
Overhead lines		400 kV	275 kV
Lagging reactive power	MVAr	42	107
Leading reaction power	MVAr	69	28
Cables			
Lagging reactive power	MVAr	25	7
Leading reactive power	MVAr	1593	1455

(b)

Transformers	
400/275 kV, 1000 MVA rating, flow	= 500 MVA
series reactive load	= 40 MVAr
shunt reactive load	= 5 MVAr
400/132 kV, 240 MVA rating, flow	= 120 MVA
series reactive load	= 12 MVAr
shunt reactive load	= 1.2 MVAr
275/132 kV, 180 MVA rating, flow	= 90 MVA
series reactive load	= 6.7 MVAr
shunt reactive load	= 0.9 MVAr
Quadrature boosters (the shunt reactive loads were not included in the data source used)	
400 kV, 2000 MVA, flow	= 1000 MVA
series reactive load	= 62.5 MVAr
275 kV, 750 MVA, flow	= 500 MVA
series reactive load	= 30 MVAr

Basic parameters assumed: transmission distance = 100 km; transmission transfer = 500 MVA.
Note: cables of this length would be very unlikely, and would be shunt reactor compensated in any event.
The nominal information in (b) was derived from the 1995 Seven Year Statement of the National Grid Company.

situation. Although specific to each utility, some of the arrangements often found will be:

(1) Delegation of the authority of the National (System) Control Centre to Regional Control Centres, each of which will then be responsible for its own area. When, at a somewhat later stage, these interconnect, one of the Regional Centres in each of the subgroups would be assigned responsibility for the operation of that subgroup.

(2) Calling senior control centre management into the centre.

(3) Calling out telecommunications and SCADA/EMS maintenance staffs, possibly some operational planning staff.

(4) Activating a System Incident Centre or equivalent; informing top management of the situation.

(5) Informing the control centres of neighbouring systems of the situation.

Precautions to observe during restoration, particularly in the early stages, will be

(1) Fault levels are likely to be low, and it may be necessary to reduce protective gear settings to ensure operation should a fault occur; potential fault currents may be lower than full load currents.

(2) At least some of the demand restored should have under frequency load shedding protection in operation.

(3) Generating plant should be operated in a frequency sensitive mode unless its integrity requires otherwise.

(4) The status of automatic switching schemes should be checked.

(5) There is a possibility that the operation of automatic switching schemes, including delayed auto-reclose, may have been halted before completion by the disturbance. If this is so, for example indicated by an 'in progress' alarm, it may be best to inhibit that particular switching sequence.

7.6 STRATEGIES FOR RESTORATION OF THE WHOLE SYSTEM

Although papers on restoration are usually system-specific, there is often some commonality between the strategies reported for large systems. It is proposed in this section to outline a representative procedure based on these, and to include references to the main alternatives, perhaps the most basic being whether total system or piecemeal restoration be adopted. In the former, the dead system would be run up to synchronous speed with very low excitation, and hence very low system voltage. Demands would be connected. The generator excitations would be increased, raising the terminal voltages and the demand powers until normal system voltages were reached, with the whole system on load. As far as is known, this has not been done in practice. It would expose the utility and consumers' plant to abnormal, practically unexplored, operating conditions. There would be no possibility of support from neighbours until virtually the end of the process. An excessive amount of co-ordination between station, transmission, distribution and demand control staffs would be required.

In the piecemeal restoration process, the system is built up gradually around one, or more likely several, generating stations, the demand being picked up gradually but at essentially normal voltage and frequency. Prior to this, however, the dead system must be prepared for restoration.

7.6.1 Preparation of the System

Once faulted plant items have been determined and isolated from the system, the possible states of the remaining plant items and suitable actions to prepare for restoration might be:

(1) for passive elements, dead and disconnected from other elements – take no action;

(2) for passive elements, dead but connected to other elements – review the merit of opening the connections;

(3) for synchronous plant elements (not generation), dead and disconnected – hold in readiness for reconnection to system;

(4) for synchronous plant elements (not generation) alive and disconnected – hold at synchronous speed ready for reconnection;

(5) for synchronous plant elements (not generation) alive and connected – probably disconnect, then hold ready for reconnection;

(6) for generators, alive and supplying local load (e.g. house load) – stabilize operating state, preferably increase load on unit;

(7) for generators, just tripped but with prime movers functioning – stabilize operating state as quickly as possible by securing house load and switching in other demand;

(8) for generator dead, but in hot or warm state – initiate the steps for starting up the unit, which may involve providing auxiliary power from other stations.

The various strategies should be reviewed, preferably practiced, by the relevant station and control staffs.

7.6.2 Rebuilding the Transmission System

There will be three main activities at the start of the rebuilding process. Stations with house supplies available should stabilize their operating state (points (5)–(7)

above). Supplies for house services should be made available to stations without these; auxiliary supplies should be provided to substations. In anticipation of such problems, 'quick start' plant and transmission routes may be nominated in the blackstart plans. The system will now consist of a number of separate power islands, each containing, say, one, two or three stations and associated demand. A nominated control room in each of these islands should instruct switching to synchronize with other islands. At about this stage, the Regional or System control centre should resume responsibility for building up the system. This may be done in accordance with a pre-defined skeleton configuration which would aim to provide a connection to all substations on the main transmission network, significant generating stations, and possibly priority demands. At this stage, the emphasis in some utilities will be on restoring the remaining unsupplied demand without necessarily meeting normal security standards; security standards will be satisfied as and after demand is reconnected. The remaining tasks will be to regularize the generation situation towards normal economic operation and complete the restoration of the transmission network.

7.7 AIDS IN THE RESTORATION PROCESS

Several of the system facilities which will contribute to rapid restoration have already been mentioned, namely that some of the generation should be capable of isolation and continued operation supplying only its own auxiliaries or these plus some local load, that a proportion should be designed for blackstart using only on-site equipment, and that major transmission substations should have on-site auxiliary power supplies. Other aids will be provided in the operational planning and control phases.

7.7.1 Operational Planning Studies

Comprehensive operational planning studies will be invaluable to control staff, including

(1) the required switching state of breakers at all substations at the start of restoration (generally open);

(2) the EHV busbar configurations to be adopted at each power station and neighbouring substations;

(3) the transmission switching schedules to connect demand to power stations;

(4) the skeleton transmission networks to be built up first;

(5) the location of system synchronizing facilities;

(6) temporary changes in protection settings;

(7) thermal and stability/voltage limited transmission capacities for circuits and selected groups of substations;

(8) essential telephone, etc. contacts.

This information will be provided conveniently by VDU. The operational planners should also be able to perform system studies at short notice.

7.7.2 Expert Systems

The problem of restoration was tackled with enthusiasm by the protagonists of Expert Systems (ES). The problem is combinatorial, and hence difficult to solve by formal mathematical methods; knowledge and data from various sources are used, and there are numerous criteria to satisfy some of which are qualitative. These problem characteristics suggest that ES approaches would be suitable. Topics studied have ranged from system restoration after a general blackout, to restoration from a partial outage, to restoration from a load disturbance with the subsidiary questions as to the objective – to achieve the pre-disturbance configuration as far as possible, or to achieve an optimum configuration using the plant remaining available for service.

Expert systems developed have often been in the range 100–1000 rules, some using forward chaining logic, some backward chaining, and some both. Most have been implemented on PCs with graphical displays.

7.7.3 Automatic Systems Switching

Some utilities have installed equipment on the system to speed up restoration switching. For instance, if voltage is detected on an incoming circuit to a busbar, this and selected outgoing circuits will be closed, establishing a live path through the network. Precautions include blocking of the closing operations to prevent closure onto a permanent fault, non-closure of breakers open before the disturbance, and manual inhibition of automatic switching from the control centre. Presumably, the risk of overvoltages would also have to be checked.

7.8 PROBLEMS FOUND IN RESTORATION

Restoration is a difficult task, during which the operators will be under pressure to restore supplies quickly, avoid actions which would damage plant, keep appropriate staff informed and not least, be able in any subsequent enquiry to justify the validity of their decisions. Coupled with the fact that the state of the system will be abnormal, and apart hopefully from training sessions, never before encountered by outstation or control staff, it is not surprising that problems occur in restoration. Some of these are discussed below.

- *Repeated failures* – there have been infrequent disturbances in which the system conditions which caused the original failure have been unknowingly repeated and a second (or more) failure has occurred. The immediate cause will be an error by the control operator, but often with an underlying cause leading to the initial and subsequent failures. A major failure in north east USA and Canada was an illustration of this. The remedy is obvious if not easy to implement or justify later – determine the cause of the failure before proceeding with restoration.
- *Overvoltages* – these are one of the most frequently encountered problems, and are an illustration of the Ferrantic effect – the voltage rise found in capacitative circuits such as lightly loaded overhead lines or cables. The consequences can be over-excitation of transformers (generating harmonic distortions and overheating), generator under-excitation, or even self-excitation and instability, and harmonic resonance. This can result in very high voltages; up to several times the sending end voltage, which may be amplified by transformer over-excitation. Flashovers and operation of surge arresters will result, and damage to these can delay restoration. Precautions to prevent this will be to deploy reactive compensation and demand, if possible, when charging circuits, and to operate to lower target voltage levels.
- *Too rapid restoration* – this occurs when control operators attempt to pick up demand too quickly; the generation is unable to supply this, frequency falls and the just re-energised subsystem again collapses. The remedy is to add demand in small increments – a figure of 5 percent of the subsystem has been suggested. The problem will also be largely solved if synchronization to a larger system is achieved.
- *Insufficient knowledge of the system* – a considerable knowledge of the state of the system and of its characteristics will be needed to achieve a trouble free restoration. One of the most important items will be to know the circumstances of the failure – was it an equipment fault (and where), overloading, human error, weather involved, problem in a neighbour, etc. Other necessary knowledge will be whether any parts of the system are still alive, are external supplies likely to be available, what is the status of the generation, what were the demand levels and distribution immediately prior to the shut down, and how will these change on re-energization. The operator will also need to estimate the load pick up capacity of subsystems as these are built up, and the effect on flows in circuits already energized as new ones are switched in. The

amount of this knowledge underlines the importance of an efficient SCADA system, and also the value of experience and training in equipping control staff to handle emergencies.

- *Liaison with distribution utilities* – control of demand is an essential component in restoration. It must be possible for the transmission, and at the early stages the generation, control staffs to instruct both amount to be restored and its location. Remote control in distribution networks is less common than at transmission voltage levels. Delays in implementing the instructions will delay the whole process.
- *Liaison between control centre staffs* – there have been cases in which misunderstandings between control centres, even control rooms in one centre, have led to extended outage times. Failures to standardize terms have been one cause, another different assumptions between two speakers on the knowledge of the other about the condition of the system.

7.9 ANALYSIS, SIMULATION AND MODELLING IN BLACKSTART

Three types of studies will be used to validate blackstart procedures. One of these will be in-depth analysis of particular aspects of restoration, such as harmonic overvoltages. The second will be routine, but complex analysis such as voltage and stability analyses, possibly load flows, and the third the approximate but fast group transfer type comparisons between expected transfers and transmission capabilities.

7.9.1 In-depth Analysis

In-depth analysis on transmission will usually be related to transient electrical problems, such as switching overvoltages and harmonic overvoltages. The range of system studies will be wide, for instance

- start up and operation of large induction motor loads, on small systems;
- post-event analysis of major disturbances;
- the effects of large and prolonged frequency and voltage variations on generation and station auxiliaries;
- the performance of system protection, such as system splitting and islanding schemes;
- checks on the start up capability of generating units from remote sources;

- the performance of voltage and frequency control systems at abnormal values;
- the prolonged operation of generating units when supplying only house load;
- the load pick-up capability of generating units;
- preferred busbar configurations at generating stations (an example of the format of information that has been provided in one utility is shown in Figure 7.1). The strategy underlying this configuration is to disconnect the local demand, and to isolate the power station from the remainder of the system.

Some utilities perform field tests to assess the validity of analyses.

7.9.2 Routine but Complex Analysis

This level of analysis will include voltage, transient stability and a.c. load flow analyses on the subsystem and system configurations that might emerge at various stages of restoration. The objectives will be to determine the amount of demand that these can support, including any difficulties such as identifying circuits prone to heavy flows or busbars prone to high/low voltages.

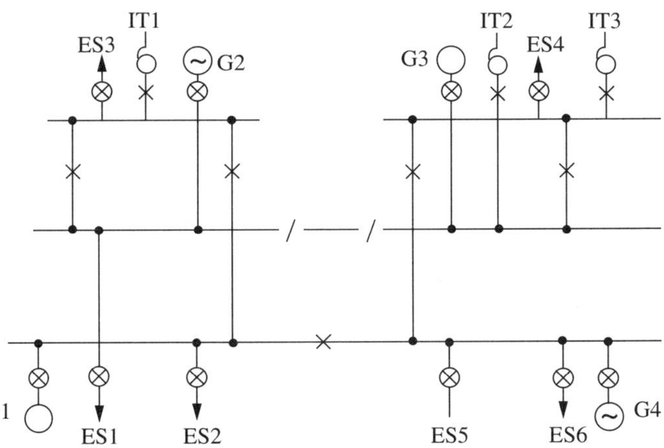

Location	Circuit identity	Switch name	Switch position	Designated control engineer	Control point
Cranby S/Sn.	ES1 ES2 Ð Ð	E105 E205	open open	Area control engineer	Cranby P/Sn control room

Figure 7.1 Example of a busbar configuration to be adopted at a major power station in preparation for restoration (also illustrating layout of information)

7.9.3 Operator Studies in the Event

Whatever training and preparations have been made, the operators will in the event be faced with the need to make quantitative assessments on power flows and voltage as switching proceeds. Experience, empiricism, simple approximations, tabulations of incremental nodal/circuit flows (coupling factors), pre-calculated values of the acceptable power flows across selected sets of circuits (cut sets) will be used. PC or remote terminal facilities have been provided in some control rooms to enable operators to make load flow and group transfer studies, including the necessary demand processing, rapidly. The value of these and their acceptability to the operators will depend upon the usefulness of the results (are they timely and give the information needed) and ease of use (simple and small data editing required, fast turn around).

7.10 RESTORATION FROM A FORESEEN DISTURBANCE

Foreseen or predictable disturbances usually develop as a consequence of shortage of essential resources, be these manpower, plant capacity, fuel or ancillary supplies. Essentially, restoration will be from a planned situation, and within reason its timing can be set by the operators. Restoration will consist of an orderly progression to normal operation as, for instance, the shortages are removed.

FURTHER READING

EPRI; 'Underfrequency operation of power systems'.

Kafka, J. *et al.*, 1981. 'System restoration plan development for a metropolitan electric system'. *IEEE Trans. PAS*, **100** (8).

Lams, J. L. *et al.*, 1986. 'Operation of generating units during system disturbances'. *Cigre Paper 39.07.*

Knight, U. G., 1986. 'System restoration following a major disturbance'. *Cigre Electra, 39.07.*

Otterberg, R., 'Restoration after disturbances in the Swedish bulk power network'. Swedish State Power Board.

Adibi, M. *et al.*, 1987. 'Power system restoration: a task force report'. *IEEE Trans. Power Systems*, **PWRS-2** (2).

Abidi, M., 1987. *IEEE Trans. Power Systems*, **PWRS-2** (4).

Marin, G., 1987. 'Service restoration following a major failure on the Hydro-Quebec power system'. *IEEE Trans. Power Delivery*, **PWRD-2** (2).

Kearsley, R., 1987. 'Restoration in Sweden and experience gained from the blackout of 1983'. *IEEE Trans. Power Systems*, **PWRS-2** (2).

Fandine, J. *et al.*, 1991. 'An expert system as a help for power system restoration after a blackout'. *Third Symposium on Expert System Applications to Power Systems.*

Lindstrom, R. R., 1990. 'Simulation and field tests of the black start of a large coal fired generating station utilising small remote hydro generation'. *IEEE Trans. Power Systems*, **PWRS-5** (1).

Cigre Study Committee 38, 1993. 'Modelling and simulation of black start and restoration of electric power systems'. *Electra*, **147**.

Abidi, M. *et al.*, 1994. 'Expert system requirements for power system restoration'. *IEEE PES Winter Power Meeting*, 94WM223-8.

Ancorra, J., 1995. 'A framework for power system restoration following a major power failure'. *IEEE Trans. Power Systems*, **10** (3).

Colloquium on Frequency Control Capability of Generating Plant, *IEE Digest No. 1995/208.*

Liou, K-L. *et al.*, 1995. 'Tie line utilisation during power system restoration'. *IEEE Trans. Power Systems*, **10** (1).

Negate, T. *et al.*, 1996. 'Power system restoration by joint usage of expert system and mathematical programming approach'. *IEEE Trans. Power Systems*, **10** (3).

Flin, D., 1998. 'Lessons from Auckland'. *Modern Power Systems.*

Germond, A. *et al.*, 1998. 'Decision aid function for restoration of transmission power systems'. *IEEE Trans.*, **13** (3).

8
Training and Simulators for Emergency Control

8.1 INTRODUCTION

It is unusual in a book dealing with engineering technology to include material on training. However, the human component in decision making is more important in real-time system operation, particularly during disturbed conditions, than in most areas, and it is thus appropriate to review the training of system operators.

Topics covered in this chapter will include the need for training, its content, alternative methods, suitable location, duration and frequency, and the resources needed. Training on simulators is acknowledged as being the most effective, and descriptions will be included of modern installations.

8.2 TRAINING IN GENERAL

An engineer's need for vocational training will depend upon the path followed and stage reached in his/her career. At first entry, the main distinction is likely to be between those with a university degree or equivalent, and those with a lower qualification (Table 8.1). The former are understood to be more usual in Europe and the latter in North America.

Large utilities may provide in-post training in management, finance, staff relations, industry organization, etc., and at the higher levels, using management and business schools, national and international aspects of these topics. In-post vocational training will also often be provided internally by large utilities. The more specific and detailed the subject, the greater will be the incentive to use in-house expertise. Small utilities will, however, often have to rely on external training facilities – university courses and seminars, teaching companies, professional groups, etc.

Table 8.1 Training at entry into a utility

Educational status on entering utility	In-company training	Job in company (assumed in engineering area)
University degree	Power system training, utility background, utility experience with some leaning towards intended work area.	Junior/more senior engineers
School	Academic work, general engineering and power systems training, utility background, utility experience, perhaps with a leaning towards intended work area.	Junior engineer Operator

8.3 THE NEED FOR OPERATOR TRAINING

As noted above, power system operators (as indeed, operators of other engineering systems) have a unique role – they must make timely, definitive and personal decisions. The reliability of supply, particularly during perturbed and non-planned conditions, will be influenced by the performance of the control personnel, noting that:

(1) Commercial and environmental imperatives will require systems to be operated closer to their technical capacities in future. Many utilities no longer operate on a cost plus basis, affecting decisions and their post-event evaluation.

(2) Restructuring in the industry requires increased competence in trading and commercial areas.

(3) Enlargement of interconnected systems (for instance, the interconnection of UCPTE and CENTREL) will change the technical, economic and political constraints and opportunities facing the operators.

(4) Innovative devices and procedures such as universal power flow controllers, supermagnetic storage, increasing use of d.c. links, real-time demand management, offer more flexibility and control, but will also require more sophisticated optimisation techniques.

(5) Newer types of generating plant are more flexible and have different output-cost characteristics (e.g. multiple slope and discontinuous input-output cost curves) compared to older plant; inertia constants may be lower, adversely affecting dynamics. The management of possibly numerous independent

power producers will increase the importance and complexity of power and energy trading. The trend towards earlier retirement increases the problems of maintaining experience and know-how in the control room.

In regard to emergency control, point (1) suggests that the number of critical situations is likely to increase, and the margins for manoeuvre will be less when these occur. In a restructured industry (point (2)), utilities may be less willing to offer mutual assistance or to release technical information to neighbours (the minimum information to be released has been specified in parts of North America). Increased interconnection (point (3)) is likely to change the operational limits for some countries. Innovative devices (point (4)), although increasing the controllability of power systems and the speed with which remedial measures can be implemented, will require operators to be familiar with the characteristics of these devices. Similarly, operators (point (5)) must be familiar with the characteristics of all generation on the system – load pick-up and rejection capabilities, overload ratings, synchronizing and desynchronizing load/time profiles etc. They should also be familiar with the impact of changes in the neighbours' systems on their own system, for instance on power flows, stability margins.

The normal process of 'father-son' tuition is shortened by early retirement, and has to be replaced by increased training. Staff replacement is not eased by the fact that in many utilities movement between the system operation and other functions seems limited.

Operators must have a comprehensive knowledge and understanding of the procedures and criteria required for operation of their own system, and also those for neighbouring systems with which they have an operational interface. At the end of the training period, a trainee control operator may be assessed formally by more senior staff. The process may be repeated as the engineer progresses up the promotion ladder.

8.4 THE CONTENT OF TRAINING

Training to handle emergencies is a subset of the total training requirement, and would include the following topics:

- principles of power system dynamics, forms of instability, recognition of these;
- assessment of security state, measures of security, recognition of insecure states;
- avoidance of emergency states;

- operating in the emergency state and recovery from this, plant behaviour and control under abnormal conditions, system switching, load shedding, black start, system synchronising and recovery.

Additionally, operators at these times should be able to use all the available communication media, remote control facilities and computer support with complete confidence, and be aware of the operating characteristics of protection systems. They should be familiar with procedures including calling out additional staff, initiating the setting up of 'incident centres' and actions in response to enquiries from the media and the public, even from their own company hierarchy.

8.5 FORMS OF TRAINING

The forms of training for handling emergencies will include father-son tuition, group discussion, training courses, seminars and simulator training.

8.5.1 Father–Son Tuition

This offers the opportunity for in-depth discussion on a one-to-one basis. Shift rotation will usually mean that the trainee will be on duty with several senior operators at different times. There is no impact on staff availability, and this form of tuition should clearly be encouraged – the senior operator will sometimes find gaps in his own knowledge!

8.5.2 Group Discussion

Structured group discussions are an excellent way to spread and exchange experience. One approach is to discuss emergency situations in depth, not only those within the utility, but also reports of disturbances in other utilities.

8.5.3 Training Courses

Most system operation courses will include material on emergency control. Their value to the professional operator will often be enhanced by lectures on plant, communications including emergency and back up facilities, data networks, trading, controls available to the operators, interaction with neighbours under emergency conditions, system control in the future, etc. Not least, these provide an opportunity for staff from different locations to exchange views and experience.

8.5.4 Organization of Training Courses

Some large utilities with several control centres will invest in a training centre where all operators are trained. The centre will contain rooms for lectures and discussions, and not least, the utilities' training simulators. There can be advantages in locating it at an operational control site.

Wide variation is found in the frequency and duration of training courses. Depending on background, a new trainee may be seconded to several operational centres – power station, district, operational planning office, trading office – before joining a control centre for some specific training in system control duties. It may be several months before a trainee is assessed as capable to take up shift duties. Refresher training will be organized at preferably not more than two yearly intervals, plus *ad hoc* courses to cover the introduction of new technology and techniques. Much of this training will be devoted to routine work, and there will finally be the need to train control staff to handle abnormal situations, done most effectively on a power system simulator.

8.5.5 Assistance in Commissioning

New computing and control systems will require acceptance tests, possibly proving and performance tests, with heavy demands on manpower. Simultaneously, shift rotas will include periods when operators are not committed for shifts, but still available for duties. These operators can then assist in the testing, easing the problems of finding staff for this work and, at the same time, familiarizing the operators with the new systems.

8.5.6 Self-tuition

The operator's job during some shift periods is characterized by periods of high activity with sometimes longish periods of relative inactivity. Many control centres then assign off-line tasks, such as administration or preparation of procedures, to the operators, maintaining their alertness and diversifying their work load. Some operators may postulate for themselves various emergencies and review what actions they would take.

In another alternative, lessons have employed a PC. This can be used, for instance, to display examples for solution by the student, the correct approach and result, plus explanations if the student is stuck.

8.6 TRAINING SIMULATORS

The broad objective of training in handling severe disturbances will be:

- to increase the operator's confidence in his ability under stress to weigh up situations and make and implement timely and correct decisions;
- to improve knowledge of the technical characteristics of the system under dynamic or degraded operating conditions;
- to improve knowledge of procedures and facilities for handling emergency situations.

The 'replica type' simulator, in which the performance of the actual system is modelled and an actual or close approximation to the operational man-machine interface is provided, will clearly have advantages over the 'generic type' of simulator, in which performance of an operational system is modelled, but there is no attempt to replicate the actual system or the man-machine interface.

8.6.1 Outline Specification for a Training Simulator

A specification for a comprehensive simulator would include the following features:

Operational and ergonomic

(1) The standard control room displays including a mimic board if used, or as close to this as possible; the audible and visual alarms should be provided.

(2) The displays should be animated with up-date response times similar to those found in practice (it should be remembered that unless the simulator is also to be used as a design tool, only 'plausible' responses are needed). The displays will receive data from a system model [8.1, 8.2], the trainer and trainee.

(3) An equivalent to the 'outside world' should be provided; this will represent the power system, transmission district and station staffs, other control rooms, even other utilities. In practice, one main component will be a comprehensive dynamic system model with capability for the trainee and trainer to inject switching and loading changes, etc. The trainer will be able to initiate alarms, and send/receive all forms of message to/from the trainee.

(4) Communication to the equivalent of the outside world.

(5) Tape decks to play back and inject into the system model demand profiles, system faults, SCADA faults, external disturbances, and possible unit commitment details; the ability to take 'snapshots' of operational conditions from the SCADA system, can provide ready made starting points for training sessions.

(6) Tape decks to record voice and data from the training session.

(7) One or more operator work stations equipped with the standard displays and communications; two or more stations will allow communication between operators to be included in the training, and a few utilities provide for interaction between two control rooms.

(8) The routine system dispatch mechanism must be modelled.

(9) Voltage control mechanism should be modelled.

(10) Generation behaviour under black start and islanded conditions.

Technical

It should be possible to model:

(11) Single, multiple coincident and sequential faults (balanced and unbalanced).

(12) Overload conditions.

(13) Protective gear operations (overcurrent, unit, impedance, demand shedding, other automatic switching schemes; in my experience, this has been one of the weakest areas of modelling in training simulators).

(14) Maloperation of protective gear.

(15) Oscillatory conditions.

(16) Voltage decay/collapse.

(17) System splitting and islanding, including the continuing dynamics within the islands.

8.6.2 Alternative Forms of Training Simulators

Stand-alone training simulators are expensive. The hardware costs will be a significant part of the cost of an operational control room, and there will also be software costs for the system model, trainer interface and displays, etc. Hence, there will be a cost incentive, if nothing more, to devise alternatives to the stand alone simulator. A summary of some of the solutions adopted is given below, in decreasing order of cost.

Stand-alone Simulator

The stand-alone simulator will comprise some six main elements (Figure 8.1): a mathematical model of the system basically as used in dynamic stability analysis; a model of the dispatch process including demand profile and generation instruction; display software; database; the trainees' interface (as a minimum one workstation and as a maximum say two control rooms); the trainer's interface and an algorithm to generate a demand profile in the form needed by the dispatch algorithm and constructed to fit the format of the demand information (for instance, demand at discrete time intervals) set in by the operator.

Stand-by Control Room and Training Simulator

A stand-by control room will contain at least the main display, communications and interfaces required in the operational control room. The addition of trainer interfaces, database and software will extend it to function as a training simulator (see Figure 8.2).

Use of Stand-by and Spare Equipment in the Operational Control Room

The work load in a control room will peak at certain times, typically when the demand is about to change rapidly or when switching for maintenance and new construction is needed. At other times, a workstation/s may be free, and with the addition of processor capacity, trainer interface, software for a system model, displays and database, will provide a training simulator; fundamentally a comprehensive stand-by CPU suite and spare workstation (or equivalent in open systems architecture) can be engineered to provide the majority of a training system (Figure 8.3). The disadvantages of this approach, and to a lesser extent of the previous one, are that the simulator facility will only be available when the components shared with operational duties are free. (I also feel that there could be some risk of confusion in an operational control room when part of it is being used to show non real-time data.) Also, it will generally not be possible to include the mimic diagram in the simulator.

Other Forms of Simulator

Simulators have been provided for training in component parts of the operators duties – for instance, switching and generation dispatch [8.3] and stability

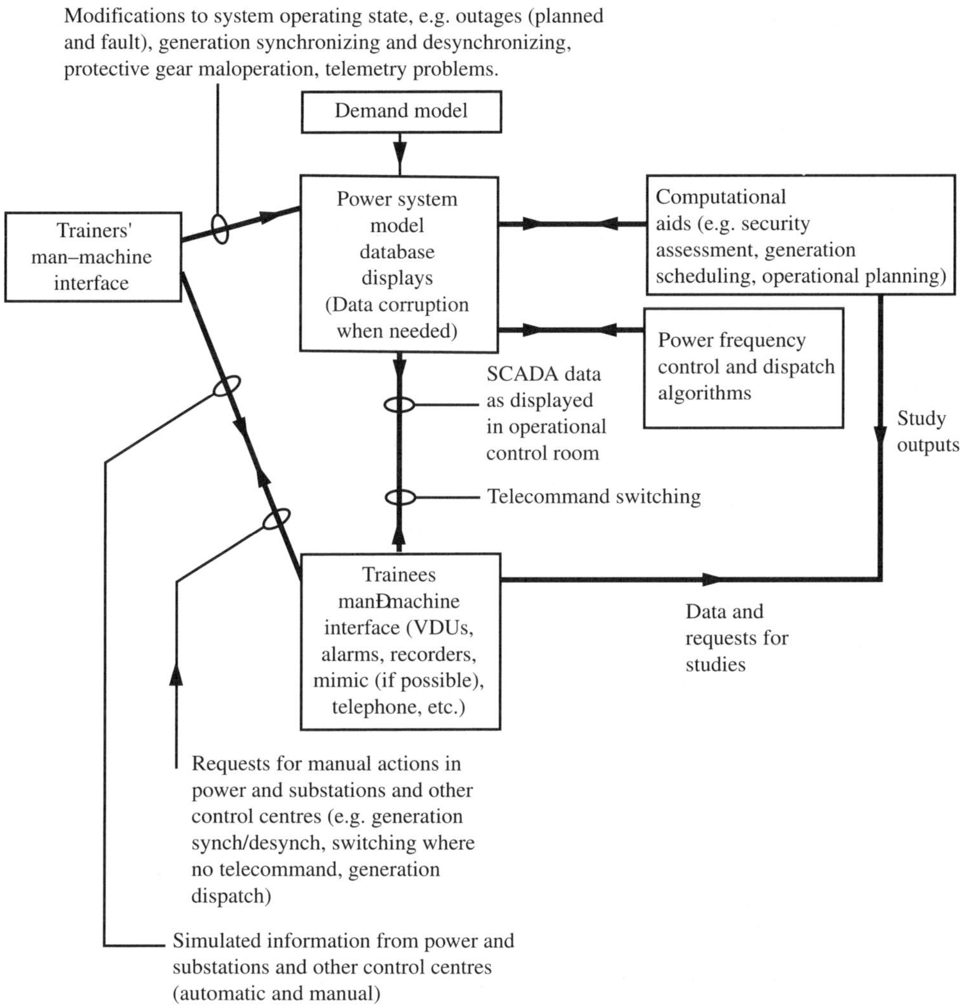

Figure 8.1 Outline of a replica simulator

phenomena. These will be much less costly, but usually only one-on-one training will be possible, and the realism of the replica simulator will be lost. Their application will be mainly in training new entrants to control room duties.

8.6.3 Some Commercial Training Simulators

Some of the simulators which have been installed by utilities or are commercially available are briefly described in this section. The information comes from papers published in the mid–late 1980s to mid 1990s.

Table 8.2 Modelling parameters of a generation dispatch training simulator (see also Figure 8.4)

Parameter steam	Quantitative
Number of steam generator units	6
Governor response	Operative between 48.5 and 50.5 Hz
No load cost	10% of full load cost
Manual loading and deloading times	40 minutes ± 8 minutes
Boiler stored energy	4% of full load output
Number of gas turbine units	6
Time zero to full load generation	2 minutes ± 30 secs
Time full load to zero generation	2 minutes ± 30 secs
Number of base generation sections	5
dead band	48 ± 0.5 Hz to 50.1 ± 0.5 Hz
output droop	2%/Hz below 48 Hz
delay in output droop	12 minutes to final value
Number of base load sections	5
frequency characteristics	2%/Hz
Grid simulator	
frequency dependence of load	2%/Hz
generation droop	2%/Hz
generation droop delay	12 minutes
reserve generation pickup rate	8 minutes
steam energy reserve	1 MWhr/200 MW of generation
Frequency accuracy	
long scale	±0.125 Hz
short scale	±0.025 Hz
Clock rate	4 × faster than real time
Outputs	Analogue instruments, chart recorder, digital clock

England and Wales (The National Grid Company)

NGC's original Dispatch Training Simulator (DTS) was provided to enable its system operators to practice controlling the system under onerous fault conditions (see Figure 8.4). A later DTS [8.5] was installed in a stand-by control room, with the facilities enhanced to cover training in commercial aspects. The DTS modelled the complete NGC transmission network with interconnections to Scotland and France, and the database was sized as follows: 250 generators, 170 supergrid substations, 970 supply point loads, 188 voltage control mechanisms, 10 000 circuit breakers and disconnectors, etc. An a.c. load flow was performed every 5–10 seconds, and the output fed to the DTS displays. The dynamic parameters of all the independent generators were modelled, and all the daily load patterns could be replicated. A session on the simulator could be based on past or present system conditions, using snapshots taken from the on-line SCADA system, or 'retro' snapshots. It incorporated an on-line demand predic-

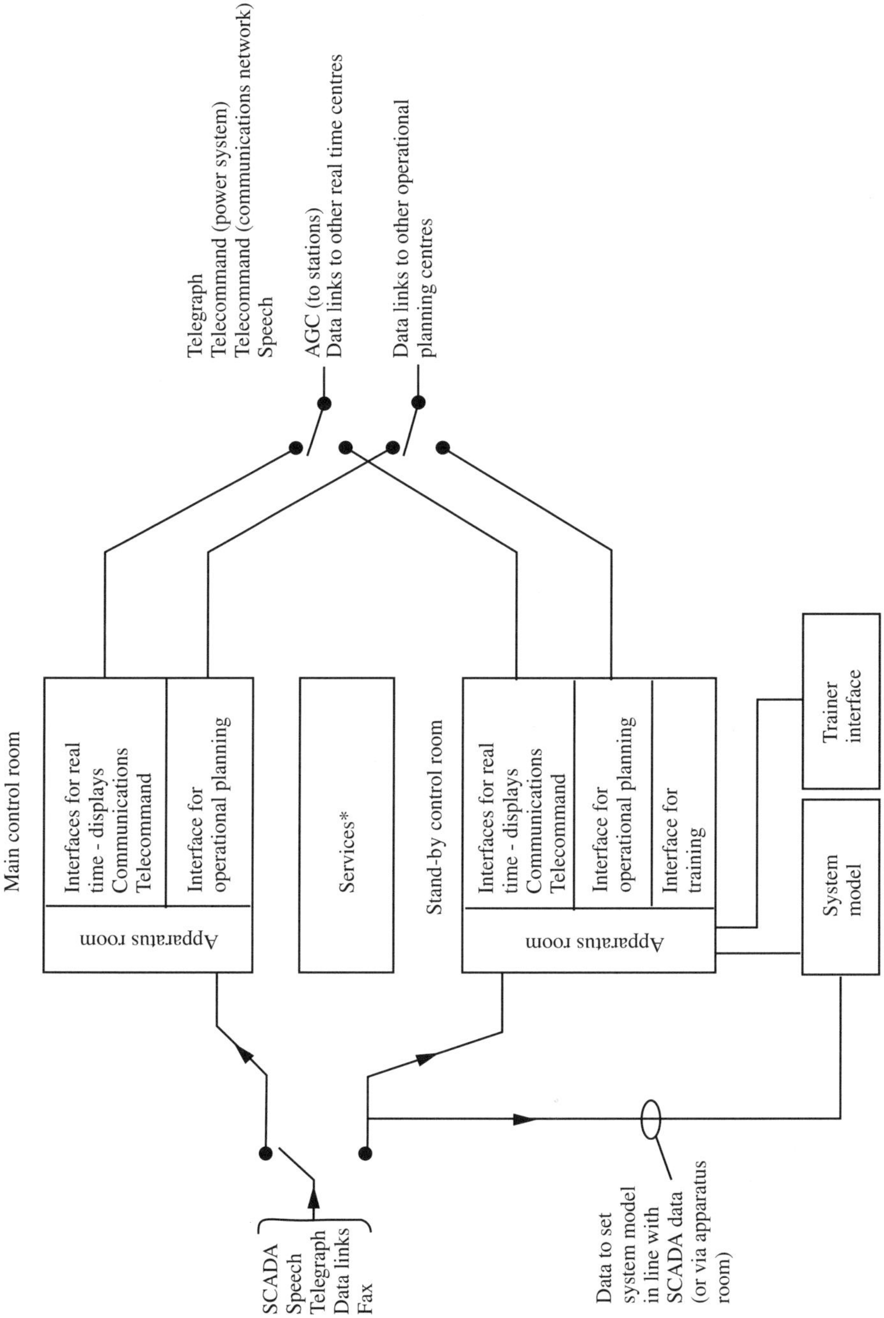

Figure 8.2 Outline of possible data flows with training facilities included in a stand-by control room. (1) Each of the data links may be duplicated. (2) 'Services' will cover power supplies, air conditioning, heating/cooling, fire prevention, security, domestic amenities

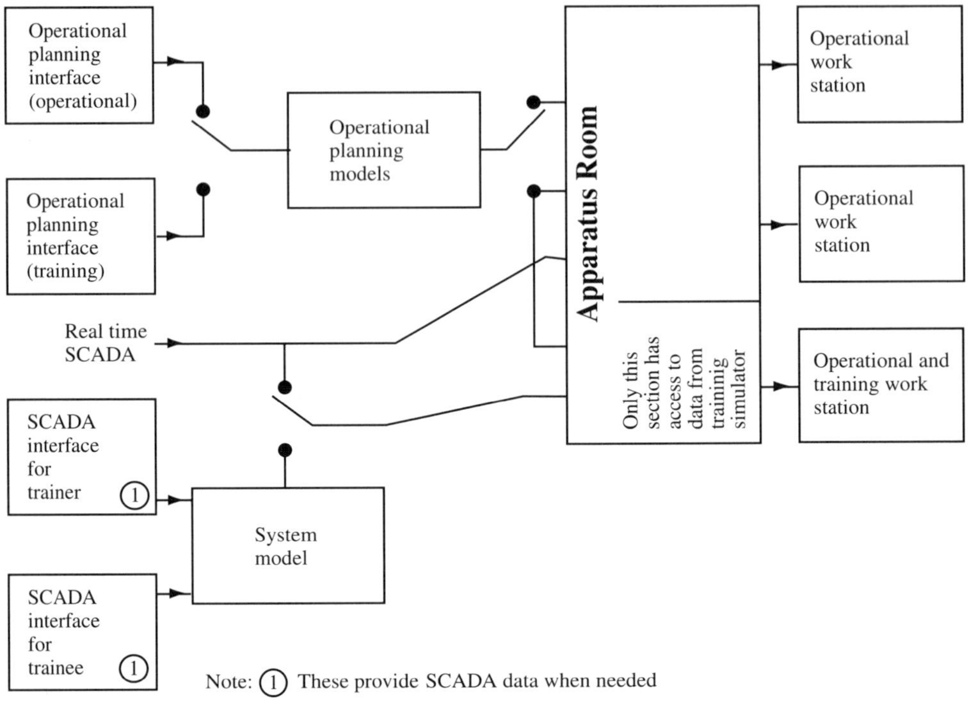

Figure 8.3 Outline of possible data flows with one workstation selectable to operational or training functions

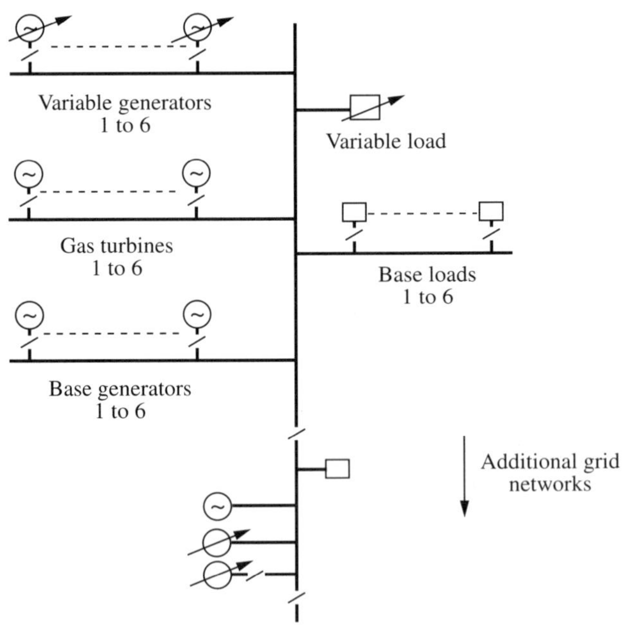

Figure 8.4 Dispatch training and emergency loading simulator. Reproduced by permission of the National Grid Company plc

tor linked to a processor that produced generation dispatch advice. The trainees also had access to an on-line real time network analysis package running a.c. and d.c. load flow and fault level studies.

Training was provided at three levels – induction, primary operational skills, and advanced operational training. Between four and ten operators attended a course, and were formed into teams, with one or two being taken from the team to act as role players (e.g. District transmission, Generators) on a rotational basis. Experienced shift teams were trained two or three times a year using the advanced training package. Each course consisted of four or five scenarios, with the trainees relocated to different functions in these. The scenarios were played in real time, and might last several hours depending on the scenario. Courses were available for external customers; sessions were also held to aid managers and staff in handling external communications. Computer-based training modules were also developed (Figure 8.5).

France (Electricité de France – EDF)

EDF installed a stand-alone replica type, training simulator at its Caen training centre in 1989 [8.6, 8.7]. It has been used to train national and regional control room operators.

There were two training rooms: one containing an exact replica of the national control centre workstation and mimic board; the other was a replica of a regional centre, with a mimic board. An artificial network was shown as it would be unrealistic to try to show seven different regional networks. Standard functionalities were provided – freeze, retro, slowdown, aids for preparing scenarios. A long-term dynamics model was included [8.6–8.8] as well as models for automatic devices, protective gear and controllers.

Figure 8.5 Computer-based training

Training is provided by a full-time eight man team, who prepare and run the scenarios. During a training session, one instructor controls the session from his workstation while a second walks round, watches and advises the trainees. These 'telephone correspondents' are also trainees, acting as power station and/or substation operators, in telephone contact with the trainees and acting as they request. The members of the team have extensive experience in system control and related work, and remain in this post for about four years.

Junior operators are given training courses, including DTS training at the beginning and end of their first year, the latter dedicated to the control of disturbed conditions. Experienced operators take one week refresher courses every 18 months. Each DTS session lasts on average six hours, with another hour for briefing and two hours for debriefing. It has also been noted [8.6] that each Regional Centre will be provided with a stand-alone simulator, but still connected to the Regional computer system to retrieve data. The simulators will be equipped with two workstations, one for the instructor and one for the trainee. It will also then be possible to train staff associated with dispatch.

The Electric Power Research Institute's (EPRI) Operator Training Simulator

North American interest in simulator training for operators stemmed largely from the blackouts of the 1960s and 1970s. Some generic, stand-alone simulators were demonstrated in the 1970s and some utilities installed simulator software in the stand-by/back up computers of their EMS's in the 1980s. These did not realistically model system behaviour over the full range of disturbed conditions, and were slow to respond to instructions.

EPRI's first operator training simulator was installed at Philadelphia Electric Company in a suite of dedicated training rooms. One room contained consoles, as used in the energy control centre, for two trainees and a mimic board. Two instructors observed the trainees from the instructors' room, one monitoring the trainees' activities, while the other conducted the training exercise and acted as the outside world. The EPRI simulator reported in 1992 comprises a dynamic power system model enabling voltage collapse, islanding and black start to be modelled [8.10–8.12]. The control centre model is the utility's own EMS software operating on a separate or back-up computer connected to the power system model through a communications interface.

8.6.4 The New Generation of Dispatch Training Simulators

There are numerous descriptions of hardware and software to be used in operator training in the literature, but instead of attempting any further review of these, the author has thought it more useful to mention recent developments. [8.8]

A Swedish Development

One of these, called 'ARISTO', has been developed by the Swedish transmission company Svenska Kraftnet, in collaboration with ABB Cap Programator [8.13–8.15]. The brochure lists system phenomena which can be simulated – transient stability, long-term dynamics with frequency control, voltage collapse, cascade tripping, island operation with any number of islands, and manually controlled restoration. Robust algorithms and models allow simulation of disturbed and collapsed systems. The user interface was based on X-windows technology.

Some proposed features of a commercial simulator included:

- advanced simulation including transient and voltage stability, long-term dynamics, large disturbances including sequential tripping of circuits, restoration;
- real-time simulation for systems for up to several hundred nodes;
- detailed substation modelling (e.g. breakers and isolators shown);
- simplified models for training and real time simulation, and detailed model for analysis;
- the unit modelling to include excitation systems, power system stabilizer, rotor and stator current limiters, turbine, boiler, turbine governor;
- automatic generation control;
- frequency and voltage dependent demand models;
- high voltage direct current;
- static var compensators;
- series and shunt compensation;
- protection models to include overcurrent, distance, over and under frequency, over and under voltage.

Other applications envisaged for the simulator include its use as a general tool for system analysis, a test bench for EMS software, study of short-term predictive operation, demonstration of power system behaviour to equipment manufacturers, and as an external system model for other types of simulation.

FAST-DTS

FAST-DTS is a dynamic simulation model whose key features are reported [8.16] as covering phenomena from rotor electrical transients to slow or quasi-stationary. The prime mover/generator model includes governor and energy or fuel systems for coal, oil, nuclear, gas turbine and hydro plant. Besides the standard control devices such as load shedding, tap change, and line overcurrent, FAST-DTS incorporates relays whose operation can only be described if system transient models are included, for example generator loss of synchronism, islanding, directional and impedance protection. FAST-DTS can be used in stand-alone mode or as a system simulator when connected to a SCADA/EMS system. The stand-alone form has been delivered to the North China Institute of Electrical Power.

8.7 THE USE OF DISPATCH TRAINING SIMULATORS IN PRACTICE

A recent Cigre questionnaire on the use of DTS was sent to 35 utilities who had installed these. The utilities ranged in size from under 7 GW to over 50 GW. Some results were [8.16]:

Proportion providing training in	
restoration	87 percent
team training	78 percent
EMS functions	74 percent
safety management	69 percent
Number of training periods per year	
less than one	7
two, three or four times	7
five times or more	5
Duration of training periods (days)	
less than one day	5
one or two days	3
three or four days	4
more than five days	3

Individual training sessions last 30 minutes and upwards, with eight hours maximum for very complex situations.

Training staff

Some 60 percent of the replies indicated that the training staff are dedicated to the use of the simulator. Otherwise, the trainers are experienced operators seconded to work as trainers for varying periods.

Team training

The control teams are rostered as teams in one-third of the utilities.

8.8 CONCLUSION

Much of industrial training is devoted to teaching the 'doing' parts of the workers' tasks. Another major component will be training in safety aspects; neither of these has been considered here. Instead, the training discussed has been in the decision making aspects involved in the operation of the system. The background necessary for this will be material on power system and plant performance, power system states, avoidance of emergency states and recovery from these. An essential component of the training syllabus will be descriptions of the system control facilities – capabilities, limitations, and how to use these facilities in all conditions.

Training can be given in several forms but there are advantages in more advanced training for emergencies in placing the trainee at some stage in the environment in which he might find himself during a serious power system disturbance. This can be done most realistically using a replica simulator or a well equipped standby control room. In each case, disturbance scenarios must be devised or selected from past incidents.

Privatization has undoubtedly increased the complexity and importance of inter-utility trading. This is a specific area of expertise and control teams will have to include or be supported by staff having a substantial knowledge of this topic.

The construction of training scenarios has sometimes been seen as difficult. A random number table plus a numbered list of circuits would seem to be a possibility to generate complex outage conditions.

REFERENCES

8.1. Waight, J. G. and Van Meeteren, H. P., 1988. 'Considerations for implementation and integration of an operator training simulator'. *Cigre Working Group 39.06.*

8.2. Yung, K., Lo, K. and Cheng, J. W., 1990. 'Operational experience on the China Light and Power Company's system operation training simulator'. *IEEE Trans. Power Systems*, 5 (2), 521–530.
8.3. *Power System Loading Simulator*, Moko Electronic Systems.
8.4. Fisher, N. C. 'Computer based training for CEGB system control staff – a pilot project'.
8.5. *The Dispatch Training Simulator*, NGC brochure.
8.6. Logeay, Y., Macrez, J. and Meyer, B., 1995. 'Training simulators for control centre operators: EDF's past experience and projects for the future. Vol. 1,' *Stockholm Power Tech. Conf.*, pp. 170–175.
8.7. Jeanbart, C., Logeay, Y. and Musart, M., 1988. 'EdF simulator for control centre operators'. *Cigre paper No. 39-12.*
8.8. Logeay, Y., Bose, A., Cukalevski, N. and Handschin, E., 1996. 'Requirements for a new generation of simulators to train dispatchers in a changing control room environment'. *Cigre Electra*, **167**, pp. 132–153.
8.9. *SCAP: Simulators for grid operation*, brochure, EDF, CORYS.
8.10. *Power System Operator Training*, EPRI Brochure, 1992.
8.11. 'Simulating the control centre'. *EPRI Journal*, 1990.
8.12. Barret, B. J. *et al.*, 1991. 'The uses of the EPRI operator training simulation for power system restoration'. *PICA*.
8.13. Edstrom, A. and Walve, K., 1994. 'A highly interactive front simulator covering transient stability and long term dynamics in large power systems'. *Cigre paper 38-2.04.*
8.14. *Aristo: the future in power system simulation.* Svenska Kraftnet/CAP Programator/ABB brochure.
8.15. *Ariadne*, advanced reactive interaction application for dynamic network simulation, brochure, Svenska Kraftnät.
8.16. *FAST-DTS: the new generation of dispatcher training simulator*, Brochure, Tractabel Energy Engineering, CORYS TESS.

FURTHER READING

Krost, G., Lutterodt, S., Logeay, Y., Konepfel, R. and Skiold, R., 1997. 'Improving human performance in the control centre'. Cigre Working Group 39.03, *Cigre Electra*, **174**, pp. 90–105.

Cukalevski, N. and Johansson, A., 1993. 'Requirements set on control room personnel'. *Cigre SC 39 Colloquium*, Sydney.

'Existing competence requirements and training for control room personnel'. *Cigre Electra*, No. 171, 1997.

Krost, G. *et al.* 1993. 'Impacts of operators' selection and training on power system performance'. *Working Group 39.03.* Cigre Colloquium, Sydney.

Carey, E. (reporter), 1993. 'Simulator primed for the real thing'. *The Grid.*

Webster, R., 1997. 'The value of training'. *National Power News* (this is an article on a station simulator).

Handschin, E. and Knight, U. G., 'System operation training simulators, Parts 1 and 2'. *Cigre Working Group 39.06*.

Wilkinson, W. 'System operation switching simulator – CEGB, mid 1970s'.

Svoen, J., Knight, U. G., Kowal, J., Marigo, L., Otterberg, R., Reilly, J., Werts, R. and Winter, W. H., 1982. 'The use of real time simulators in operator training and power system control'. *Cigre Electra*, **84** (includes a copy of the questionnaire used), pp. 85–103.

Handschin, E., 1989. 'Status and trends of dispatch training simulators'. *Cigre Symposium on Operation of Electric Power Systems in Developing Countries*, Bangkok.

Elder, E. and Metcalfe, M. J., 1981. 'An efficient method for real time simulation of large power disturbances'. *Cigre paper No. 81 TR 02 SC 32*, Rio de Janerio (this is one of several papers by these authors on this subject).

Kosonen, H., Solberg, A. *et al.*, 1992. 'Computer based training for power system operators'. *Cigre*.

Cigre Study Committee 38, Working Group 02. 'Modelling and simulation of black start and restoration of an electric power system: results of a questionnaire'. *Electra*, **131**.

Schaffer, G. *et al.*, 1992. 'Scenarios for dispatcher training simulators'. *Cigre Electra*, **141**.

Necar, W., 1996. 'Old desks make way for world showpiece'. *NGC Network*, May.

9

Plant Characteristics and Control Facilities for Emergency Control, and Benefits to be Obtained

9.1 INTRODUCTION

I had intended originally to call this chapter 'The Costs and Benefits of Emergency Control', and to provide a simple comparison of the cost of emergency control facilities against an estimate of the benefits to be achieved by their installation. Several factors persuaded me otherwise:

- Numerical values of costs and benefits are likely to be system-specific; qualitative rather than quantitative comparisons are more appropriate in a general review.
- The costs to the economy at large of a major disruption of supply will heavily outweigh the costs of emergency control facilities, and hence will not be a deciding factor in judgements on whether or not to provide emergency control facilities; rather, these costs may provide a basis to select the better alternatives if capital expenditure has to be limited.
- It is difficult to draw a line between equipment provided for emergency control and that for normal control, and if done, how to deal with shared costs, the voltage and current transformers for instance; from this point of view, it would seem best to attribute to emergency control only the equipment and costs provided to handle emergencies, underfrequency relays would be an example.
- Even with this distinction clear, the costing can still present problems; work will be involved at many substations involving access to current and voltage transformer circuits, tripping circuits and relays, relay and control panels, displays, etc.
- The estimation of potential savings to the economy will also be difficult; this will require the cost per unit of unsupplied energy and the quantity of

unsupplied energy to be assessed, both being judgmental and possibly contentious.

- Utility managements will be very conscious of the bad press which major disturbances will attract, and of the need to be able to demonstrate due diligence in minimizing the incidence and impact of disturbances; the qualitative and quantitative factors to be discussed here will be essential background to these arguments.

In view of these factors, the first part of this chapter will review the total facilities and characteristics for emergency control which should be considered by a utility, with emphasis on functions and relationships to normal control rather than costs. The second part will discuss benefits in qualitative and quantitative terms. The author should add that the whole topic of costs and benefits is difficult, and hopes that the approach of considering benefits as avoided costs is found acceptable.

9.2 THE CHARACTERISTICS AND FACILITIES REQUIRED FOR EMERGENCY CONTROL

The characteristics and facilities needed for comprehensive handling of emergencies are outlined below for the various parts of the power system, including problems within the system control structure itself. Brief comments have been added on how these relate to normal operational requirements – in summary, it can often be said that 'emergency control is the same as normal control, only a bit more so'.

9.2.1 Generating Plant

The characteristics of generating plant with added importance for emergency control will be overload capacity, the changes in active power output possible over the loading range, and the times in which these can be achieved, the ranges of reactive power available over the active power range, the times to load the plant from various at rest states (cold, hot, etc.), and the times to deload it.

9.2.2 Transmission Plant

As with generation, an important characteristic of transmission in the context of emergency control will be its overload capacities. This is discussed below for individual components – overhead lines, cables, etc. – as influenced by thermal ratings. Other factors such as voltage regulation and stability will influence the system capability, and any scope for exceeding planned limits in respect to these is discussed in Section 9.3.

9.2.3 Overhead Lines

The thermal rating of overhead lines is a well researched subject. Continuous ratings will depend upon ambient conditions (temperature, solar radiation, conductor surface, wind speed), and equations are available to estimate current carrying capacities of a conductor in terms of these parameters and its permissible core temperature, the annealing temperature of which will be the limiting factor. Short term ratings are heavily influenced by the duration of any overload, and by the value of the line flow prior to the overload, as illustrated in Table 9.1 (based on data in *Modern Power Station Practice Vol. L*).

9.2.4 Cables

Cable ratings will depend upon the maximum permissible conductor temperature, soil temperature and soil thermal resistivities. Together these will fix the insulation temperature, its rate of deterioration, and hence the expected life of the cable. Installation factors affecting the rating will be the cable cooling, trench backfill, proximity to other cables and their loadings. A cable circuit will usually be required to transmit its rated power only when onerous and fairly infrequent loading and configuration conditions occur, often in periods when a system reinforcement is imminent. Thus, on the rare occasions when the full continuous rating is required, a conductor operating temperature (and hence a rating higher than that applicable through the life of the circuit) may often be assumed.

Cable systems may be specified to match the ratings of standard overhead lines over a range of core temperatures.

9.3 THE SYSTEM AND DEMAND

The control mechanisms brought together in the system are principally the network configuration, the demand, and the mechanisms for the adjustment of active power flows in series elements and reactive power flows in shunt elements, respectively (there will be second order effects from these mechanisms, such as the impact of power changes in shunt reactance elements on power flows in series elements).

Table 9.1 Comparative thermal ratings with respect to winter continuous rating (PL = protection limit)

Season – Winter

Duration mins \ Pre-fault flow	85%	60%	30%
20	1.23	1.32	1.38
10	1.32	1.55	PL
5	1.51	PL	PL
3	1.63	PL	PL

Season – Spring/Autumn

Duration mins \ Pre-fault flow	85%	60%	30%
20	1.14	1.21	1.27
10	1.22	1.42	1.55
5	1.39	PL	PL
3	1.58	PL	PL

Season – Summer

Duration mins \ Pre-fault flow	85%	60%	30%
20	0.98	1.04	1.09
10	1.04	1.21	1.31
5	1.18	1.51	
3	1.33	PL	

9.3.1 Configuration

However detailed the operational planning work, it is very likely that network configurations which have not been studied previously will occur during emergency conditions. Hence, changing the configuration, one of the quickest ways to adjust power flows, may be worth while, although it would be prudent to check the viability of the change before making it if suitable real-time load flow facilities exist. Other alternatives might be to lower the security standard, say from a double circuit to a single circuit criterion, or to shed demand; provided there was no undue risk of faults in the extant ambient conditions, the operator might well judge the former to be the better option. Switching of series capacitors could provide a further mechanism.

The facilities provided for switching in normal operation will generally be adequate to handle emergency conditions, particularly with the increasing use of remote control.

9.3.2 Demand

Except in very abnormal situations when major transmission substations may be deliberately disconnected, control of the demand is exercised at the distribution voltage levels either by disconnection (load shedding) or by reducing the voltage (tap locking). Some supply tariffs include agreements whereby the utility can disconnect consumers, perhaps up to agreed durations or numbers of times per year, to suite the utilities' plant demand conditions or operating costs.

Although load shedding is a coarse measure for control, the mechanisms should provide flexibility in the location and amount of demand shed, the signal(s) which initiate the shed, the intentional time delays, and the system frequencies (or other operational variable) at which stages should be shed. As a general rule, the requirement will usually be to shed demand as quickly as possible after the shedding frequency threshold/s are reached. The equipment needed to detect low frequencies will only be required for emergency control.

'Tap locking' is a mechanism to hold transformer taps at a fixed position, irrespective of terminal voltages. The author judges it to be a worth while adjunct in normal operation, and hence not chargeable to emergency control.

9.3.3 Adjustment of Active Power Flow

The ability to adjust active power flows is also a valuable feature in normal operation, and should not generally by chargeable to emergency control. The mechanisms available within a fixed configuration will be by changing nodal transfers, or by injecting voltages into the series elements of the network by a quadrature booster, whether as a single device or as part of a universal power flow controller.

9.3.4 Adjustment of Reactive Power Infeeds

Nodal reactive power infeeds are the main factors in the control of network voltages. The requirement in normal operation will be to keep the voltages within the operational standards. In some circumstances, there will be merit in changing the target operating voltages, usually to meet difficult weather conditions:

- One utility considered reducing the transmission voltage possibly by up to 10 percent to reduce flashovers on overhead line and substation insulators in adverse conditions; these could be rain, snow or a thaw following several weeks of dry weather, during which pollution or ice would accumulate on insulators; whichever occurred, conducting films could form leading to flashovers.
- Conversely, overhead line capabilities could be increased by operating the system at a higher target voltage, still keeping the maximum voltage within the equipment voltage ratings; this would be particularly useful during very hot weather.
- Generators are often the main source of controllable reactive power infeed in the UK. The high speed automatic voltage regulator installed on every generator acts continuously to provide constant voltage, normally the design figure, at the stator terminals. The required reactive power is then obtained by changing the tap position on the generator transformer. For instance, when the station operator is given the target voltage at the h.v. side of the generator transformer, he will tap change on this transformer to achieve the voltage. Procedures will differ between utilities, since not all provide tap changers on generator transformers. An interesting example is the use of 'simultaneous tap changing' on generator transformers to adjust transmission voltage levels system-wide on instruction from the system control centre; the instruction will give the number of taps and direction needed. The problem if generating stations tap change independently is that action taken at one station to adjust its reactive output will be compensated for by neighbouring stations as they detect and react to the network voltage change.
- A practical problem is the tap change situation when a system fault has caused a voltage depression, perhaps prior to a loss of supply. Unless safeguards have been fitted, automatic tap changers on distribution transformers will then move to the extreme position to raise local voltage levels. If restoration were then started by providing supplies at nominal or higher voltage level, damage to consumer equipment might be caused by excessive voltage levels in the short period before automatic tap change systems respond. It is essential that, in such circumstances, distribution control engineers undertake network charging at the lowest possible voltages, the transmission and distribution control centres working closely together.

9.4 SYSTEM CONTROL COSTS FOR EMERGENCIES

Most of the literature on normal and emergency control deals with technical aspects, and published information for the general assessment of system control costs is hard to find. The difficulty is increased in this case by the need to separate out the proportion to be allocated to emergency control. Hence, in the following paragraphs, the facilities which may need to be enhanced for emergency control have been itemized.

The main items needed for emergency control will be:

- disconnection and reconnection of demand by telecommand in selectable groups of substations – the usual facility is to be able to switch demand at one substation at a time;
- mimic diagram – the necessity for mimic diagrams is contentious, but a recent Cigre survey on graphical interfaces showed some two-thirds of the respondents in favour of their provision. Reasons given included: better overview of the system; information available to all dispatchers simultaneously; better monitoring of disturbed conditions; the basic system is always available even if all the on-line systems are not functioning;
- extended ranges on instrumentation to cover abnormal operating conditions – this should present no problems with digital displays, provided instrument transformer ratios are suitable;
- simultaneous transformer tap changing for selectable groups of transformers;
- broadcast telegraph messages to selectable groups of substations;
- broadcast speech;
- software enhancement: identification of system splits; provide alarms for high levels of overload and other abnormalities in operating conditions; flexible selection of groups of substations and parameters associated with these (e.g. summated generation, demand and transfer, available generation margins); computational assistance to cover a wider range of contingencies with user-friendly input to define and compute the end results of these contingencies;
- the system control structure and telecommunications – turning to emergencies within the system control facilities, the security provided for normal operation such as triangulation of data links to outstations, and duplication or equivalent of EMS systems should be adequate. The additional hazard will be the total loss of the normal control centre, and to cover this, an emergency centre with limited facilities can be provided. [1,2]

In round terms, the cost of equipment and buildings for modern system control, excluding controls within substations, is thought by the author to be in the order of 1 percent or less of the net generation and transmission assets of a utility. Excluding energy auxiliary supplies, the facilities specifically for emergency control may be some 10 percent of the total system control costs, or some 0.1 percent of the net generation and transmission assets of the utility.

9.5 INDIRECT COSTS

Some of the expenditure, mainly revenue, which will be incurred to minimize the impact of possible disturbances will include:

- insurance against the costs of damage to plant;
- insurance against costs resulting from loss of supply to consumers;
- the organizational structures to provide the necessary additional services (police, fire fighting, etc.) which might be needed.

Utilities can insure against the costs of repairing or replacing damaged plant, or may choose to carry their own risk, although only the large ones are likely to do this. Presumably, the premiums will depend upon the hazards to which the plant is exposed, and annual reports and accounts may give some indication of the sums involved (probably a few percentage points of the total expenditure).

Although the direct costs to the utility involved will only be a small percentage of its budget, the costs to the whole economy will be many times greater, as discussed in Section 9.7. Utilities may be potentially liable for damage claims covering, for instance, personal injury, loss of business, perishable merchandise, loss of wages, etc. It is not known whether insurance against such items is possible, or whether a utility must rely for its defence on due diligence and acts of God.

An essential feature when dealing with emergencies is to respond rapidly, and this cannot be done unless preparations have been made. In organizational terms, this will involve setting up procedures to call in extra staff at control centres, generation stations and transmission districts, to activate emergency control rooms and incident reporting centres, to alert telecommunications staff and to enhance liaison with distribution utilities.

9.6 THE BENEFITS OF EMERGENCY CONTROL

Some of the benefits to be obtained from the installation of emergency control will be in the shape of avoided costs to the utility, and to the economy at large, whilst others will be qualitative. The latter are discussed first.

9.6.1 Qualitative Aspects

The societal impact of a loss of supply will be determined by (a) the nature of the area affected (terrain, population, economic activities and types of industry and commerce, employment and income levels, transport, ethnic diversity), (b) the day, the time and the duration, and (c) the weather conditions. The security of supply provided by the utility, sometimes influenced by connection charges paid by the consumer, will take account of the factors in (a), but otherwise the impact will be random; precautions should anticipate the worst in regard to 'when' and the weather conditions. Some of the general aspects found in all developed countries will be food spoilage, interruptions to transport through terminals closing (all forms) and loss of motive power (electric trains), failure of staff to move between home and work, and for those at work loss of power in office and factory, loss of patient support systems in hospitals, traffic congestion due to commuters taking to their cars and worsened by failure of traffic control devices, and problems with essential supplies such as water (power for pumping) and fuel (forecourt pumps) [3]. There is no doubt that the potential problems from supply failure are growing as dependence on electricity increases, for instance more air conditioning, and lifts and escalators in high rise buildings. Even the trend towards working from home does not eliminate these problems, as seen by the increasing use of personal computers in the home office.

Overloading of telephone systems is frequently reported, mainly from domestic traffic as people try to inform family and friends where they are and why they may be delayed. It is interesting to note that telephone companies often install backup supplies at their exchanges, and will be capable of operating for many hours on stand-by power.

The problems will be most severe in city areas, and the worst time will be when people are travelling to and from work, particularly when dark with precipitation or fog. The concentration of commerce found in cities could also mean that city blackouts will have the most serious effect on the economy, although the direct impact on manufactured output will be less than for blackouts in industrial parks. Having once experienced the public reaction in the London conurbation to threatened shortages, I feel that the main problems caused by suburban blackouts would be stockpiling of food and fuel. There would also be greatly increased telephone traffic, and possibly severe road congestion as people transferred from suburban electric trains to car and bus. Apart from the general inconvenience, supply failures in country areas will obviously affect farming activities (milking at any time of the year, processing of crops at harvest time, animal husbandry), and because of difficulty in obtaining petrol and diesel, the operation of transport and farm machinery.

Remembering that, in many cases, loss of electricity supply will delay an activity rather than terminate it, an 'adverse impact' factor/duration of outage relationships can be postulated as in Figure 9.1 for supply losses, starting mid-morning and extending for periods up to the early hours of the morning, and for losses starting about midnight and extending several hours into the working day. As far as is known, such graphs of inconvenience against the duration of outage

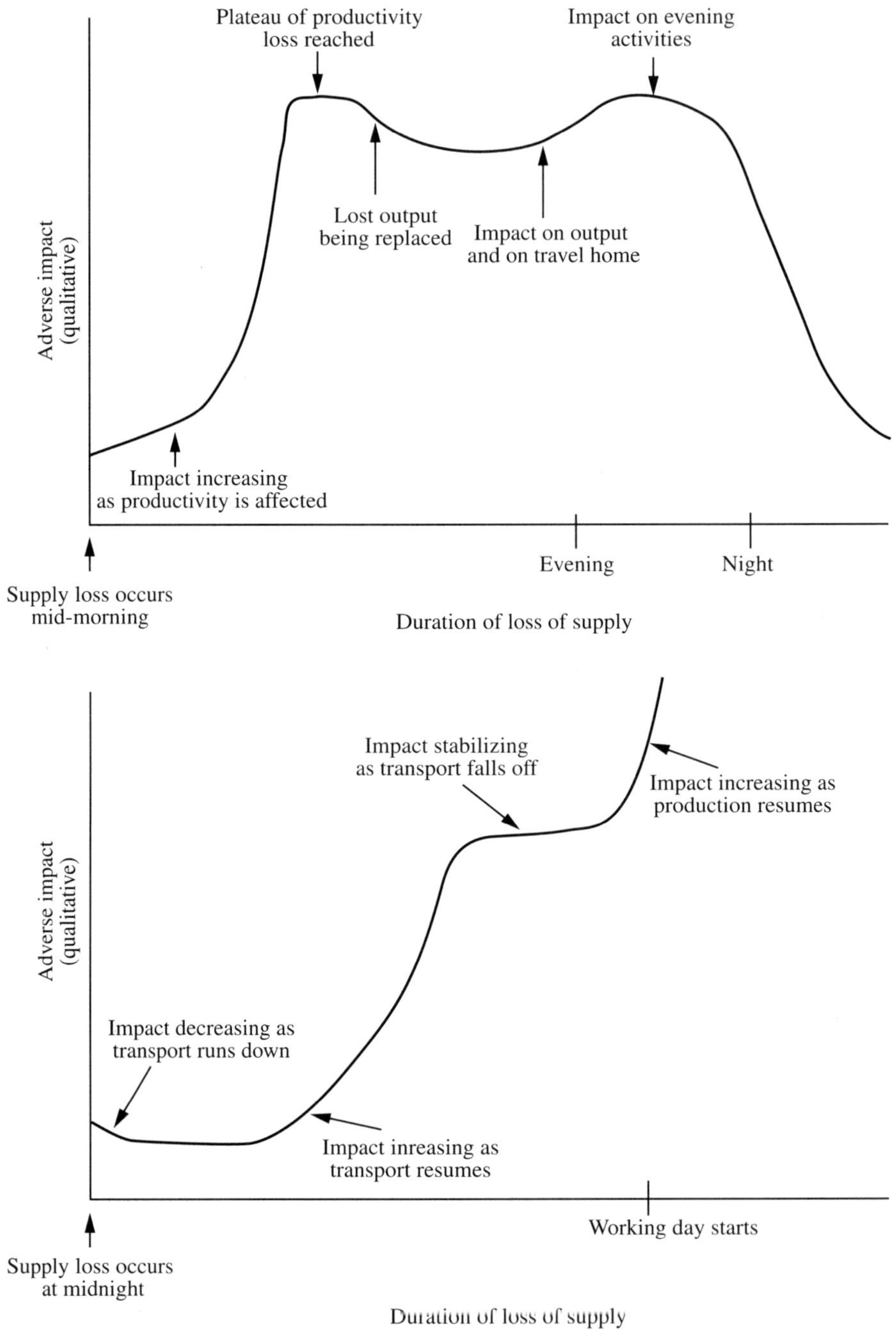

Figure 9.1 Adverse impact on communities

graphs have not been proposed before; their construction makes for clear thinking into the effects of losses of supply.

The importance of minimizing the duration of interruptions is increasing as the amount of electronic and computer equipment installed by consumers increases.

9.7 QUANTITATIVE ASPECTS

Whilst qualitative assessments can only be expressed in general terms as above, quantitative assessments can be made for applications ranging from enhancing supply to critical loads to system-wide installations. In all cases, however, the simplest concept of costing the unsupplied energy at or just above the normal supply tariff completely neglects the value added by the processing enabled through the use of electricity. It would be equivalent to the consumer setting his product price at just the cost of the raw materials used in its production.

Some economic assessments of reinforcement schemes include on the benefit side a term for the reduction in unserved energy expected from their implementation. This has often been put at some 50 times the cost of energy in the supply tariff. A reliable although resource-intensive technique is to obtain the consumers' estimates of the value of electricity by means of a questionnaire. This can address such questions as the value assigned by the customer to outages of various durations, would there be damage to equipment, and how would this vary with outage duration, would loss of production be permanent or could it be made up, would outages be likely to lead to vandalism, theft or arson, would outages affect the consumer's competitive position, etc? A more direct approach would be to ask each consumer what had been spent on safeguarding electricity supplies against failures of its normal sources, were there any plans in this area, what events would initiate such expenditure and what level of expenditure would be acceptable? One problem with this approach would be to weed out replies based on a 'what can I lose, why not inflate problems?' attitude. Further questions could ask what benefits would be obtained from warning of supply interruptions (surprisingly, in one Canadian survey, just over half the industrial consumers surveyed thought this would have no effect on the cost of an interruption), and what would be the impact of voltage reductions?

Although there seems no reason why this technique should not be applied to any size of outage, apart from the data processing involved, a simpler approach is to use the quotient Gross National Product (GNP)/total electrical energy consumption as a measure of the cost per unit of unserved energy. This can be criticized on several counts:

- it assumes the economy is only producing products which appear in the GNP;
- it assumes an average consumption of electricity for all products;

Table 9.2 Estimates for costs of unserved energy

Geographical area	Date of estimate	Purpose	Cost of unserved supply energy $/kWh	£/kWh	power $/kW	£/kW
Sweden	Published 1994	To assess value of busbar protection	13.3	8.3	3.4	2.1
Canada	1977	Survey of large users	15	9.4		
New York	1978	Cost of New York shutdown in 1977	4–11	2.6		

- does it in fact count the productivity of electricity more than once – for instance, first into the basic material (iron, steel, plastic, etc.), and then into the finished consumer product?

Whatever measure is used for benefits, it is suggested that in any comparative study, this needs to be in terms of benefit per consumer item rather than absolute terms such as total benefit.

Several figures for cost per unit of unserved energy are listed in Table 9.2. These have either been published as such, or obtained from the basic GNP and energy cost figures. The figures are quoted in dollars and pounds at exchange rates for late May 1999.

9.8 IS EMERGENCY CONTROL WORTHWHILE?

The ability to handle emergencies efficiently will have repercussions on the whole structure and facilities required for system control. The main impacts will be on the amount and speed of response needed, telemetry speeds, and additional displays and processing of telemetered data. The author has no doubt that any cost benefit analysis will justify extending normal control facilities to handle emergencies. A more useful study is into the priority to be assigned to the different components of emergency control when capital expenditure is limited.

Assessing return to normal as more important than detailed diagnosis, the priority components of emergency control will be:

(1) Instrumentation, telemetry and display for basic monitoring, i.e. frequency, total generation, total transfer.

(2) Telecommand for demand disconnection.

(3) Speech communication with outstations in the utility, with neighbours and with distribution utilities and directly served consumers.

REFERENCES

1. Couchman, M., 1990. 'Survey of back-up control centres, requirements and options', *EPRI*, Report 4000-4.
2. Power, M. 'Report on backup control procedures in emergency back up control centres', *Cigre*.
3. Lamaare, L., 1998. 'When disaster strikes', *EPRI Journal*, **23** (5).

FURTHER READING

Massiello, R. D., 1981. 'Cost benefit justification of an energy control centre', *IEEE Trans. PAS*. **100** (5).

Arnott, I. A., Bergstrom, W. 'The impact of control centre performance on power system reliability'.

10
Systems and Emergency Control in the Future

10.1 INTRODUCTION

Clearly, population growth and improved living standards will lead to increasing demands for electricity, only partially compensated by more efficient conversion processes. Expectations in regard to quality – freedom from interruptions and, increasingly, from transients and waveform distortions – will also rise as consumers introduce more sophisticated equipment into their home and workplace. Utilities will themselves be exposed to political and economic pressures as politicians, investors, consumers and society at large hear about, and perhaps experience directly, developments around the world. Hence system control, including emergency control, will be driven by the following imperatives:

- the need to meet system growth;
- the need to accommodate the consequential changes in the system, including developments in plant;
- the need to avoid catastrophic losses of supply as a result of abnormal environmental or human induced conditions;
- the need to accommodate changes in the organization and structure of utilities, many of which will be consequential upon privatization and electricity markets;
- the need to provide better quality of supply in terms of continuity, reliability, purity of waveform and freedom from transients;
- the need to minimize the use of resources – capital, land, wayleaves – through better use of system capabilities.

Some of these issues will be discussed in this chapter under the headings of organizational changes, the regulatory environment, and facilities (hardware and software).

10.2 CHANGES IN ORGANIZATION

Before considering the impact of possible organizational developments on emergency control in the future, it is appropriate to review briefly current utility structures as these have evolved into the later years of the 20th century.

Public ownership (central government, regional and local authorities) has been commonest, with government the major owner of generation and transmission, and regional/local authorities the main owners of distribution. Ownership tends to follow the perceived political creeds, for instance, public ownership in eastern Europe. The transmission system often has one owner, USA and Germany being major exceptions. A few utilities usually dominate the generation scene. There are often significantly more distribution owners. Some utilities have interests in other businesses, such as steam and heat supply, gas sales and distribution, high capacity telecommunications, provision of engineering services, and planning and operational support to utilities in other areas and countries.

Many variations can be found for the internal organization within a vertically integrated utility. A large one might be self-sufficient in virtually all aspects of its activities, comprising, for instance:

- Generation and Transmission Divisions, providing development, construction and operations services in these areas, and including supervision of power stations and transmission districts.
- Technological Division, covering system and station planning, research and development, metering and telecommunications developments.
- Services Division, covering secretarial, legal, computer, administrative buildings, general staff issues, administration of external liaison, finance and commercial activities. The general policy could be to centralize administration requirements in this division, contracting the engineering and technical work to the appropriate division.

Several functions important in the context of emergencies have been deliberately left for extended discussion below.

System operation may be combined with operational planning functions (fuel supplies, plant outage programming, documentation) in an Operations Division. It requires a range of support activities (relay settings, fault analysis, specification or checking of enhancements to system control and system operation facilities). Its main *raison d'être* will, of course, be to manage the day-to-day operation of the power system.

Fuel supplies is a vital function at all times, but becomes particularly important in day-to-day operation when there is the threat of shortages.

Outage/Maintenance Programming is also essential. It will have major impacts on the security and operating cost of the system.

System Analysis, Simulation and Forecasting techniques are the basis of all the more specific tasks in system work, ranging from on-line security assessment to dynamic analysis and loading simulation covering hours or days.

Inter-utility trading has usually been either 'opportunity' trading, in which incremental costs for say an hour or two ahead are compared between neighbours and power exchanges agreed in a direction to equalize their incremental costs, or 'contractual' trading, in which longer term contracts for days to months ahead are made for stated powers, sometimes at specific times. Differences between actual and scheduled energy transfers over specified time periods will be made up at agreed times in the future.

Ways in which inter-utility trading have been implemented in the past are summarized in Figure 10.1. In Figure 10.1(a), trades within the Pool are agreed between neighbours: there is no System Control Centre or System Operator; security is monitored by each of the participating utilities; transfers to utilities outside the Pool may be supervised by one of its members, or by those utilities having links to external utilities. Strategies and requirements for handling emergencies will have been agreed at the planning/operational planning stage, and will cover issues such as under-frequency protection, generating plant response, operator action in the event of extreme frequency changes, voltage drops or plant overloads.

Some Pools have established System Centres in which decisions on economic operation are taken (Figure 10.1(b)). Typically, each utility will provide costs C_k for incremental power changes ΔP_k to the Pool Centre, who will select and instruct the increments necessary to balance the Pool load. A flowchart of the typical sequence of computations involved in short-term operation, including unit commitment, is shown in Figure 10.2. The extensions to software and hardware to include security assessment are less than in Figure 10.1(a).

Electricity markets spurred initially by developments in the UK and in South America, there is now worldwide interest in introducing a more market orientated approach to trading between utilities. (This resulted at one stage in a plethora of new names for what were often old concepts.) It involves segregation of the different operations of a vertically integrated utility into separate businesses (unbundling), and will often be a functional rather than organizational separation.

This process of organizational and perhaps of ownership separation is sometimes called *restructuring*, in which for example a vertically integrated utility would be split into a number of separate companies, some of which would have the same function but cover different areas of the originally vertically integrated utility, whilst others would have different functions and sometimes geographical areas.

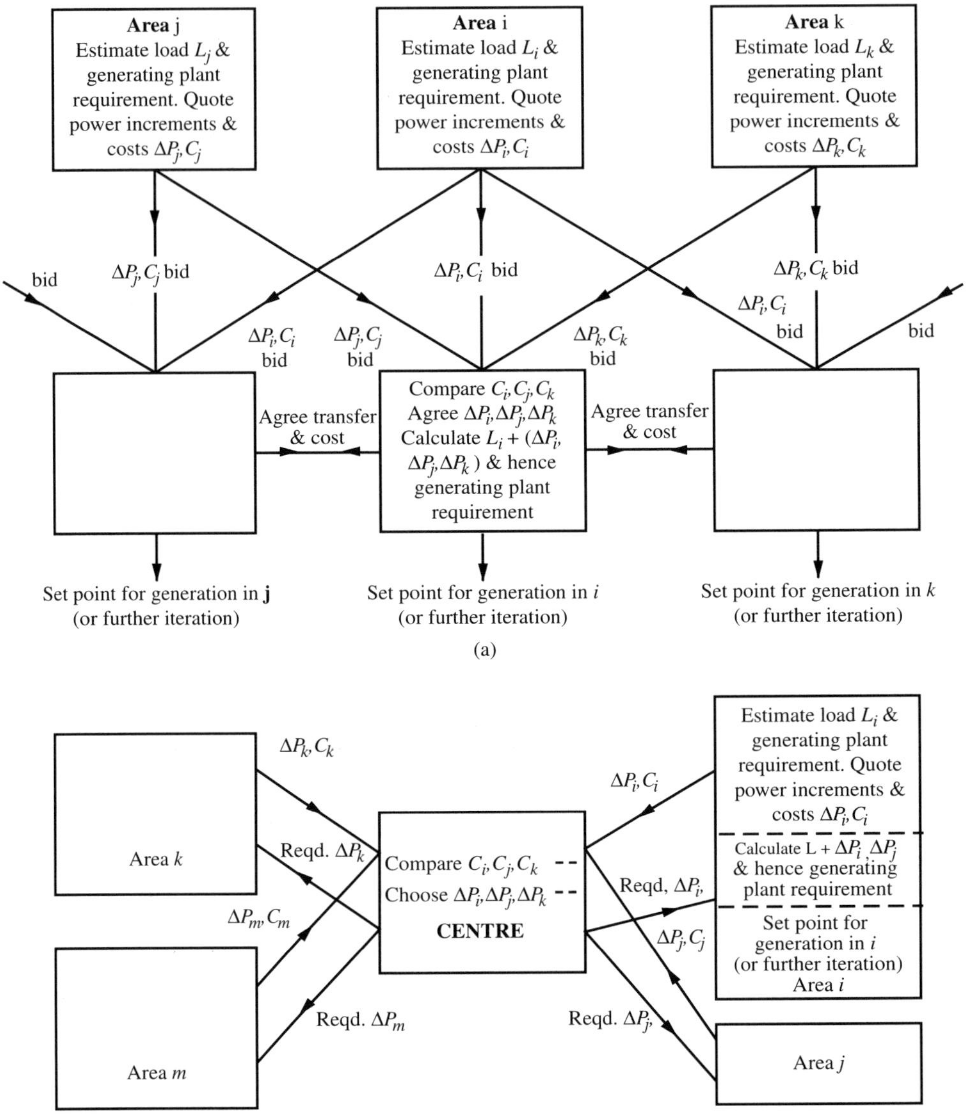

Figure 10.1 Inter-area bidding with (a) no central control, (b) with central control

The structural components of an electricity market are illustrated in Figure 10.3. Decision lead times are approximately decreasing from left to right. Separation of the functions in the one organization is achieved by so called 'Chinese Walls', devised to prevent all but legitimate and authorized information passing between the activities.

Power Exchange this is the market for future energy trades, from an hour up to any length of time ahead. It may be in several parts, e.g. one or two

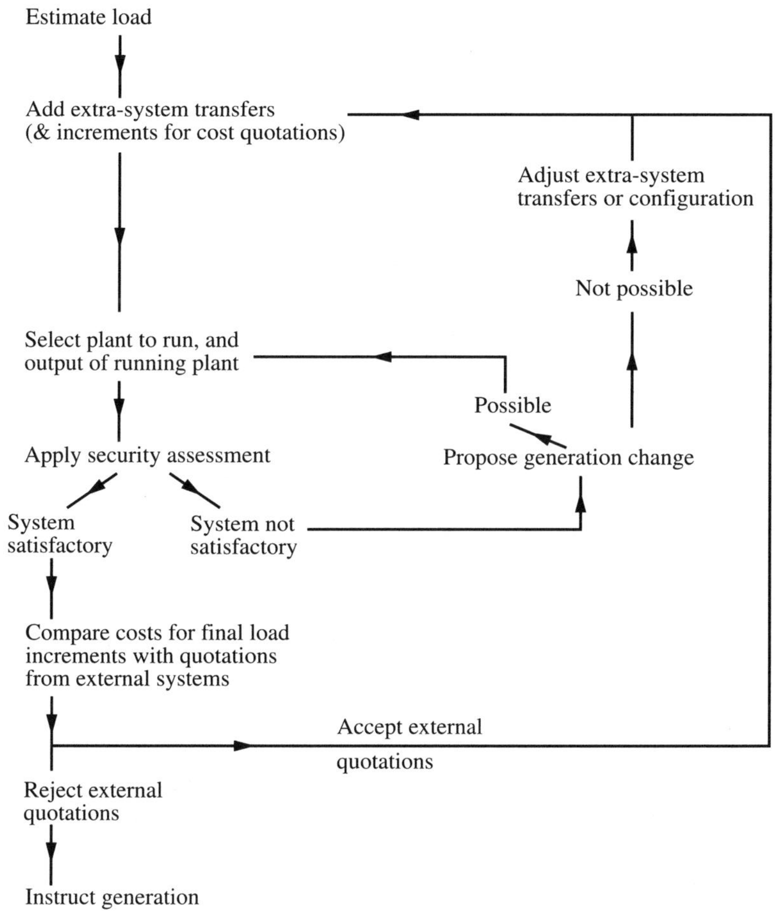

Figure 10.2 Outline of the sequence of computations in determining inter-system transfers

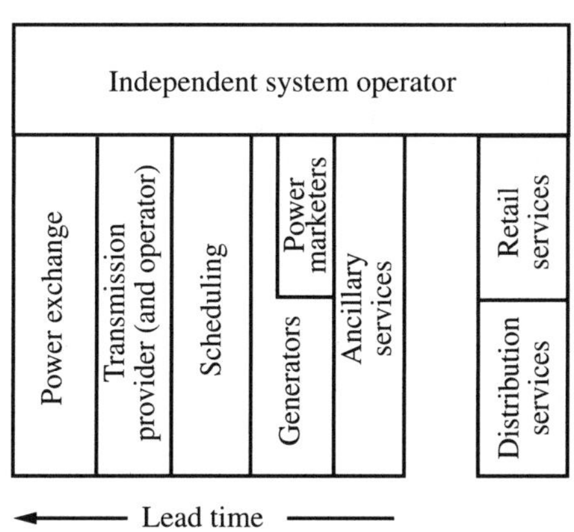

Figure 10.3 Components of an electricity market

hours ahead, day ahead, longer term. A single component bid will cover just one level of transfer, but to improve scope and flexibility, iterative bidding may be used.

Transmission Provider the organization that is responsible for transmission services will frequently own the transmission. To ensure a level playing field, the Provider should offer open and non-discriminatory access to all users. This requires that the Provider should have no financial interest in generation or supply. The transmission network operator should provide objective and impartial operation of the various grid functions. System Operation will sometimes be part of the transmission group.

Generators perhaps because of their high capital value or because large power stations tend to be 'stand alone' operations, the generation assets have often been hived off first from vertically integrated structures and also into more separate businesses (thereby offering more competition), some comprising only one station. Generators may interact directly with the Independent System Operator (ISO) or through Power Marketers. Even in the days of merit order operation when staff salaries depended upon station outputs, the station objective would be to achieve long periods of operation at high steady loads. This remains the case, with the outputs set by the power marketing contracts.

Power Marketers these are commercial organizations which market generation and other resources to the ISO/Pool to ensure viable operation.

System Operation and Control Centre operational objectives will include the usual reliability and minimum cost (or whatever is relevant) criteria. However, the constraints placed by the plant operator on plant operation may be more severe, not least to meet contractual obligations.

Wholesale electricity markets will often be administered by the ISO. This function will have a pivotal position in the electricity market, as will be seen from the list of its key tasks in Table 10.1, which also indicates the other members of the market for whom the results are of main interest. These tasks will differ between organizations, but in some will cover the whole timescale from long-term planning to post-event analysis, and include the provision of operational facilities. Some of the ISO tasks may be delegated to the system operation groups in the individual utilities making up the market.

Although the facilities required within an ISO control centre will be similar to those provided in a vertically integrated structure, it is suggested that some of the changes which might be found will be:

- less information available to the ISO on the internal costs and characteristics of the members of the market;
- more emphasis on security of information;
- more emphasis on justification of ISO decisions to the market members;

Table 10.1 Independent system operator – key tasks

Generation planning (margins (a), siting (a) (b) (d))
Transmission planning (standards (a) (b), configuration (a) (b) (c) (d))
Generation outage planning (timing and location ((a) (b) (?d) (?c))
Transmission outage planning (timing and location (a) (b) (c) (?d) (?e))
External connections (timing and location (a) (b) (?d) (?e))
Short term scheduling (unit commitment (a) (b), network configuration (a) (b) (c) (d) (?e))
Demand prediction (for short term scheduling, possibly dispatch (a) (b) (?d))
Dispatch (generation output (a) (b) (?c) (?d), network configuration (a) (b) (?c) (?d))
Emergency control (measures and operation to contain disturbances and restore normal conditions on ehv system (a) (b) (c) (?d))
Security of supply (planning and operating to provide a secure supply (a) (b) (?c) (?d) (?e))
Quality of supply (planning and operating to provide adequate quality in supply (a) (b) (c) (?d) (?e))
Administration of electricity market (setting up and operating the electricity market (a) (b) (c) (d) (?e))
Provision of infrastructure for operational planning and real time control (specification of facilities and checking proposals (?c) (?d) (f)
Preparation of operational memoranda (a) (b) (c) (?d) (?f)
Costing of transmission services (determining and collecting fees for the use of the transmission system, and of ancillary services (a) (b) (?c) (d))

Notes
(1) Taking the first line of Table 10.1 as an example, the key tasks of the ISO in relation to generation planning will be to advise on or determine generation margins and sites. This information will be passed for information and action to utilities (a), (b) and (d).
(2) The various utilities and other organizations are identified by function as follows:
(a) generation utilities (one or more of planning, design, construction, operation of generation)
(b) transmission utilities (one or more of planning, design, construction, operation of transmission)
(c) distribution utilities
(d) power marketer (utility or organisation managing trading between utilities (see next section)
(e) utilities or marketers trading with the Pool but not part of it
(f) external consultants, manufacturers.
(3) A question mark indicates that the organization may only have a peripheral interest.
It may be felt that privatization has less impact on system operation than might be expected.

- more emphasis on rapid reporting of the operation of the market to its members.

Scheduling any differences in scheduling between the old and new markets seem to be in degree rather than in kind, with the utility schedules now more constrained by meeting external contracts rather than internal needs. The tasks will be to balance supply and demand energy schedules, having regard to optimum use of pertinent contracts, as well as covering bidding and unit commitment.

Settlement money transfers in a market environment are likely to be more numerous (more players) and more complex (wider range of separately-buyable services) than in the past. The process of determining and making the money transfers generally starts immediately after the end of the operational day, and will require metered data from all the trading partners. The process and data should be auditable, with high integrity. It is not new within the industry, but rather on a larger scale than experienced before.

Ancillary services have been defined [10.1] as 'those services necessary to support the transmission of energy from resources to loads while maintaining reliable operation of the transmission provider's transmission system in accordance with good utility practice.' The market concepts have probably made more impact in this area of operation than elsewhere as economists and other protagonists of market trading have come to realize that some of the resources available almost automatically within an electricity Pool can be traded.

Resources often included within an ancillary services package are listed in Table 10.2. Active power reserves will be classified in terms of speed of response – regulating reserves to correct second to second imbalance between generation and demand, spinning reserves to compensate for sudden losses of

Table 10.2 Resources provided in an Ancillary Services Package

Quantity	Function
Active power margin (generation)	Frequency control; maintenance of system integrity following sudden loss of power infeed.
Active power margin (demand)	Frequency control; maintenance of system integrity following sudden loss of active power demand.
Reactive power margin (magnetising)	Voltage control; maintenance of system integrity following loss of reactive infeed or increase in reactive demand.
Reactive power margin (demagnetising)	Voltage control; maintenance of system integrity following loss of reactive demand or reactive outfeed.
Black-start capability	Provision of start-up capability from internal resources in the event of black out conditions occurring.
Fast-start capability	Provision of capability from internal resources to increase generation rapidly in the event of sudeen loss of active power generation or infeed.
Scheduling, system control and dispatch	Provision of the operational planning and real-time control functions necessary for economic and secure operation of the utility either independently or as part of an interconnected system.

The costs of providing these resources may be recovered from the Generators, or Distributors through "uplift" charges on, for instance the energy charges.

The costs of ancillary services will be small in comparison to the total energy costs. They are nevertheless essential–"the oil that keeps the mechanism running"

generation and supplemental reserves to replace spinning reserve which has been committed. Spinning reserve will include quick start plant such as aerotype gas turbines. It may be necessary to supplement the reactive power available from the active power generation; it will be in the commercial interests of the Generators to operate their plant at as near unity power factor as stability and voltage constraints permit, leading possibly to a market in reactive power and reactive power reserve. 'Energy balance' is the mechanism whereby inadvertent, generally small, deviations from controlled exchanges are compensated. A utility which does not own the appropriate mix of plant to start up its system from a completely dead state will need to purchase blackstart supplies. Scheduling and system control and dispatch have been discussed in the previous section.

Fink [10.2] has noted that many alternative structures can be constructed from the main elements found in a traditional vertically integrated utility, these being generation, transmission, distribution, system planning, system operations, bulk power markets and retail sales. For instance, the numbers of possible organizations covering seven, six, five, four and three of the seven functions will be:

providing seven functions $= {}^7C_7 = 1$

providing six functions $= {}^7C_6 = 7$

providing five functions $= {}^7C_5 = 21$

providing four functions $= {}^7C_4 = 35$

providing three functions $= {}^7C_3 = 35$

Although many of these will be impractical, there will be ample opportunity for politicians, economists, consultants and utility managements to explore alternatives, not least in regard to the secondary functions such as R&D. Some of the structures found in practice are shown in Table 10.3. Numerous alternative organizations are also possible within the Transmission function. As an example, the Independent System Operator, Transmission Owner and Operator, and Power Exchange are in one business group, and Ancillary Services in a second in the National Grid Company of England and Wales [10.3].

10.3 RESTRUCTURING, UNBUNDLING AND EMERGENCY CONTROL

Restructuring and unbundling has led to the establishment in the last 10–15 years in some countries of many more utilities based on geographical boundaries, or on function, or even on task. One may ask how far this process will continue, for instance whether countries in which utilities are geographically defined, such as Germany and North America, will introduce functional separation on a large

Table 10.3 Some alternative organizations in the engineering and operations areas of utilites

Basic structure	Main departmental areas	Departments	Sections
A-4 Main areas	Transmission	Design	Main plant specification/design: secondary plant specification and design (e.g. telemetry, metering, control, protection): R & D etc
		Construction	Main plant: Secondary plant
	Generation	Design	Main plant specification/design: secondary plant specification/ design (e.g. station auxilaries, fuel & water supplies, coal handling etc): R & D etc
		Construction	Main plant: secondary plant
	Planning	Station	Sites: wayleaves: fuel, water, ancillary supplies: waste disposal etc
		System	Sites: wayleaves: system design & analysis: secondary plant facilities: R & D etc
	Operations	Station	Fuel, water, ancillary supplies: waste disposal: maintenance: staffing: etc
		System	Operational planning & outage programming real time control: monitoring & post-event analysis: system control facilities (maintenance & perhaps development): communications (maintenance & perhaps development): etc
		Transmission	Network maintenance: network operation
B-3 Main areas	Transmission	Design	Largely as A above
		Construction	Largely as A above
		System planning	As listed for system planning in A
	Generation	Design	Largely as A above
		Construction	Largely as A above
		Station planning	As listed for station planning in A
	Operations	Station	Largely as A above
		System	Largely as A above
		Transmission	As A above
C-3 Main areas	Transmission	Design	Largely as A above
		Construction	Largely as A above
		Operation	Largely as transmission operation in A
	Generation	Design	Largely as A above
		Construction	Largely as A above
		Operation	Largely as Station Operation in A
	Planning	Station	Largely as A above
		System	As A above
		System operation	As system operation in A above

Notes: 1. Commercial & trading activities are not included above
2. The work involved will be done by utility staff, consultants, manufacturers or contractors.

scale into the geographically defined utilities. How will this affect security and quality of supply, what will be the impact on system planning, system operation and control in general and emergency control in particular?

10.3.1 Regulatory Aspects

Regulatory aspects will be illustrated by reference to practice in several parts of the world – the Americas, Western Europe, Asia and Australasia*. The role of some of the international organizations will also be mentioned. As general comments, the advent of deregulation has seen the breakdown of the vertically integrated utilities into functional units – generation, transmission, distribution, supply – and often further splits into geographically differentiated areas. The ethos of public service has been overtaken by a more mercenary attitude at all levels (or is it an instinct for survival?). Calculations of benefits do not usually extend into the total social benefits from reliable supply, and so some regulation has to be imposed even in a deregulated industry to maintain socially acceptable levels of continuity and quality. Other factors likely to require regulation concern tariffs and environmental effects. The large financial resources required to participate in the wholesale electricity market will limit the credible competition, and to counter this it may be necessary to regulate wholesale tariffs. Environmental issues may restrict wayleave and site availabilities sometimes needing compulsory powers to acquire the only suitable locations.

Latin America

South America has been a crucible for much of the regulatory reform in the electricity supply industry worldwide. Developments in Argentina, Brazil, Chile and Venezuela are outlined below.

Argentina Prior to the early 1990s, the industry consisted of state owned companies, and deregulation started with a reformation law in 1989. A subsequent law defined the organizational features of the transformed power sector based on the following main aspects [10.4,10.5]:

- creation of a National Regulatory Authority;
- vertical unbundling of the existing state companies and creation of business units for generation, transmission and distribution;
- generation to be a free, competitive activity, having open access to the transmission network, with dispatch based on declared marginal costs and payment for capacity as a function of system failure risks;

*A valuable source of information on utilities worldwide is available from ABS Publications.

- transmission to be monopolistic, with one concessionary, Transener, for the national 500 kV network and six for the regional 132 kV to 330 kV networks;
- transmission tariffs to be based on the nodal cost of losses and network unavailability plus charges for operation and maintenance;
- penalties for transmission failures to be a function of the transmission charges;
- extensions to be paid for by the interested parties;
- Transener does not buy or sell electricity; it charges for use of its network;
- distribution charges in federally regulated areas to include failure penalties based on unserved energy;
- the wholesale electric market to be managed by a private company (Cammesa), whose shareholders comprise the associations of Generators, Transporters and Distributors and the state. Each holds equal shares;
- long-term contracts in the wholesale market to be freely negotiated between the Generators on one side and large users on the other, plus a spot market for power surpluses.

Brazil For many years the majority of the generation and transmission in Brazil was owned by the Federal and the individual State authorities. Eletrobras owned the Federal Authority, and was responsible for the overall planning and operation of the interconnected systems. The power sector was reorganized in the mid 1990s, including unbundling the Generation, Transmission and Distribution activities, and an organization called Sintrel made up of the transmission installations of the earlier companies was established to be responsible for the national transmission system [10.6,10.7]. The objectives of restructuring included stimulation of competition between Generators and obtaining further private investment.

Guidelines for the wholesale market will include obligations for it to buy and sell all energy, registration of contracts, rules and/or procedures for 'commercialisation'‡, accountancy, settlement, mediation, independent auditing, financial warranties and the treatment of hydrological risks. A mechanism was devised to share these risks between the Generators. The new Transmission companies (Transcos) owning the transmission system will make these assets available to an Independent System Operator (ISO) on payment of fees. This operator will plan the system, operate it in line with a hydrothermal optimization programme, and administer transmission services. Open access to the transmission grid will be guaranteed with transmission users paying a wheeling fee.

‡Commercialization is an independent business activity located between end users and the Generators, Transporters and Distributors of electricity.

An independent regulatory agency was set up with a staff of over 300. Its duties included implementing the federal government's guidelines, promoting bidding for hydro-power concessions and defining their optimal use, trouble-shooting between players, approving approaches for the evaluation of transmission losses and, with the Fuels Regulatory Agency, regulating transport costs for fuels destined for electricity generation. It should define standards for quality, reliability, costs and safety of service and facilities, but above all, guarantee the maintenance of free competition in the energy market.

Central and Southern Chile Following a period in the early 1970s when state control increased, the private sector was given a greater role, and deregulation started in 1978, a new electricity law being approved in 1982. Some of the main elements of this were:

- prices based on short-term marginal costs, but with no constraints on those to large consumers;
- separation between generation/transmission and distribution activities;
- competition between Generators to supply large consumers and the distribution companies, to be based on cost;
- provision of an independent load dispatch centre;
- an open access system for the shared use of transmission;
- establishment of a procedure for generation planning.

The emphasis on competition in generation has stimulated interest in more efficient technologies and cheaper energy resources. There is a central pooling organization, the CDEC, with guiding principles to ensure open, competitive access by any Generator to the system and, above all, to ensure that this competition is compatible with secure and economical operation. Reference [10.8] suggests that regulations on security and quality have been lax, but the National Energy Commission is trying to enforce some standards for example on the number of times and hours per year a load may be shed.

Unlike most countries, the national grid in central/southern Chile on privatization was vested with the largest Generator and this, plus pricing difficulties, has resulted in other Generators deciding to build their own transmission. Two power suppliers have given place to seven generating companies. The system in the north has been sparse, with the main public supply centred on the small city of Antofagusta, although an interconnection between Chile and Argentina across the Andes mountains has been discussed. As of the late 1990s, there is no connection between the northern and central/southern systems.

Venezuela Venezuela is very well endowed with natural energy resources – hydro, gas and oil. The supply industry consists principally of a large company serving the capital city of Caracas and its metropolitan area, plus three minor private utilities and four medium to large publicly owned companies. One of these (Edelca) is in charge of developing and operating large hydroelectric projects on the principal river system, and also owns and operates most of the bulk transmission network in the country. Venezuela has undertaken a large reform of the electric power sector to improve competition. Some regulating reforms, a consistent tariff system and basic rules for the electricity business were introduced.

Central America

Mexico The Federal Electricity Commission (CFE) of Mexico was given the responsibility of providing electricity as a public service in 1960, when the government began acquiring the stock of the then investor owned utilities [10.9]. The Electric Public Service Law was modified in 1993 to allow participation of the private sector in the generation of electricity, with the anticipation that most generation additions in the future would be met by non-utility Generators, variously via schemes for self supply, cogeneration, small or independent power production. Import and export by private utilities would also be allowed. The general duties of the CFE are to ensure a supply of electricity at acceptable levels of quality, quantity and price, provide good service, protect the environment and promote social development. The Commission was also required *inter alia* to provide transmission services. It was anticipated that the new law would provide incentives to build stronger links between the Mexican and US systems.

The Mexican system is longitudinal in structure, and CFE anticipates having nine SVCs and several series capacitors in service by the end of 1999.

Columbia Prior to the mid-1990s, there were six major generation companies, about 23 distribution companies and two vertically integrated generation, transmission and distribution companies in Columbia, all publicly owned. Problems experienced and attributed to centralized planning included lack of efficiency incentives and of accountability, monopoly power over regional markets and deterioration of financial viability [10.10, 10.11]. A sector reform strategy was introduced in the 1990s including:

- establishment of a competitive wholesale market;
- independent and transparent regulation;
- participation by the private sector;
- financial viability of state owned enterprises.

Following a financial crash in the country, there were severe electricity shortages in 1991/92. A decree in 1991 set up rules for the introduction of private capital into the generation business and empowered the government to make decisions about new generation. Rules were introduced to limit the vertical and horizontal integration between businesses. A wholesale market was introduced in 1995, with compulsory participation for the larger Generators, and commercial dealers assisting end users. Transactions can be made either through bilateral energy contracts or transactions in an energy stock market. Prices have been volatile, and because of this a capacity charge was introduced to maintain long-term price signals at a level corresponding to the supply reliabilities needed. The plant composition (60 percent hydro) makes the system sensitive to weather conditions, including the El Nino effect, and because of this a Statute of Rationing was introduced better to control the energy situation.

Some 75 percent of the transmission network is owned by one company. Open access is available to the network on payment of connection and use of network charges, the latter being based on the user's contribution to maximum flows in a minimum network. As of early 1998, there were virtually equal numbers of private and public Generators (14 and 15, respectively), eight public and three private transmission companies, and 55 commercialization dealers.

North America

USA The utilities in the USA are regulated by two, sometimes three authorities. The Federal Energy Regulatory Commission (FERC) is best known, and regulates hydroelectric facilities, electricity transmission, wholesale electric transactions (and natural gas transmission) and service terms and conditions. The rates must be 'just and reasonable', which 'translates to as low as possible for consumers and high enough to attract capital for investors' [10.12,10.13, 10.14]. There are also state regulators who set rates that are just and reasonable, and are responsible for approving new plant, setting local reliability requirements, approving terms and conditions for local retail service, approving purchase contracts for electricity sales from suppliers to utilities and, in some cases determining the sites of new transmission facilities.

The Public Utilities Regulatory Policy Act (PURPA) of 1978 opened the generation market allowing non-utilities to build, own and operate generation, and requiring utilities to buy the output of these plants at administratively set 'avoided costs'. *Inter alia*, it resulted in non-investor owned utilities dominating construction of new generation, and it is reported that considerable over-capacity developed. The Energy Policy Act (1992) exempted generators who sold only in the wholesale market from price regulation, allowing wholesale buyers to shop for power and requiring transmission owning utilities to provide a transmission service. All transmission users, including the owner, pay the same prices for using the transmission. Nevertheless, some Generators were suspicious that transmission owning Generators had an unfair advantage, and FERC introduced the ISO concept. Each ISO is to be responsible for planning and operation, and is governed by a totally independent body representing all interested parties and dedicated to system reliability and the free movement of power within and across the system. The more detailed information below is based on [10.15] and the original includes reference to FERC publications. Although based on one Coordinating Council, the author judges the criteria will give a general guide for other areas.

ISO Responsibilities (see also Table 10.1)

- *Operation* – co-ordinate short-term operations.
- *Reliability* – ensure reliability while supporting the competitive spot market.
- *Independence* – no subset of the market should be able to control criteria or operating procedures.
- *Non-discrimination* – access to and pricing of services should apply to all participants without distinction on identity or affiliation.
- *Unbundling* – services should be unbundled.
- *Efficiency* – operating procedures and pricing should support an efficient and competitive market with fair attribution and division of costs.

ISO operating rules
The operating rules the ISO should follow when meeting these responsibilities are in summary as follows:

Financial

Manage transmission congestion contracts and payment
Have no financial interests in the power exchange or any generation or load
Determine the marginal costs for locations

Development

Provide the communications systems for the scheduling networks

Operations
Bear the main responsibility for final operation and despatch to
(1) maintain reliability, at the lowest total cost for all users
(2) maintain the standards for frequency and voltage
(3) co-ordinate any necessary redespatch and implementation of final schedules in order to achieve reliability, least costs and system balance
Co-ordinate scheduling and balancing for the day ahead
Procure the necessary ancillary services
Provide open and non discriminatory services for the use of the transmission grid.

As regards payment for use of the network, the multiple wheeling fees which were previously incurred when moving power across several utilities will be replaced by a single lower ISO wheeling fee. Electric bulletin boards [10.16] will be provided to make the market transparent, displaying rates for use of the system, operating constraints, scheduling, interruption criteria, etc. This seems to contrast significantly with earlier market procedures, as well as with other industries. Electricity pools in this new environment will concentrate on receiving generation and demand bids, providing a visible market clearing price. It will be responsible for providing a preferred dispatch to the ISO (the operational planner's merit order in the old system?). In engineering terms, the markets are often based on the Power Pools that have been extant for many years (see Table 10.4). These differ in detail, but the core function is dispatch, transmission constraints are often included and power exchange is incorporated.

Looking to operational aspects in evolving markets, in the USA there will be more classes of players with more, sometimes considerably more, players in each. It has been said that electricity is becoming a commodity, and judging from much of the literature on the evolving markets, it seems that many of the prospective players have been mainly familiar with its financial aspects. The first experience has been obtained in California, where the California electricity market (CalPX) went into operation at the end of March 1998. Some of the judgements based on this short experience [10.17] were that 'competitive markets function better and more efficiently than regulated markets... and a full changeover to retail and wholesale direct access achieves competitive markets more quickly than a phased in approach... temporary intervention in the market place is appropriate during the transition to obtain a fully competitive market... deep and liquid spot markets are essential... accurate and timely settlements are complex but very important to induce more market participants.' The near future (1999) CalPX goals with general application included establishing firm transmission rights, introducing elasticity into demand-side links, and starting a forwards contracts market.

Canada (Ontario Hydro)

Ontario Hydro is a publicly owned utility providing generation, transmission, distribution and some related services in the Canadian province of Ontario. The company supplies municipal utilities and large direct industrial consumers. It is a member of the North American Reliability Council (NERC) and North East Power Co-ordinating Council (NPCC), and as such voluntarily follows the reliability guidelines of these organizations. Its statutory obligations include ensuring generation adequacy and transmission security in the province. A new structure is to be introduced, and will include an independent Central Market Operator to operate the electricity spot market and ensure reliability, a transmission

Table 10.4 Features of ISO and market developments in the US

Name	Western Power Exchange (WEPEX)	Pennsylvania-New Jersey, Maryland	New York Power Pool	ISO New England (New England Power Pool)	ERCOT (Electricity Reliability Council of Texas)	IndeGo-Independent Grid Operator (in Pacific North West and West)	MISO (Midwest ISO)	Alliance Group	Desert Star (Desert Southwest Transmission and Reliability Operator)
Start date	31 March 1998	1 April 1998	As at April 1998 FERC decision awaited on NYPP proposals	1998	1997	The proposals, which fell between WEPEX and PJM views, were dropped in March 1998 with prospective members concerned about geographical size.	Proposals appeared to collapse just before filing them with FERC. It could make the largest ISO considered up to that time.	This would be a subset of the MISO group, but as at mid-1998, details had not been published.	
Functions	ISO and Power Exchange providing dispatch, transmission access and ancillary services	ISO co-ordinates short term operation through multi-part bid phased economic dispatch	ISO, Power Exchange and Reliability Council with ISO administering dispatch and spot markets for the Power Exchange	ISO will administer a bid based dispatch system with no separate power exchange	ISO schedules transmissions use and administers cost sharing scheme to deal with congestion and transmission expansion				

Features	Includes several markets for energy and services. 'Must run contracts' for generators. Transmission congestion pricing with payments to Generators to ease congestion	Locational marginal cost pricing for spot market energy transactions and fixed transmission rights. Generator bid prices can be limited to prevent units in constrained areas influencing market prices.	Locational marginal cost pricing and transmission congestion contracts. Day ahead forward market with multi-part bids, settlement system, transmission congestion contract auction. Separate generation capacity market for long term reserve margins	Seven separate markets for energy services. Transmission congestion costs spread across all users.	Described as a 'working, minimalist ISO'
Future		Day ahead forward market, market hubs, auctions of fixed transmission rights			
Governance	Independent Boards and multiple advisory committees	Independent Board and committee to advise the Board	ISO Board of independent directors, ISO management committee, State Reliability Council	Includes independent Board of directors. Retail competition planned for later in 1998	

Note: A main source of the information in this table is the short article 'ISO developments in the United States' by Professor W.W. Hogan, *Power Economics*, 1998. This is extracted from the paper 'Independent System Operator: pricing and flexibility in a competitive electricity market; available at http://www.ksg.harvard.edu/people/whogan

services company, agents or brokers to act on behalf of Generators and customers and direct retail access by customer to Generators [10.18,10.19]. The Generators will sell ancillary services such as reserve, reactive power and regulation. Transmission will be a regulated monopoly whose responsibilities include maintenance, operation and extension of the grid.

The business processes and functions of the Central Market Operator will include:

- providing the data transmission and information system between the Operator and the Market participants;
- planning and scheduling the operations of the market, e.g. evaluating system capability and transfer limits (power and energy as appropriate), assessing system security, scheduling generation and ancillary services based on spot bids and bilateral contracts;
- real-time operation of the Ontario Hydro system in accordance with NERC and NPCC criteria and the Market code;
- system restoration;
- reconciliation of bids/contracts with the event;
- calculation of market clearing prices;
- billing;
- disseminating data, and statistics to the Market operators.

In preparation for operating in an Electricity Market, Ontario Hydro restructured and instituted internal practices which would enable the company, within itself, to emulate the operation of an electricity market [10.18]. This included an 'Electricity Exchange', some of whose responsibilities were to develop and operate commercial activities, manage real-time operation, and manage the information and settlement process (Figure 10.4). The system was designed to be flexible, and would seem to be a powerful tool in the market operations of the utility.

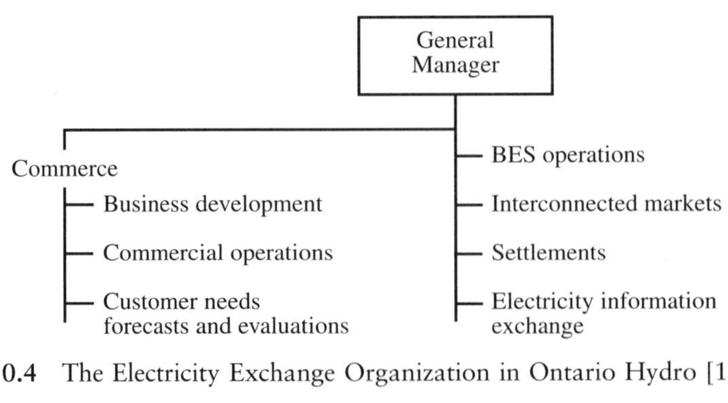

Figure 10.4 The Electricity Exchange Organization in Ontario Hydro [10.18,10.19]

Western Europe, The European Union and UCPTE, Individual Countries

Utilities in western Europe are subject to mandatory regulation and/or codes of practice from both international and national bodies. The requirements set by some of these in respect to security of supply are outlined below.

The European Union (EU) The Commission of the European Union first presented its proposals for an Electricity Directive in 1992. The Directive, to come into force in 1999, establishes common rules for the generation, transmission and distribution of electricity with the objective of introducing competition into electricity markets throughout the Community. The Directive requires the owners of transmission systems to maintain and develop these systems, make information on their operations available, and avoid discrimination. The owners should require transmission operators to develop and publish technical rules. The operators should dispatch units within their supply areas using an economic merit order (with priority to small renewable and similar plant) and be able to give priority to specific units for security of supply.

UCPTE (Union for the Co-ordination of Generation and Transmission of Electricity) UCPTE provides a 'system operators' club' for its members which, in 1998, comprised ten member states of the European Union, Switzerland and the states of the former Yugoslavia [10.20]. It also liaises with utilities responsible for system operation in CENTREL (Poland, the Czech and Slovak republics, and Hungary), in NORDEL, the UK, Iceland and North Africa. All these countries (with the exception of Ireland) are synchronously or asynchronously interconnected. It has been suggested that connections might extend across North Africa to complete a Mediterranean ring post-2000.

It is involved with several other international organizations:

- MEDELEC – a liaison committee of associations of utilities in the Mediterranean region;
- SUDEL – an association of utilities in southern Europe;
- EURELECTRIC;
- NORDEL – the association of utilities in Scandinavia.

The main function of UCPTE is to facilitate electricity exchanges between its members, and to ensure secure operation of the electricity networks. The Union then defines technical rules to ensure a fair division of resources required from each member [10.21], including the following:

Primary control reserve

- This should be 3000 MW, with each control area contributing a specified amount
- The maximum times for deployment should not exceed the values shown in Figure 10.5
- A frequency fall of 200 mHz should result in its total deployment. Conversely, generation should be reduced by the total primary control reserve if the frequency rises by 200 mHz
- The dead band of controllers should not exceed ±10 mHz

Frequency performance

- With the system in an undisturbed state, a sudden loss of 3 GW of capacity must be offset by primary control alone, without the need for frequency activated customer load shedding
- Assuming the frequency characteristics of the demand to be 1 percent/Hz, the loss of 3 GW of generation should not result in a quasi steady-state frequency rise exceeding 180 mHz
- In order to maintain "network time" in agreement with astronomical time, the frequency may be varied within the range 50 ± 0.1 Hz over a period of 24 hours
- The frequency should not fall outside a range ±20 mHz in order to avoid activating the primary reserve

System operation

- In order to acheive these targets, the system must be operated so that its power frequency characteristic falls within a relatively narrow band, e.g:

Demand frequency characteristic (percent/Hz)	Network power GW	Network power frequency characteristic
1	150	16 500
	300	18 000
2	150	18 000
	300	21 000

The required value for the network power frequency characteristic is 18 000 MW/Hz.

Turning to individual countries within UCPTE, information relating to structure and responsibilities for some of these are given in Table 10.5. This is believed correct for the later 1990s.

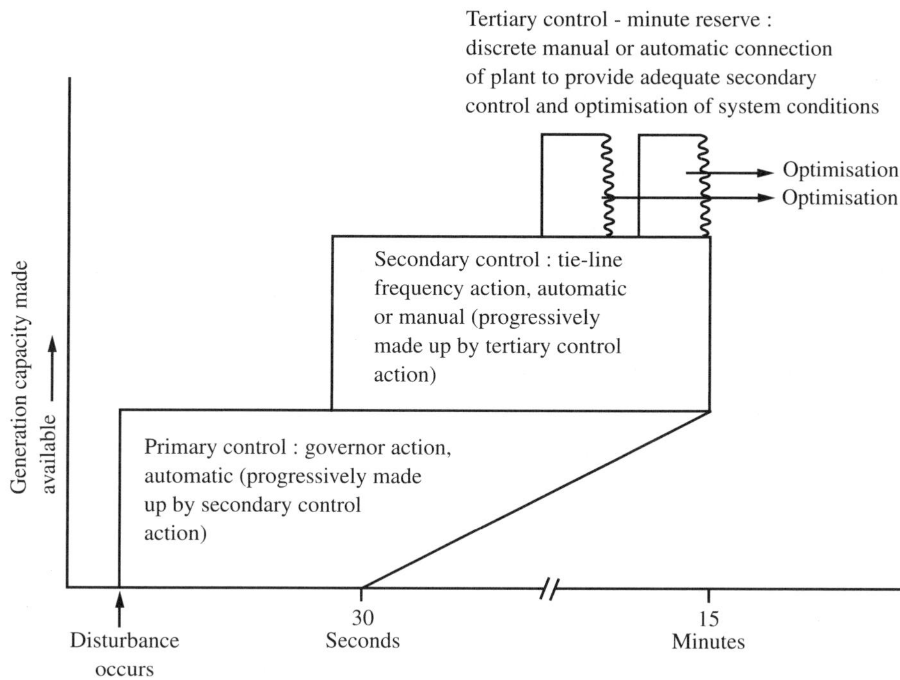

Figure 10.5 Response times for control actions

England and Wales The UK was the second country in the world to restructure its supply industry in the interests of greater commercial freedom. In the ten years since the various companies were created, the Electricity Pool and its attendant commercial and operational rules were developed, the latter being based substantially on CEGB practice. The general duties of the Director General of Electricity Supply were set out in a Parliamentary Act (1988), and included: ensuring demand is satisfied; promoting efficiency; economy; research and development; taking account of the effect on the environment; protecting the public from dangers arising from generation and transmission; and protecting consumers' interests on prices and quality of service. The same Act provided for the granting of Licences for transmission, generation and supply as follows:

(1) *Transmission Licence (to NGC)* – places obligations in respect of merit order operation, security standards, code of practice, statement of charges and terms for connection and use of system; in addition, it requires NGC to develop and maintain an efficient, co-ordinated and economical transmission system, and to facilitate competition in generation and supply.

Table 10.5 System responsibilites in some countries of Western Europe (mid-late 1990's)

Country	Main utilities	Organisation (1)	Responsibility for generation	Responsibility for transmission	Responsibility for real time control	External connections
France	EdeF	State owned, vertically integrated	Mainly EdeF	EdeF	EdeF (national dispatch center)	Belgium Germany Italy Spain Switzerland U.K.
The 'old' Western Germany	Some 10 major generation/ transmission companies	Generally investor owned	The individual companies	The individual companies	Decentralised with companies having own system control centres	Austria Belgium Checkoslovakia Denmark France Holland
Italy	ENEL	State owned, vertically integrated	Mainly ENEL	Mainly ENEL	ENEL (national dispatch centre)	Austria France, Yugoslavia
Belgium	Electrabel & subsidary CPTE (company for coordination of generation & transmission of electrical energy)	Investor owned	Mainly Electrabel	CPTE	CPTE (national dispatch center)	Germany Holland France
Netherlands (as of the late 1990's discussions on reorganisation of the industry were in hand	SEP (transmssion) 4 major regional generating companies	SEP is owned by the generating companies	The generating companies	SEP	SEP	Belgium Germany (& proposal for d.c. link to Norway)

Spain	RED Electrica (transmission) 4 major generation/ distribution groups	Favoured structure is operation within a regulated framework with a mixture of government & investor ownership. UNESA is the industry trade organisation. RED Electrica is owned by the generation/ distribution groups	The individual companies	RED Electrica	RED Electrica. A control dispatching organisation optimises the use of facilities	France Portugal
United Kingdom	National Grid Company A number of generating companies, some large. Some 12 Regional distribution companies. Two generation/ transmission/ distribution companies	Nearly all investor owned, vertically integrated	The generating companies	N.G.C.	N.G.C. for the main transmission & main generation	France (there are connections within the country between the English & Scottish utilities. A submarine link is being constructed between Northern Ireland and Scotland
Eire	Eire Supply Board (ESB)	State owned; vertically integrated	ESB	ESB	ESB	A small interconnection with Northern Ireland. It is understood that a submarine link between Eire and Wales is being considered

(2) *Generation Licences* – requires each holder to operate its plant as instructed by NGC, be party to agreements on the operation of the Pool and comply with relevant Codes of Practice.

(3) *Public Electricity Supply Licence* – requires each Regional Electricity Company to develop and maintain an efficient, co-ordinated and economical system of electricity supply, publish a Distribution Code, offer terms for supply and provide the necessary lines and plant. A new type of business called 'supply' has been introduced, which allows anyone who can satisfy the Regulator of their capability, to provide electricity to specified groups of consumers. Since 1998, consumers have been able to purchase electricity from any such 'second tier' supplier.

The Pool has been described as a 'compromise between the free market expectations of the [right wing] government of the day and the public sector, engineering-led sensitivities of the power sector. It largely met the needs of engineers while not quite satisfying the market purists [10.22]. The net benefit of the initial restructuring has been put at 3 percent of the final sales of electricity, mainly to shareholders. Criticisms made of the Pool were that 'gaming' would occur in the market, with Generators able to extract artificially high price for their output. OFFER calculated that in 1998, UK customers paid £90m over the odds for their electricity [10.23]‡.

A new system, RETA [10.24], is planned to replace the Pool soon. This will provide a balancing market to the contracted exchanges of power, and will deal with the practicalities that on any day both the Generators' outputs and the customers' demands will differ from expected values. Key tasks in introducing this will be preparing the balancing and settlement code, establishing the mechanism for its governance, establishing a Market operator and setting up a 'Power Exchange'. The National Grid will buy services such as reserve power and auxiliary services in this market. The balancing power will be only a fraction of the power trade. To encourage the striking of contracts rather than operating entirely in the balancing market, buyers in that market will pay a surcharge on the commodity price for the short-term flexibility needed, the market retaining the surcharge.

Scandinavia

Sweden The Swedish State Power Board (Vattenfall) was made responsible for the planning, development, operation and maintenance of the transmission network in 1946. Transmission and distribution was governed by an Act of 1962, whereby the contribution and operation of transmission and distribution

‡Although a large sum, it is nevertheless small in comparison to the turnover of the industry.

was based on Government concessions assigned on a line (electrical) or area basis. A new state authority, Svenska Kraftnet, succeeding Vattenfall was established in 1992 to run and operate the grid, including security and load frequency control. An Act in 1994 provided for open access to the networks on payment of a connection fee, unbundling of grid services from generation and distribution and a regulatory authority to supervise grid services [10.25].

The Pooled Operations Group currently runs the Swedish Power Exchange consisting (mid-1990s) of two main blocks of members (Vattenfall and KGS). Subject to satisfying size conditions, including sufficient reserve capacity, these have transmission rights in the grid.

The system is planned and operated in accordance with a number of 'dimensioning criteria', which include the ability to withstand without loss of load the sudden loss of any generator or production unit, the tripping of any line or system transformer, a three-phase transient fault on any line with auto-reclose, a single phase permanent fault on any line with a single reclosure and any three-phase busbar fault. Various types of reserve are specified.

Finland The Finnish Electricity Market Act was introduced in 1995 to ensure preconditions for the efficient functioning of an electricity market. The market should secure a 'sufficient supply of high standard electricity at reasonable prices'. This would be achieved through competition in production and sales, and reasonable and equitable service principles in operation. The main electricity production groups in the Finnish market are the municipalities, the wood pulp and paper industry and the state-owned company IVO. The last two are consolidating their grid assets into a single company which will also operate the international lines. There will be open access to the whole network on payment of a fee. The Finnish Power Balance (FPB) organization will manage the balance between production and consumption of electricity, and act as a trading centre for the balance power, that is the difference between purchase and delivery of the 'regulation responsible parties' (those producers who have sufficient capacity to enable them to regulate their output in the very short term).

Far East and Australasia

Taiwan As of the late 1990s, electricity supply in Taiwan is provided by the state owned monopoly Taipower [10.26]. An electricity act is being considered to liberalize the electricity market by privatising Taipower, with the objectives of reducing tariffs, and easing problems experienced in obtaining sites and planning permission. Taipower will retain the transmission and distribution functions, with the generation split into private generation companies, each of these and any other independent power producers being limited to a maximum of 20 percent of the island's generation capacity. The act may establish an electricity market, based on the British pool model, and a regulatory agency.

New Zealand Transpower, the New Zealand transmission authority, became a state owned enterprise in 1994, when it was separated from the New Zealand Electricity Co-operative. Owning the main grid, its responsibilities included transmitting power for any potential customer across the grid, and system operation. An Electricity Market Company, formed in 1993, had already established an electricity exchange, including a spot market and forward markets for short and long term contracts in tradable electricity. The industry is subject to the same antitrust and commercial legislation as the private sector and, of course, the national supply standards and supply legislation [10.27]. Several of the electricity markets, including those for wholesale generation and ownership, and operation of new distribution networks are routinely evaluated.

Direct access was introduced as part of the restructuring, allowing since 1994 free entry into all aspects of the industry. The comment is made that allowing direct access without imposing operational standards on direct access suppliers does not seem to have had a detrimental effect on retail competition.

In conclusion, the New Zealand government seems to have been very active in pursuit of reforms of the supply industry. One reference suggested that in the mid-1990s, the political context was such that the form of wholesale market reform in the future was uncertain and controversial. It also noted that interim views on the major failures of supply in Auckland in 1998 were that the way the new industry was privatized had no bearing on these failures.

International Organizations

International Conference on Large High Voltage Electric Systems (CIGRE) The International Conference on Large High Voltage Electric Systems (CIGRE) was founded in 1921 in France to promote the development of technical knowledge and exchange of information between all countries in the fields of generation and high voltage transmission of electricity. Its scope covers station and transmission high voltage equipment and the development (planning, construction and operation) of transmission systems. It is made up of 800 collective members (administrative bodies, scientific and technical organizations, research institutes and public or private companies in the field), 80 educational institutions, and some 3400 individual members. In all these are drawn from more than 80 countries [10.28].

CIGRE has three governing bodies: an Administrative Council with decision making powers; an Executive Committee which makes recommendations to the Council; and a Technical Committee. This is composed of the chairmen of the 15 Study Committees, and sets the technical direction and work of CIGRE via a

Strategic Plan. The Study Committees promote and exchange information in their fields of activity, guided by the Strategic Plan, and organize and carry out studies in these fields, often through *ad hoc* working groups and task forces set up by the Study Committees for specific tasks. The members of the committees, groups and task forces must be either individual or collective members of CIGRE, and be in a position to collaborate actively. A Study Committee is made up of a chairman, a secretary and up to 24 regular members of CIGRE, each from a different country. 'Experts' can be appointed as observer members. It therefore has access to a considerable body of expertise. Members of the Study Committees are appointed for a period of six years. In addition, National Committees have been set up in 50 countries and will, for instance, provide a mechanism to acquaint more organizations and engineers within individual countries with CIGRE activities. In the UK, the British National Committee issues a BNC Newsletter and holds an annual UK Liaison Meeting.

The Study Committees cover the full range of power system technology – equipment (static, rotating, a.c. and d.c., power electronics), transmission plant, systems (insulation co-ordination, protection, communications, and of particular interest in the context of this book, system planning and development, system analysis techniques and system operation and control).

CIGRE plenary conferences lasting several days are held every two years in Paris. Typically, some 2500 members attended the 1998 session, and most of the committees, working groups and task forces find it convenient to arrange meetings during the conference periods. Study Committees will adopt topics to be the main subjects for discussion at the next conference (Preferred Subjects) and papers submitted for that conference should relate to these topics. Countries are allocated a number of papers (the largest allocation is ten). Competition to have a paper selected is keen, and this often results in multiple authorship. Proposals are normally vetted by the National Committee before acceptance as one of a country's quota and submission to the CIGRE Paris headquarters. Study Committees may provide reports on their activities. Sometimes, committees and task forces will prepare papers on specific topics. 'Special Reporters' are nominated to summarize the papers relating to the preferred subjects. The papers are published.

Increasingly, Regional meetings are being held, often in the non-conference years, and organized by one or more of the local Study Committees. These may be arranged to coincide with technical conferences or symposia, and provide an excellent opportunity to assemble an international audience in particular subjects; for instance, a colloquium was held in Brazil in 1999 on regulation and privatization.

CIGRE activities are published in a bimonthly, bilingual (English/French) journal *Electra*. Conference and symposium proceedings are also published as technical brochures, reproducing study committee papers on particular topics. These are not as widely available in libraries as, say IEEE or IEE publications, but they can be obtained from the CIGRE Central Office in Paris; copies are often on sale at CIGRE arranged/supported symposia. They provide an excellent source of information on up-to-date practice and developments.

Other organizations There are several other organizations which, although based in one country, have an international membership, for instance the Institution of Electrical and Electronic Engineers (IEEE), USA, the Institute of Electrical Engineers (IEE), UK, the International Federation of Automatic Control (IFAC), World Energy Conference, etc. The Study Committees, Working Parties and individual members of these organizations will have authority and knowledge to influence criteria and standards, and to collate international experience.

Increasingly, co-operation is being strengthened between the learned societies, for instance the IEEE, IEE and CIGRE.

UNIPEDE UNIPEDE (International Union of Producers and Distributors of Electric Supply) was founded in 1925, and was a professional organization whose membership consisted of companies or groups concerned with generation and distribution. It has been described as a mouthpiece through which electricity companies can voice their common position with public authorities and international bodies. Questions referred to it have included safety, environmental issues, new uses of electricity, and the study of large systems and international interconnections. Its proposals are often adopted by participating countries.

Its work was conducted through a series of Study Committees and, for specific limited areas of work, Task Forces whose members were proposed by the utilities and professional associations making up UNIPEDE. Triennial international conferences, called Congresses were held at various venues. It operated an information office to keep members informed of its activities, keeping in touch with national and international bodies dealing in its own areas of interest. It is understood that it has now combined with EURELECTRIC.

10.4 FACILITIES FOR EMERGENCY CONTROL IN THE FUTURE

An appropriate starting point for reviewing future facilities for emergency control is from the system need and application viewpoints. On the assumption that present trends towards privatization with multiple ownership of generation, but retention of an integrated transmission network and a system-wide control organization continue, plus an acceptance by all generators and users that in emergency situations preservation of system integrity should override commercial considerations, the mechanisms for response to emergency situations will essentially be the same as those that have evolved since large scale systems and interconnections developed, namely:

- adjust (a) generation and/or (b) demand to eliminate imbalances between them, both system-wide and, as dictated by transmission capacities, within the system;
- adjust the distribution of active power generation primarily and, to a lesser extent, demand and reactive power generation to reduce circuit overloads and/or too high angles between nodes;
- change extra-system transfers to eliminate power imbalances, abnormal voltages, and excessive power flows within the system;
- adjust reactive power sources to maintain nodal voltages within the prescribed limits;
- examine the possibility of changing the network configuration or parameters to reduce circuit overloads, abnormal voltages, inter-nodal phase angles and fault levels.

This unchanging operational need must be met in the context of several technical and non-technical developments:

- system changes, including those needed to fit in with organizational changes;
- main plant developments;
- manpower attitudes and changes;
- quality of supply issues;
- control plant developments.

Organizational Changes

The major driving force for change worldwide will be to meet organizational developments, including changes in ownership. As a consequence of multiple ownership of generation and contracts to supply consumers, there may be significant proportions of both generation and demand which, under normal conditions, will be outside the jurisdiction of the ISO. Parts of the transmission network may be associated with these pockets of generation and demand. Hence, the task of the ISO could be to accommodate prescribed operating conditions within these pockets with optimal operation of the remainder of the system, in modelling terms a potentially highly constrained optimization problem. The ISO may, for commercial reasons, have to be satisfied with minimal information on the characteristics and operating states of the system within these pockets. It should, however, be able to rely on the co-operation of all participants in the system to maintain viable operation during an emergency.

System Changes

It seems likely that the proportion of smaller capacity generating plant with low operating costs will increase in future. Nevertheless, for many years, large coal, oil and nuclear generating units dominated plant programmes, which means that the plant mix of many utilities will contain such large units well into the 21st century. The siting requirements of the smaller plant are likely to be less onerous than those of the large conventional stations, so that it may be possible to site these nearer to load centres. In turn, this could imply less reliance on transmission, and as transmission failure is a significant cause of large-scale failures of supply, less risk of these occurring.

Generating Plant Developments

These changes could lead into the 'distributed utility', that is the use of small, affordable generation and storage units in the power range of, say, 1 kW to 10 MW [10.29, 10.30]. These may include fuel cells, photovoltaics and micro-turbines. Features of such sources are their low voltage and low output, often d.c. It would seem worthwhile to evaluate the use of local d.c. distribution loops to supply consumers from such power sources. The impact on consumers could be substantial (the d.c to a.c. change of the early days of the supply industry in reverse!), and it would be necessary to maintain a.c. and d.c. distribution systems in parallel, perhaps by using one phase of a 3-phase and the neutral cable for d.c.

Transmission Developments

The future for transmission seems fairly well defined:

- for high power very long distance (e.g. several hundred kilometres): ehv direct current (e.g. ± 500 kV); series and shunt compensated ehv a.c. (up to $1000\pm$ kV).
- for meshed networks with medium and short transmission distances: ehv a.c. (up to say 400 kV) with shunt, perhaps series, compensation; ehv and d.c. for identified point to point transfers; FACTS.

Superconductivity has been studied for use in cables, transformers, machines and energy storage devices. The promise for transmission is reduced energy losses and higher ratings, of particular value in the reinforcement or refurbishment of supplies to urban areas.

Manpower Attitudes and Issues

The attitudes of staff reflect the underlying aims of an organization, an affect likely to be reinforced by the trend that 'like hires like' when hiring and training staff. With privatization and more emphasis on shareholder returns, staff will be predisposed to opt for minimum operating cost rather than security and quality of supply, if the option exists, as might have been the case in the past. Other effects are that utilities are likely to be more conscious of commercial secrecy and less willing to exchange experience or information on developments and future plans.

Quality of Supply

Quality of supply in the past has usually been quantified by the number of interruptions (very short interruptions being excluded), voltage magnitude and consistency and waveform. The higher quality supply needed for computer installations will be provided from rotating or static rectifier/inverter units, etc. With the encroachment of microprocessors for control and data management into most aspects of life, customers at all levels will need supplies free from voltage disturbances and interruptions; so called 'premium power'. Two approaches to an advanced distribution system are being studied by EPRI. One uses low cost sensors and software to detect and correct system problems, including incipient faults and momentary line contacts, providing automated isolation and restoration. The other is the 'Custom Power' family of electronic controllers, which include devices to protect sensitive customer equipment from system disturbances, to protect consumer problems from affecting the supply quality and to extend these concepts to groups of consumers [10.31]. Combining such devices with energy storage devices will provide capability to ride through outages.

Control Plant Development

One of the main areas of innovation in recent years has been the introduction of power electronics into very high power applications. This has made possible the development of plant whose characteristics are specifically tailored to control the operating parameters of the system, variously through adjustment of circuit impedances, nodal reactive powers and nodal active powers as described below.
Static var compensators The Static Var Compensator (SVC) is a static implementation of the much older rotating synchronous condenser. A modern version consists of banks of capacitors and reactors connected via

thyristors and step-up transformer to the transmission system (Figure 10.6), at typical power levels (late 1990s) of 100 MVA [10.32]. The combination of switchable fixed capacitors and adjustable (via the firing angle of the thyristors) reactors enables the continuous voltage-current characteristic of Figure 10.7 to be obtained. Their response times will be less than those of mechanically switched compensation. In some utilities, mechanically switched capacitors are installed, switched in response to the local reactive demand in a load following mode or in response to the ehv voltage, to improve system security following disturbances.

The use of Gate-Turn Off (GTO) thyristors in FACTS devices has been explored in recent years. Such devices are expected to be physically much smaller than conventional FACTS controllers, but to have higher losses. In one line of development, the basic element is a voltage source of adjustable magnitude and phase produced by a three-phase bridge circuit made up of six (or a multiple thereof) GTOs, each paralleled by a reverse diode, Figure 10.8. The a.c. output waveform will be improved by increasing the number of GTOs in the converter (e.g. to twelve). The application of this device in an SVC is illustrated in Figure 10.9. The operating characteristics of such an advanced SVC will be superior to those of a conventional SVC, as indicated in Figure 10.9, capacitative support down to lower voltages being the most valuable addition.

Static series compensator Series capacitors have been used for many years to compensate the inductance of long transmission lines, thereby increasing the power transmission limits of stability limited interconnections. Mechanical switching of these can be replaced by thyristor switching either of the series capacitors or of reactors in parallel with fixed capacitors. Several

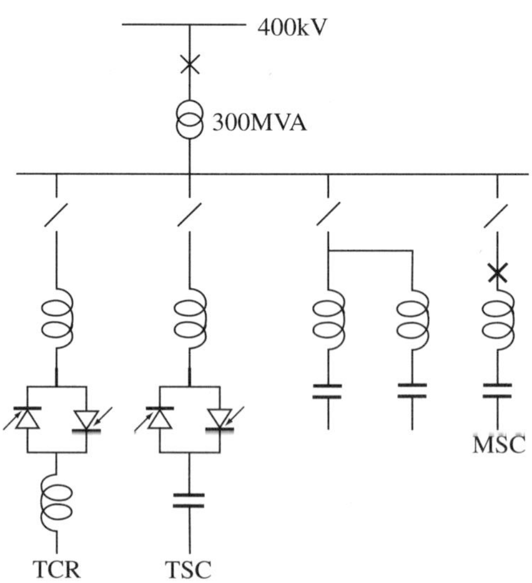

Figure 10.6 Static var compensator using GTOs

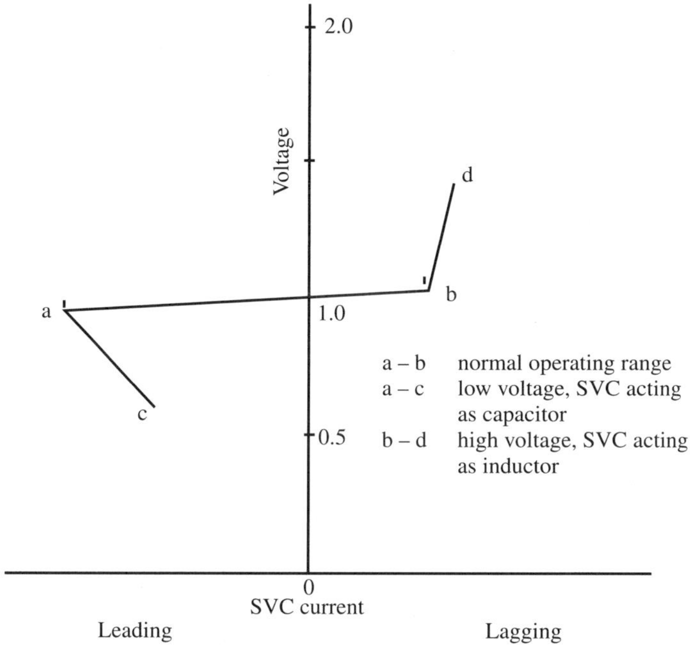

Figure 10.7 Voltage–current characteristic of SVC. Reproduced by permission of IEE

alternatives are shown in Figure 10.10, and are described briefly below (from [10.33]).

- *Thyristor Controlled Series Capacitor (TCSC)*: this would provide fast control of series compensation. A thyristor controlled inductor is connected in parallel with the capacitor (Figure 10.10(a)). Reference 10.33 suggests that the TCSC

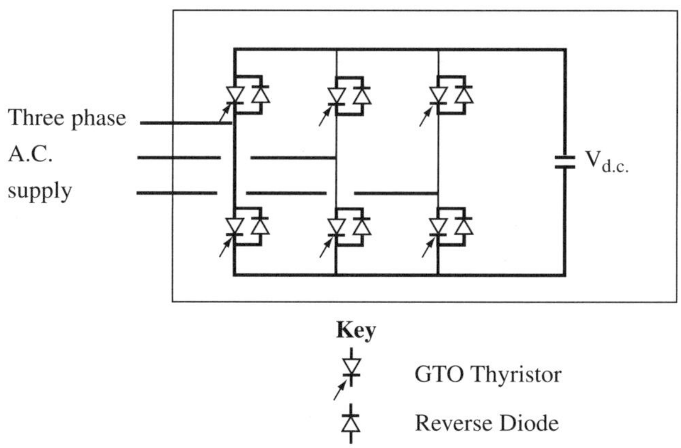

Figure 10.8 Voltage source converter

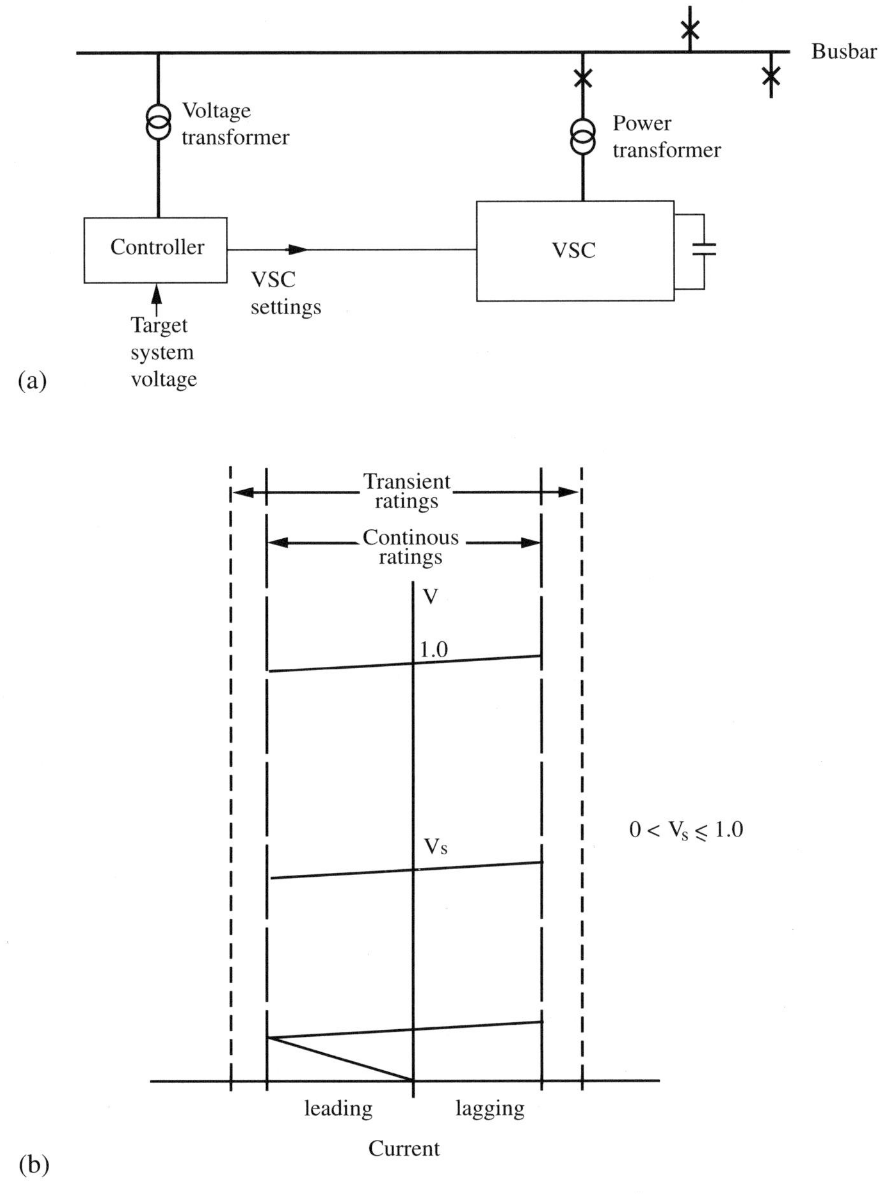

Figure 10.9 (a) Advanced SVC. (b) Voltage–current characteristic of advanced SVC

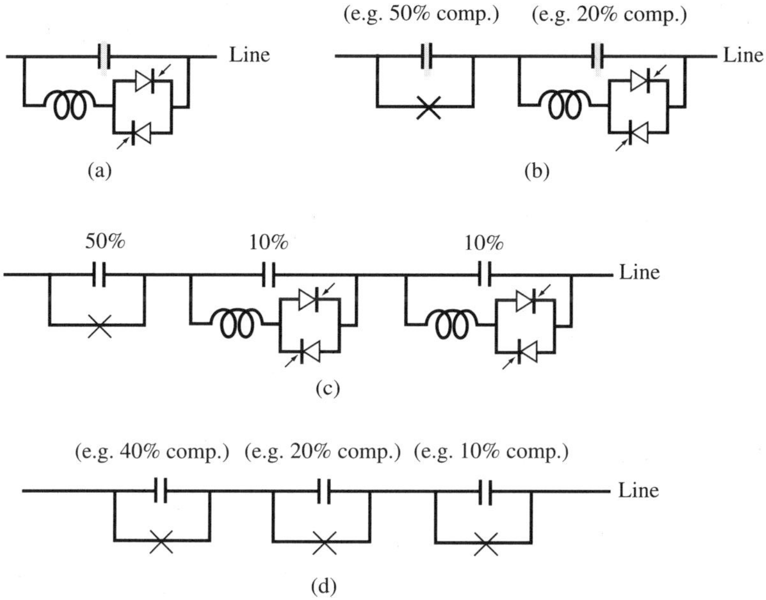

Figure 10.10 Series compensation of a line using switched capacitors. (a) Basic TCSC, (b) outline of typical TCSC application, (c) outline of TSSC application, (d) outline of MSSC application (this would give compensation up to 70% in 10% steps). Reproduced by permission of ABB Power T&D Limited

would only be implemented on part of the capacitor installation (Figure 10.10(b)).

- *Thyristor Switched Series Capacitor (TSSC)*: in this device, the series reactance of the circuit is varied by controlling the proportion of the total series capacitance bank that is switched into circuit (Figure 10.10(c)). It would be used for flow control and oscillation damping.
- *Mechanically Switched Series Capacitor (MSSC)*: the arrangement shown in Figure 10.10(d) would provide line compensation up to 70 percent in 10 percent steps, with a response speed under, say, 100 msecs.

As a comment, it is judged that series capacitor installations will grow rapidly to improve transmission capability, improve load sharing and improve system damping.

Transformer tap changers Although practice differs between utilities, tap changers are installed on transformers between system voltage levels so that target voltages can be achieved, reactive power flows and, to some extent, reactive demands controlled using the two transformer tap stagger principle. This can result in several tap change operations daily, with consequential wear on contacts. There have thus been incentives to replace the conventional electromechanical tap changer by a thyristor switching device. A further advantage would be higher operating speed.

Phase angle regulator This resembles the conventional quadrature booster, except that the magnitude of the injected quadrature voltage is controlled by thyristors.

Unified Power Flow Controllers (UPFC) The UPFC [10.34] consists of two voltage sourced converters, one connected in shunt and one in series (Figure 10.11). A variable phase voltage can be injected in series with the phase voltage, and hence provides control of the real and reactive power flows through the line; in system terms, it should provide power control, voltage control, phase angle control, reactive compensation, oscillation damping and an aid to transient stability. A typical power level (late 1990s) of 200 MVA has been suggested.

Braking resistor [10.40] This is a shunt connected thyristor-switched resistor which, when connected say at the generator terminals or generation busbar, can be used to improve stability or to maximize power fluctuations on a generator (Figure 10.12).

Series reactor This can be a fixed series reactor shunted by a thyristor switched reactor.

NGH damping device [10.40] This device, named after its inventor N.G. Hingorani, effectively adds resistance into a circuit to inhibit low frequency oscillations on a long heavily loaded line (Figure 10.13).

Fault current limiters Fault currents are directly related to network power densities, and hence will increase as demands and generation rise (see Chapter 2). Measures to contain these can be incorporated in the system development, and others can be implemented in operation, but these may have drawbacks – cost, introduction of new ranges of plant (voltage levels, current ratings), reduced flexibility and increased complexity in operation. Fault current limiters are network series elements which limit maximum currents during faults to values which will not damage other parts of the network. Other desirable features are that they should be resettable, capable of construction in such sizes that the system capability is not decreased, have low losses, cause no overvoltages or harmonics and operate at sufficient speed both to protect other equipment and themselves. Both non-super-

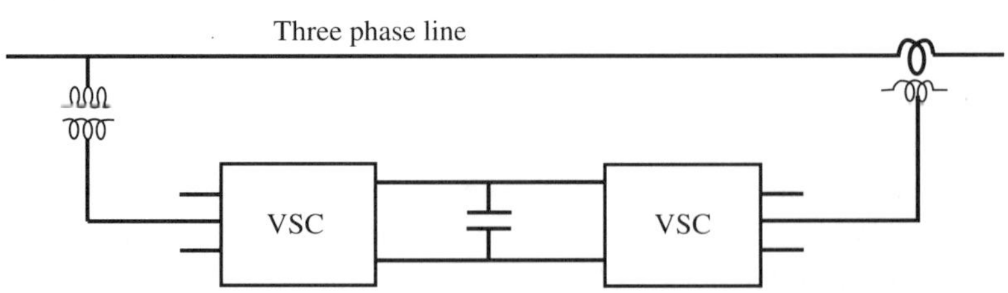

Figure 10.11 Unified power flow controller (Reproduced by permission of IEE from [10.34])

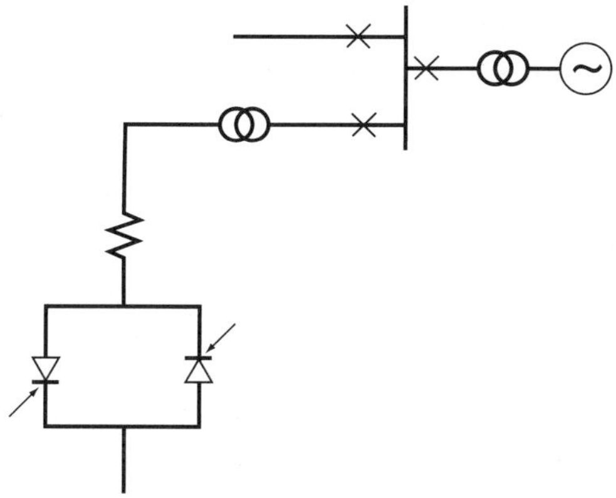

Figure 10.12 Braking resistor (Reproduced by permission of IEE from [10.40])

conducting and superconducting approaches have been studied.

The long standing and simplest method is to install a series reactor/s, generally at the point of excessive fault level. These can be by-passed (short circuited) when system conditions permit; fault levels are not strongly related to the capacity of plant running, but fortunately, the excess of fault level over switchgear rating is usually quite small in practice, and a small reduction should often be sufficient. A parallel capacitor/variable inductor scheme (Figure 10.14) has been installed in the USA, which provided both reactive compensation of a circuit and fault level reduction.

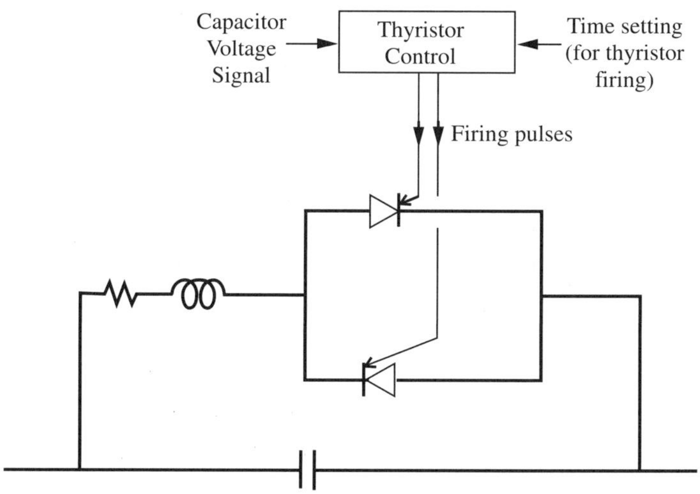

Figure 10.13 NGH damping device (Reproduced by permission of IEE from [10.40])

Several superconducting fault current limiters have been proposed [10.30]. Figure 10.15(a) is the simplest and potentially the cheapest. The superconducting series element carries the normal load current in a superconducting state. It returns to its normal resistive state when the critical current density of the material is exceeded on the occurrence of a fault. It would be necessary to ensure that the heat generated in the superconducting element can be dissipated before it is again placed in the circuit, which would dictate the return to service time. In Figure 10.15(b), the primary winding of a transformer with a superconducting secondary winding is connected into the line. During normal operation, the ampere turns of the primary and secondary windings would balance, and there would be no leakage flux. On a fault, the current in the superconductor would exceed the critical value, there would be some leakage flux providing the needed

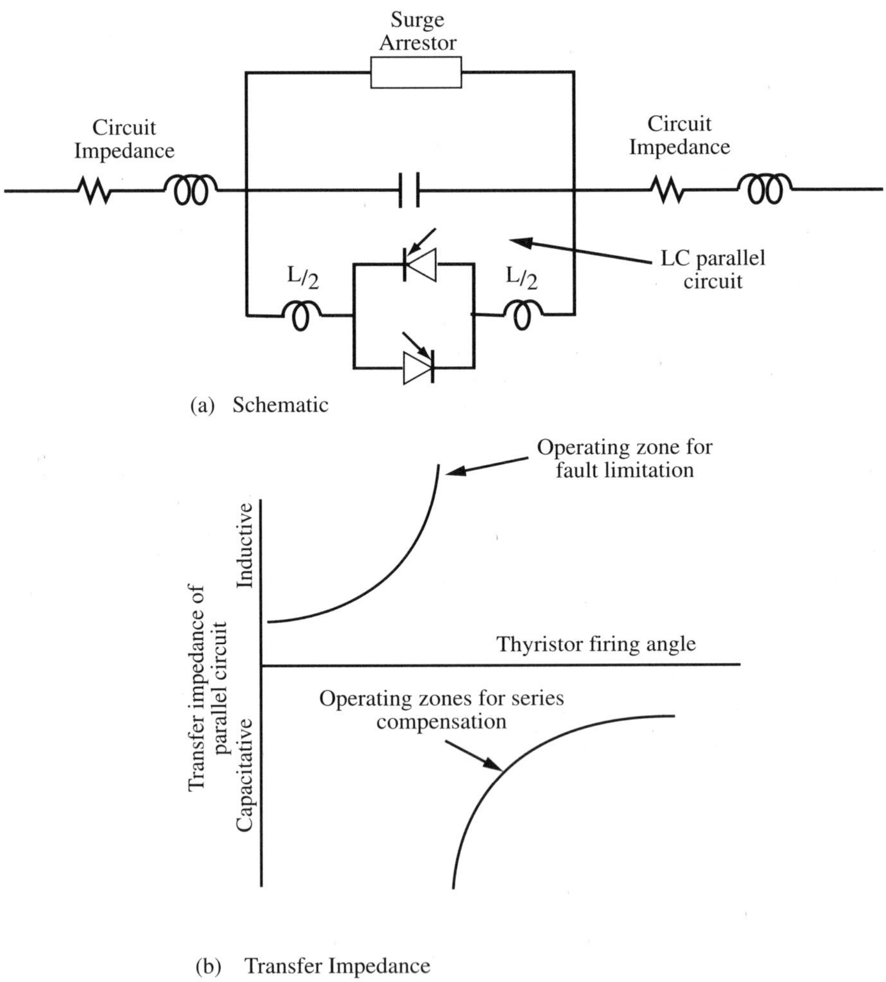

Figure 10.14 Fault current limiter (Reproduced by permission of IEE from IEE Digest No. 1995/026)

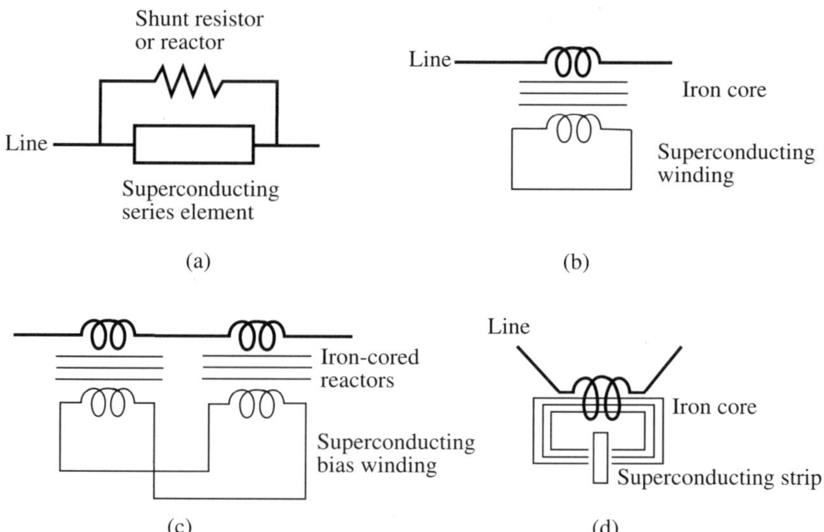

Figure 10.15 Possible implementations of superconducting fault current limiters. (a) Series resistance, (b) shielded inductance, (c) saturated inductance, (d) Ain-gap (Reproduced by permission of the IEE)

emf, and the transformer would provide an impedance in the line which would limit the flow of fault current. The iron cored reactor in Figure 10.15(c) is held in saturation by the superconducting winding. When a fault occurs, the fault current takes the core out of saturation and the impedance of the reactor increases greatly; the two devices provide the current limitation for both polarities of current flow. The feasibility of this device was demonstrated in the 1980s. Figure 10.15(d) is somewhat similar to 10.15(b); the strip of superconducting material operating in a similar fashion to the superconducting winding of 10.15(b). The conclusions were that (as of the mid-90s) no power device was available for use at the ratings of the NGC supergrid system.

Supermagnetic energy storage systems (SMES) SMESs have been applied to protect sensitive demands such as paper mills against voltage drops.

Regenerative power systems Although new methods to provide and manage running spare may be found, the problems of providing substantial energy outputs are much greater, entailing new methods to store energy or, alternatively, provide it in a changeable form (e.g. chemical $\leftrightarrow$ electrical) on a large scale and continuous basis. In the UK, National Power have recently announced an energy storage technology 'Regensys', which may accomplish this [10.36]. Its principle is shown in Figure 10.16(a), and combines the features of a battery and a fuel cell. A reversible chemical reaction is used which takes place across a membrane (regenerative fuel cell modules in Figure 10.16(b)). The electrolytes, sodium bromide and sodium polysulphide, are stored in separate tanks and pumped through the fuel cell modules when the unit is charging or discharging (on charging, the sodium bromide is reduced to bromine and the sodium in the electrolyte which dissolves, and *vice versa* on discharge). Since the electrodes in the module do not take part in the reaction, the electrical storage capacity is increased merely by increasing the electrolyte storage. The process is self-contained with no emission. Each cell produces about 1.5 volts, so a large number have to be connected in series to produce the 1000–1500 volts needed for efficient inverter/rectifier action. It can be built in modules ranging from 5–500 MW output. As of July 1999, full size electrochemical modules were being manufactured in a dedicated assembly plant. Following successful pilot plant trials, National

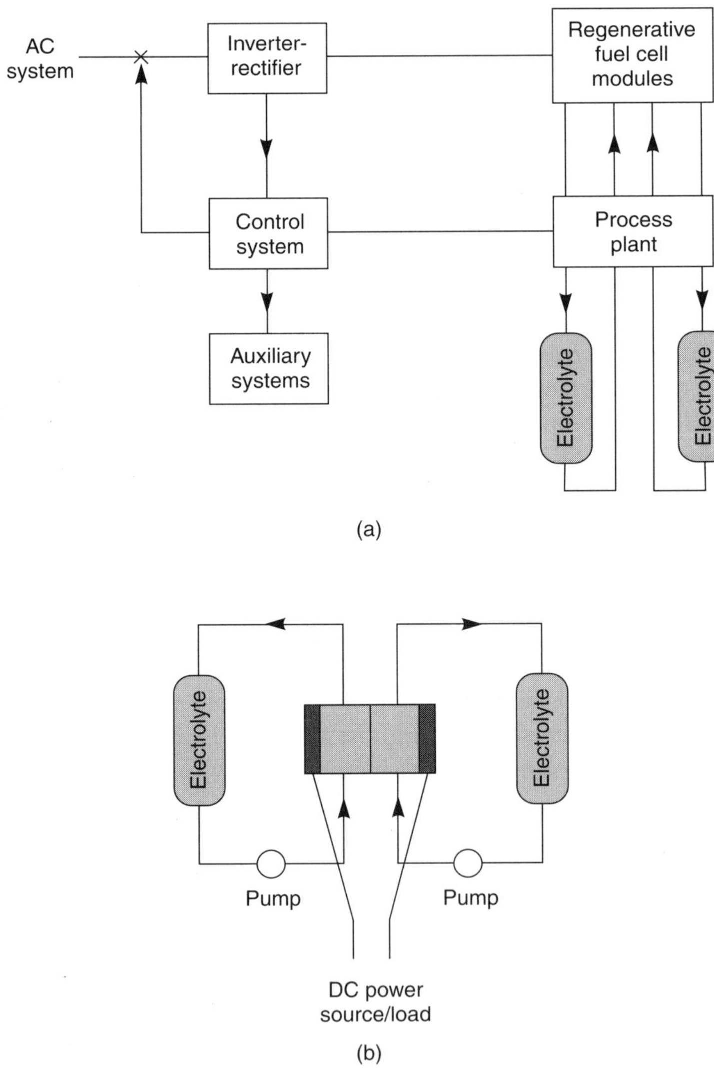

Figure 10.16 The Regenesys system. (a) System, (b) regenerative fuel cell (reversible chemical reaction: sodium bromide + sodium sulphate $\rightleftarrows$ sodium sulphide + sodium tribromide) (Figure reproduced by permission of National Power)

Power planned to build a commercial 120 MWh plant. Capital costs of £100/kWh were considered achievable, falling significantly with volume production and further development.

Another large scale reversible battery (200 kW × 4 hours) uses a vanadium process, using vanadium recovered from flue gases [10.37]. This has operated successfully over hundreds of cycles. Other storage systems described in this reference with capacities large enough to be of interest for supply applications are sodium-sulphur, zinc-bromine and vanadium-redox, whilst flywheels have been studied for providing spinning spare.

Presumably, the search will continue for electrochemical reactions with optimum characteristics for such applications (the redux process used in Regenesys was discovered in the last century), but otherwise, the future will be with the already available forms of rapid response plant and with transmission as typified in the flexible a.c. transmission system developments.

FACTS and emergency control In some 10 years FACTS devices have developed from mechanically switched equipment – quadrature and in phase control, on-off states – to the variety of control mechanisms outlined in the previous sections. This has been made possible by the convergence of high power electronics and possible system control applications. Many of the applications proposed so far, however, have been to improve the steady state performance. Significant improvements can be achieved.

NGC in the mid-1990s installed several conventional quadrature boosters to improve load sharing on the 400 kV network. Well over 2000 MVAr of SVCs and mechanically switched capacitors were installed in the south west, midlands and south coast areas of the UK in the 1980s and 1990s to improve voltage conditions.

Studies into the impact of FACTS on the control of power systems during emergencies have been limited. The impact will, in my view, be found mainly in the restoration phase. Because of the high speed with which it should be possible to adjust network conditions independently of generation, it may be possible to disassociate the possible network post-fault states from the pre-fault state more completely than in the past. This should provide greater freedom in planning and operation and better system utilization. Apart from the capital cost, the use of silicon-based and related devices will incur significant power losses, suggesting that the development of devices and circuits in which the device is only placed in circuit when needed would be profitable.

10.5 SUPERCONDUCTIVITY

Several references have already been made to superconductivity. Although discovered in 1911 by K. Onnes at the University of Leiden, the potential application of superconductivity in electrical power transmission applications was limited by two factors: the near absolute zero temperatures needed to achieve the superconducting state and the disappearance of this state in the presence of external magnetic fields above a certain strength, including those caused by current flow through the conductor itself. However, in 1986 'high temperature superconductivity' (HTS) was discovered, the special interest being that the transition temperature from superconductivity was above 77°K, the saturation temperature of liquid nitrogen at one atmosphere pressure which is a readily available commercial refrigerant. Studies into commercial applications have included small and medium sized superconducting magnetic energy storage systems (SMES), fault current limiters, transformers, cables, generators and motors. For instance Reference [10.38] describes an analysis of the economic benefits from incorporating SMES into the reinforcement of two long parallel transmission lines with capacity limited by voltage collapse.

A demonstration of the reinforcement of an urban network is planned for 2000 [10.39]. Three HTS cables each 120 m long and rated at 99 MVA are to be drawn into existing 10 cm ducts, replacing nine conventional cables (the cable runs include several right angle bends). Each cable will carry 2400 amp at 24 kV and over 8200 kg of copper cabling is replaced by 110 kg of HTS conductors.

Electricity supply in conurbations is characterized by high power densities, expense and difficulty to obtain sites/wayleaves and high security and quality standards, problems which can at least in part be addressed by superconductivity. Apart from the promise of a major increase in the capability for power transmission of existing rights of way, the lower voltages at which HTS cables can operate means that power could be delivered into urban areas at the subtransmission level avoiding the environmental impact and possibly some of the cost (depending on the voltages available at the out-of-city power source) of conventional networks.

Additionally urban networks are likely to be designed for high power infeeds from the high voltage networks such as could be found in HTS applications.

Various design and hardware development groups have examined the potential for HTS transformers, expected advantages being smaller size, greatly extended capability, and freedom from the pollution, fire and toxicity problems associated with many insulating media. In one concept, the windings would be within a liquid nitrogen enclosure, the remainder of the transformer being at normal temperatures.

Summing up, and in relation to the core subject of this book, emergency control, it seems likely that it will be some time before superconducting devices will warrant special attention, from a system point of view, when reviewing security criteria and actions in the event of disturbances.

10.6 CONTINGENCY PLANNING AND CRISIS MANAGEMENT

Concluding this book on its central theme, managing the power system to counter the 'slings and arrows of outrageous fortune', these can take two forms – on the one hand, the short, unexpected and often acute problem, and on the other, the long unveiling of events, often foreseen and often with progressive impact. The approach to these is quite different. The first, typically a system fault or sudden bad weather, is met by providing redundancy on the system at the various timescales. The second is covered by holding stocks of fuels, plant spares, etc. Inter-utility agreements will be valuable in both areas. In the past, the first has received more attention, but this bias has in recent years been eliminated.

Although automation has allowed some aspects of emergency control to be taken over by automatic (including some very important) mechanisms, human operators still have a vital role. Training to handle emergency conditions remains most important. This tends to be in the real-time aspects, the acquisition of operational planning expertise being left to normal development and tuition as the operator progresses up the staff hierarchy.

The operator must be supported by adequate control aids – SCADA, real-time contingency analysis, unit commitment, tie line and frequency control/economic dispatch for the first, and network capability/loading simulation/fuel purchase and stocking for the second. These aids should include comprehensive guides on procedures to be followed in normal and emergency conditions, and also schedules of necessary studies with time scales extending from the immediate future to months and years ahead.

The experienced operator is the backstay for dependable system operation, competent to get the most from the resources built into the power system whatever the environment, the ambient conditions and the state of the system.

REFERENCES

10.1. Rahimi, F.A., Vojdani, A., 1999. 'Meet the emerging transmission market segments', *IEEE Computer Applications in Power*, January.

10.2. Fink, L. H., 1995. 'Impact of electric utility restructuring on energy management and generation control', *APPA Engineering and Operations Workshop*, Philadelphia.

10.3. Ford, R., 1996. 'Ancillary services in England and Wales', *IEE Colloquium*, June.

10.4. Sbertoli, L. V., 1994. 'Reorganisation of the electric transmission system in Argentina', *IEEE Power Engineering Review*.

10.5. Caruso, L.M., 'Transformation of the Argentine wholesale electricity market', *ibid*.

10.6. Antloga do Nascimento, J.G., Marangon Lima, J. W., 1998. 'New regulatory framework in the Brasilian power industry', *IEEE Power Engineering Review*.

10.7. Alqueres, J.L., 1994. 'Brasil and the electrical interconnections in the Mercosul region', *IEEE Power Engineering Review*.

10.8. Rudnick, H., 1994. 'Chile, pioneer in deregulation of the electric power sector', *IEEE Power Engineering Review*.

10.9. Arriola, E., 'Electric power sector in Mexico: past, present and future developments', *ibid*.

10.10. Dussan, M.I., 'Restructuring the electric power sector in Colombia', *ibid*.

10.11. Chahin, C., 1998. 'New regulatory framework in the Colombian electrical sector', *IEEE Power Engineering Review*.

10.12. Denton, D.M., 1997. 'Deregulation risks and opportunities', *IEEE Power Engineering Review* (one of several papers on energy market environments in Europe and the United States).

10.13. Hyman, L.S., 1999. 'Transmission, congestion, pricing and incentives', *IEEE Power Engineering Review* (this reference includes draft of the proposed FERC notice of proposed rulemaking on Regional Transmission Organisations).

10.14. Perl, L.S., 1997. 'Regulatory restructuring in the United States', *Utilities Policy*, **6** (1).

10.15. Albuyeh, F., Alaywn, Z., 1999. 'Implementation of the California independent system operator', *IEEE Proc 21st International Conference on Power Industry Computer Applications*.

10.16. Tian, Y., Gross, G., 1998. 'Oasisnet: an Oasis network simulator', *IEEE Trans. Power Systems*, **13** (4).

10.17. Sparks, D., 1999. 'The California electricity market', *IEEE Power Engineering Review*.

10.18. Goulding, D., Fraser, N., 'Preparations for a competitive environment and the impact on power system operation: Ontario Hydro's experience'.

10.19. Goulding, D., *et al*., 1997. 'Evolution from system control to central market operator', *Cigre Symposium*, Tours.

10.20. UCPTE Annual Report 1998.

10.21. Ground Rules Covering Primary and Secondary Control of Frequency and Active Power within UCPTE, 1998 (UCPTE also publishes annual reports).

10.22. Newbery, D.M., 1998. 'The regulator's view of the English electricity pool', *Utilities Polity*, 7.

10.23. Office of Electricity Regulation: five papers reviewing trading in England and Wales and overseas, 1998.

10.24. Coleman, D., 1999. 'Impact of new market/new power game', *National Power News*.

10.25. Olsson, A., Engstrom, L., 1997. 'Pricing for transmission access in Sweden', *Cigre Symposium*, Tours

10.26. Hayes, D., 1999. 'Taiwan restructures', *Power Economics*.

10.27. Bergara, M.E., Spiller, P.T., 1997. 'The introduction of direct access in New Zealand's electricity market', *Utilities Policy*, **6** (2).

10.28. Professional networking on a worldwide scale, brochure International Conference on Large High Voltage Electric Systems (Cigre); e-mail: secretary-general@cigre.org.

10.29. Moore, T., 1998. 'Emerging markets for distributed resources', *EPRI Journal*, April/May.
10.30. Rastler, D., *et al.*, 1993. 'The vision of distributed generation'. *EPRI Journal*, April/May.
10.31. Sundaram, A., 1996. 'The EPRL distribution system power quality project' *EPRI*, (http:www.epri.com).
10.32. Edris, A., 2000. 'FACTS technology developments: an update', *IEEE Power Eng. Review*, March 2000.
10.33. ABB Power Systems, 1994. 'Controllable series capacitors', *Cigre Expo-94*.
10.34. Gyugyi, L., 1992. 'Unified power flow control concept for flexible a.c. transmission systems', *IEE Proc. C*, **139** (4).
10.35. IEE Colloquium on Fault Current Limiters–a look at tomorrow. *IEE Digest No. 1995/026*, 1995.
10.36. National Power. '*Regenesys, the flexible solution to energy storage*', (brochure).
10.37. Altimari, J., 1994. 'Venezuelan energy resources and electric power systems', *IEEE Power Engineering Review*.
10.38. Mann, T. L., Zeigler, J. C., Young, T. R., 1997. 'Opportunities for superconductivity in the electric power industry'. *Trans on Applied Superconductivity*, 7 (2).
10.39. Kiyotaka Ueda, Takasake Ageta, Shinichi Nakayama, 1997. 'Super G-M and other superconductivity projects in Japanese electric power sector', *IEEE Trans in Applied Superconductivity*, 7 (2).
10.40. Hingorani, N. G., 1994. 'Facts, Technology and Opportunities', *IEE Digest No. 1994/005*.

ADDITIONAL READING

Persoz, H., 1998. 'International interconnections towards the year 2000', *Cigre Electra*, **177**.

Perez-Arriaga, I.J., 1995. 'International power system transmission open access experience', *IEEE Trans. Power Systems*, **10** (1).

Holmberg, D., *et al.*, 1997. 'Transmission planning in the joint Norwegian-Swedish power market'. *Cigre Symposium*, Tours.

Brunekreeft, G., 1997. 'The 1996 reform of the electricity supply industry in the Netherlands', *Utilities Policy*, **6** (2).

Puttgen, H.B., *et al.*, 1997. 'Energy market environments in Europe and the United States', *IEEE Power Engineering Review*.

Clough, M., Hughes, M., 1998. 'The EC electricity directive–light at the end of the tunnel?', *Power Economics*.

Objects of World Energy Conference (WEC) e.g. in 1985 WEC annual report.

Williams, J.W., 1996. 'Open transmission access', *IEEE Power Eng. Review*.

de Jong, H., 1999. 'Going Dutch on mergers and acquisitions', *Power Economics*.

Pricing of Ancillary Services: an international perspective, IEE Colloquium, Digest No. 1996/164.

Janssens, N. (Cigre Task Force 38.02.14): 'Analysis and modelling needs of power systems under major frequency disturbances', *Cigre Electra*, **185** 1999 (summary of brochure).

Szechtman, L., Long, W.F., 1999, (Cigre Task Force 14.29): 'Co-ordination of controls of multiple FACTS/HVDC links in the same system', *Cigre Electra*, **187** (summary of brochure).

Moore, P., Ashmole, P., 'Flexible A.C. transmission systems', (tutorial in four parts) *IEE Power Engineering Journal*, December 1995, December 1996, August 1997, April 1998.

London Symposium, 1999. 'Working plant and systems harder enhancing the managment and performance of plant and power systems', *Cigre Electra*, **187**.

Appendix 1
Some Major Interconnected Systems Around the World: Existing and Possible Developments

While preparing material for this book the author has been impressed by the growth of interconnections between separate utilities which are geographically adjacent, sometimes within national boundaries (e.g. USA and the Commonwealth of Independent Power Systems (IPS)), and sometimes crossing these (e.g. UCPTE). Interconnections with neighbours will have a significant impact on operation during emergency situations, and membership of interconnections may place operational and sometimes planning, obligations on participants.

The 1990s have been a period of rapid organizational change, and it is more difficult to keep up to date with these than it is with technical developments. The notes which follow outline some of the major interconnections that existed in the mid–late 1990s. The information provided is patchy and sometimes approximate, and it is suggested that any reader who requires definitive and up to date information should approach the Secretariat or a principal utility of the interconnection concerned.

WESTERN EUROPE

Organization	Union for the Co-ordination of Production and Transmission of Electricity (UCPTE), founded 1951.
Countries/Utilities represented	Belgium, Germany, Spain, France, Greece, Italy, Yugoslavia, Luxembourg, Netherlands, Austria, Portugal, Switzerland, Slovenia, Croatia, Bosnia-Herzegovina, Serbia, Montenegro.

Structure	Loose pool.
Installed generating capacity	390 GW (1992).
Energy production	1533 TWh (1992).
Main voltage levels	400/380 kV, 225 kV.
Interconnections	Nordel, CENTREL, UK, a 600 MW a.c. link to Morocco.
Operation and control	No central control of network or generation dispatch; each member must maintain adequate frequency correction reserve (2–5 percent) and ensure single outage security in the operation of its tie lines. Each partner will normally provide a secondary regulation capacity at least equal to that of his largest unit in service, the target being to restore normal conditions within five minutes.

ENGLAND, WALES AND SCOTLAND (as at the mid-late 1990s)

Organizations (exercising control of the transmission networks)	National Grid Company (NGC) – covers England and Wales. Scottish Power (SP) – covers southern part of Scotland. Hydro-Electric (HP) – covers northern part of Scotland. OFFER, the Office of Electricity Regulation, and the Electricity Pool have had a major impact on the electricity supply industry, but have now (1999) been superseded by OFGEN, an organization covering the electricity and gas industry. The responsibilities of the Director General of OFFER were to ensure that reasonable demands for electricity were met, licence holders had adequate finance to promote competition, to promote consumers interests, to promote research and development, and to take account of the effects of generation, supply and transmission on the environment. The Pool is the trading organization. Any party wishing to trade electricity in England and Wales must do so through the Pool. Generators must sell into it and suppli-

	ers out of it. It defines the market trading rules, sets the half hourly Pool selling price and settles payments.
Utilities	Including subsidiaries of NGC, SP and HE, licences to build and operate power stations had been granted to 40 companies at October 1997. NGC, SP and HE had been granted transmission licences to transmit power from stations to the companies responsible for distributing it to end users. Twelve Regional Electricity Supply Companies are responsible for distribution in England and Wales. Both SP and HE are vertically integrated companies.
Geographical area	229 000 km^2
Structure	Each company has its own transmission network with 400 and 275 kV interconnections between NGC and SP, 275 kV and 132 kV interconnections between SP and HE.
Maximum demand	England and Wales 1997/8, approx. 50 GW.
Registered generating capacity	England and Wales 1997/8, approx. 72 GW.
Main voltage levels	400 kV, 275 kV, 132 kV.
Interconnections	D.C. link to France. A.C. link under construction to Northern Ireland.
Operation and control	Each transmission licencee must provide an agreed amount of reserve. SP provides the operational interface with NGC on behalf of HE and British Nuclear Fuel. NGC undertakes several functions for the Pool in England and Wales – it produces demand forecasts, generation schedules, dispatches plant, and owns and operates the transmission system within the agreed security criteria. It provides the ancillary services (frequency and voltage control, black start), and monitors reserve holding in real time for the whole system. Programmes for transfers for longer term (two months), short term (weekly, daily), and control room timescales are agreed between the utilities. Generation schedules are deter-

mined by the companies in accordance with these programmes.

SCANDINAVIA

Organization	NORDEL, founded 1963.
Countries/Utilities	Part Denmark, Finland, Iceland (but no connection), Norway, Sweden; those sections of the Danish System located on the European mainland are part of UCPTE.
Structure	Loose pool.
Maximum demand	≈55 GW (1997).
Installed generating capacity	86 GW (1992); hydro, nuclear, coal/gas.
Energy production	347 TWh.
Main voltage levels	400/380 kV.
Interconnections with	UCPTE, Russia (d.c. links)
Operation and control	Power and energy balances for each country are computed up to three years ahead. Meetings are held a few times per year to discuss power exchanges, generation outages, transmission limits, fuel prices. Information is exchanged on a weekly basis to determine marginal costs and power exchanges for use in real time.

PART CENTRAL AND EASTERN EUROPE

Organization	Unified Power System (UPS) of Russia (first interconnections in the 1950)s.
Countries/Utilities	Part of former USSR.
Structure	The UPS is divided into 13 power pools.
Maximum demand	144 GW (1993).
Installed generating capacity	213 GW (1991).
Energy production	937 TWh (1993).

Main voltage levels	1150 kV, 750 kV, 500 kV, a.c.; d.c.
Interconnections	Nordel (small), Finland, the Independent Power Systems of East Europe, Ukraine, Belarus and Baltics.
Operation and control	The structure of system operation within UPS is hierarchical – central dispatching office (Moscow), 12 dispatching offices of power pools, 104 Regional control centres, 600 distribution control centres and 1000 power station control centres. The primary reserve is usually not less than 2 percent of generation. Combined load frequency control is exercised through the central dispatching office, power pool dispatchers and the regulating stations. Two systems are responsible for frequency control.

Suggestions have been made, backed at Government level, to construct a European power link with a capacity of 400 MW from Germany across Poland, Belarus and into Russia (Smolensk). The alternatives considered were

(i) d.c. (± 500 kV or ± 600 kV) with four or five converter stations;

(ii) a 750 kV a.c. link from Russia to Belarus, with an HVDC back to back converter station at the Belarus-Polish border and connected into the Polish/UCPTE 400 kV system.

The a.c./d.c./a.c. stage is necessary to overcome differences in power quality between the western and eastern European networks.

A BALTIC RING

A Baltic Ring project was included in a European Union programme 'Transeuropean networks'. In this the Baltic Sea states would be interconnected by a.c. or d.c. links as appropriate from the German terminal of the East–West link up into the Nordel system, then Nordel to Finland, Finland to Russia direct, and also via the Baltic States to substations on the eastern sections of the East–West link.

CENTRAL EUROPE

Organization	CENTREL (formed 1992).
Countries/Utilities	Poland, Czech Republic, Slovakia, Hungary.
Installed generating capacity	Some 62 GW (1993), including part Ukraine and VEAG (Eastern Germany).
Energy production	250 GWh.
Main voltage levels	750 kV, 400 kV, 220 kV.
Interconnections	UCPTE, IPS of Ukraine.
Operation and control	Central dispatch from control centre in Prague. The intention (1995) was to operate synchronously with UCPTE and asynchronously with eastern Europe.

NORTH AMERICA

Organization	The North American Reliability Council (NERC), established in 1968, is made up of nine Regional Reliability Councils. It covers the whole of North America and the northern part of Baja California, Mexico.
Countries/Utilities	USA, Canada, part Mexico.
Geographical area	19.1 m km^2
Structure	There are four distinct interconnections – Western, Eastern, Texas and Quebec. These interconnections contain several power pools, most operating as loose pools.
Maximum demand (1993)	Some 600 GW
Interconnections	None external.
Operation and control	The major utilities will often have a control centre for the on-line dispatch of generation. The minimum telemetry will be generation outputs, external tie line flows, some voltages, and system frequency. Some power pools have established Pool Control Centres.

INDIA

As at mid-1996, there were five Regions in India, only two of which (Eastern and North Eastern) now operate in parallel. Some engineers see value in interconnecting all the regions, but several face acute power shortages, and most of the time the frequency is low. Hence, the practical solution will be to develop d.c. links (either 500 MW or 1000 MW capacities) between contiguous regions, with the exception of the link between the Western and Eastern regions for which an a.c. link is considered. In the longer term, a.c. links would be proposed, but the d.c. links would still have a role in facilitating power flow control.

MIDDLE EAST AND NORTH AFRICA

Studies have been carried out for widespread interconnections in the Middle East extending into North Africa. The countries involved would be Egypt, Syria, Iraq, Turkey, Lebanon, Israel and Iran (totalling some 70 GW of generation in 1993): Saudi Arabia, Kuwait, Bahrain, Qatar, United Arab Emirates, Oman and the Yemen (totalling some 31 GW in 1993: Saudi Arabia is a 60 Hz system, the others 50 Hz); and Algeria, Morocco and Tunisia (totalling nearly 9 GW in 1993). These developments are seen as ultimately forming part of a grid stretching around the Mediterranean.

PEOPLES' REPUBLIC OF CHINA

The installed generating capacity was 199 GW at the end of 1994, growing at several gigawatts a year. A nationwide network should be achieved quite early in the 21st century based on six regional power systems, with capacities varying between some 25 GW and 60 GW. The network will be divided into three interconnected sections. The main transmission voltages are 500 kV and 200 kV, with some 330 kV.

The Three Gorges hydro project on the Yangtze river in eastern China is the largest generation development at present. The generation capacity will be 17.7 GW (26 units of 680 MW). Another major project is construction of two hydro stations at Tiangshengquia, capacity 2500 MW with a 500 kV hvdc link some 1000 km long to a terminal station near Hong Kong.

AFRICA

There are some 50 separate countries in Africa, including off-shore islands, covering 20 percent of the world's land area. The installed generating capacity (1992) was some 74 GW, mainly concentrated in the south (South Africa) and along the Mediterranean seaboard (Egypt, Libya, Tunisia, Algeria and Morocco). These countries are, or soon will be, interconnected at various voltages up to 200 kV to form the Northern Africa power pool, with a generation capacity of some 26 GW. A 500/400 kV interconnection may be superimposed.

The Southern Africa power pool stretches from the southern tip of Africa north for some 3000 km. The installed generation capacity was some 26 GW (1992). Transmission voltages up to 750 kV are used. The Western Africa power pool includes Nigeria. It had an installed capacity of 8 GW (1993). The growth rate is somewhat lower than in the other pools. The Central Africa power pool contains the Inga generation project, on the Zaire river basin, already developed to some 2800 MW, although with much greater potential. The potential capacity of the Zaire river is estimated to be 800 TWh per year. A 500 kV d.c. link transmits power some 1700 km from Inga to Shaba in the southern power pool. The Eastern Africa pool is the smallest with a generation capacity of about 2–5 GW.

Very ambitious development has been discussed for the African continent. Inga has a potential of 30–40 GW, 240 Twh, and it has been suggested that it could be the hub of four d.c. links:

- Inga to Western Europe (Spain) roughly along the western coast of Africa – some 7200 km;
- Inga to Western Europe (Spain) through Congo, Nigeria, Morocco – some 5200 km;
- Inga to Western Europe (Italy) through Congo, Morocco, Tunisia – some 5100 km;
- Inga to the Middle East (Turkey) through Sudan, Egypt, Jordan, Syria – some 7400 km.

Some 1 percent of the routes would be added. With conservative design, the operating voltage could be 600 kV.

Another scheme for a final capacity of 20 GW proposed 800 kV links from Inga to Morocco, and then Spain, and from Inga to Turkey via Egypt; there would also be a link from the Egyptian terminal through Libya to Tunisia, and thence to Italy. It is interesting that transmission losses on this proposal would approach 25 percent.

Overall, the cost per MWh of the Inga generation plus transmission to Europe is put at about half the cost of a coal plant development in Europe over a lifetime of 60 years.

SOUTH AMERICA

Nearly 25 percent of the world's hydro potential is located in South America, with one third of this in Brazil, some 1120 GWh annually equivalent to an installed capacity of, say, 260 GW. The Itaipu hydro station (12 600 MW from 18 700 MW generators) was built on the Parana river in Brazil in the late 1980's/early 1990s. Half the generators operate at 50 Hz, and are connected by two $\pm$600 kV d.c. bipoles to the load areas of Sao Paulo and Rio de Janeiro. A small supply is also provided at Itaipu to Paraguay. The other generators operate at 60 kV, and are connected by 750 kV a.c. lines nearly 900 km long to the Brazilian transmission network.

The potential hydro capacity in the Amazon Region is some 130 GW. Transmission distances to the Rio–Sao Paulo–Belo Horizonte demand centres vary between 2200 and 2800 km, and to the Recife–Salvador areas 2100–2400 km. Eletrobras, the Brazilian authority responsible for managing transmission and interconnection has considered a.c. alternatives at 800 kV, 1050 kV and 1200 kV and d.c. alternatives at 600 kV and 800 kV. Half wave and six-phase transmission were also studied. Transmission losses would be about 10 percent, considerably less than in the Inga scheme. The transmission cost, including losses is put at 11–13 mils/kWhr, about 1/3 the cost of developing the generation.

CENTRAL AMERICAN POWER GRID

The six countries of Central America are, with support from the Interamerican Development Bank, developing plans for an interconnected grid stretching from Panama to Guatemala. Objectives would be to improve efficiency and security. An international power market would be established.

INFORMATION SOURCES

The information in this appendix has been obtained from various sources. Particular reference should however be made to the papers in the *IEEE Power Engineering Review*, often collated by T.J. Hammons (University of Glasgow), papers by Professor H. Rudnick (Catholic University of Chile), reviews in the journal *Modern Power Systems*, and the review, of power systems available from ABS Publishing, Woking, Surrey, GU22 7PX.

Appendix 2
Glossary of Useful Terms

Some of the terms often found in the literature of emergency control and related subjects have been described below. These are mainly the generally accepted understanding of terms rather than strict definitions. The author has drawn on his own knowledge, various of the papers included in the references and, in particular, an IEEE publication, *Glossary of Terms and Definitions Concerning Electric Power Transmission System Access and Wheeling*, and also *Modern Power Station Practice.*

Active power model – this is a system model in which demands and generation, or equivalent net transfers, are represented by active power nodal transfers, and system series elements by reactances; it is commonly referred to as a d.c. (direct current) model (not to be confused with direct current transmission).

Adaptive control – a form of control in which the set values and actions of control equipment are changed (adapted) to suit the current system conditions optimally (an example would be to adjust load shedding so that more demand would be disconnected in areas short of generation than in those with adequate generation).

Alarms, alarm handling, alarm rationalization – 'alarms' are the visual or audible indication within a control room that an unexpected event has occurred: thus, if a circuit breaker opens to clear a fault, the following indications may be given in the relevant control room/s:

- a chime ('ding dong')
- a flashing lamp within the room;
- a flashing 'general alarm' lamp on the mimic diagram;
- the lamp indicating that particular circuit breaker on the substation diagram may flash, alternatively a 'turn-over aspirin' operate;
- an indication (lamp or turn-over aspirin) on a wall mounted or desk mounted substation diagram may operate; alternatively, if CRT displays are in use, the appropriate circuit breaker on the relevant substation display may flash together with a message on a general alarm or incident list.

Typically, alarms will be restricted to events which are not the direct result of operator actions, for instance the automatic opening of a circuit breaker to clear a fault, indications of operating parameters outside limits or of abnormal configurations such as a system split, unauthorized entry to a substation.

Ampacity – the thermal current rating of a circuit.

Amp overload alarm – an audible or visual alarm given when the current in a circuit exceeds the setting value.

Ancillary services – the services, additional to the main transmission, necessary to ensure reliable operation of the system. These will typically include frequency response, forms of reserve, reactive power supplies, and black start.

Artificial intelligence (AI) – a much quoted definition is "artificial intelligence is the science of making machines do the things that would require intelligence if done by man." Intelligence can perhaps be taken as an ability to study and determine solutions to problems which cannot be solved by rote in an acceptable period of time. AI is commonly taken to include some techniques which would normally be classified as emulating thought whilst others would be essential to provide an interface between the outside world and intelligence, e.g.

Intelligence aspects	**Interface aspects**
Expert systems	Perception (visual, auditory, tactile)
Heuristics	Artificial speech
Optimization	Language understanding
Automated reasoning	
Guided search	
Knowledge acquisition	
Automatic programming	

Asynchronous operation – some part of a system normally operating in synchronism is out of synchronism.

Authorized person – this is a term used in the UK, and no doubt utilities elsewhere will have equivalent concepts. It denotes a *competent person* (see below) who has been nominated by an appropriate officer of the utility to carry out duties specified in writing. A 'senior' authorized person is someone who additionally is nominated to prepare, issue and cancel safety documents.

Automatic Generation Control (AGC) – automatic adjustment of the output of generation within a specified part of the system to maintain frequency and active power transfers to the remainder of the system at target values (sometimes called tie-line frequency control; a related term is Economic Dispatch Control (EDC)).

Automatic reclosure – the automatic closing of a circuit breaker which has opened by operation of a protective relay; reclosure may be delayed (e.g. reclosure after, say, 20 seconds), high speed (e.g. reclosure under one second), one shot or multi-shot (e.g. one attempt or more than one attempt is made to close the circuit), single phase or multi-phase (generally single phase is only found at the highest voltages when in order to stand the best chance of maintaining stability individual tripping of the phases of a faulted circuit rather than tripping of all three phases is used).

Black start – the process of bringing generation from zero load, usually disconnected from the system, to load; it can be applied to a single generator, a power station, or a whole system.

Boiler following turbine – a method of plant control in which the turbine of a boiler turbine-generator unit receives the signal to change output and the firing rate of the boiler is adjusted to provide the steam demanded at the target steam conditions. In 'turbine following boiler', the output change is signalled to the boiler.

Boiler stored energy – the energy stored in the boiler which if released, say, by a drop in external pressure will provide additional output over a limited time.

Bucholz relay – a relay fitted to a transformer, reactor or other plant item to detect abnormal conditions in the cooling medium. Frequently, operational rules will require the main plant item on which a Bucholz relay has operated to be switched out of service rapidly. Gas and oil pressure alarms on cables will have a similar purpose.

BSP (Bulk Supply Point) – a point of connection to large consumer or to lower voltage network from a transmission network.

Bus, busbar, node, switching point – alternative terms for the connection point of two or more circuits, usually through circuit breaker/s or disconnector/s.

Cascade tripping – a general and now rather old-fashioned term used to describe a sequence of automatic switchgear operations often not able to be explained fully.

Code of practice – obligatory or recommended procedures in a specific area of work.

Communication channels – facilities used to transmit data (analogue and/or digital) and/or speech; these include

- power line carrier (plc) – signals and speech;
- high frequency radio – speech;
- low band VHF, mid band VHF – speech;

- microwave links – data and speech;
- hired circuits (e.g. from British Telecom in the UK) – all data and speech; very high capacities can be provided, tariff likely to depend upon capacity available and point to point distance);
- pilot wires (cables) – generally refers to cables for data and/or speech layed with main cables (the term 'private wires' has been used to denote cables owned by a utility);
- optical fibre – data (in large quantities), speech;
- satellite – signals are beamed to a communications satellite in stationary orbit and thence back to a receiving point on earth.

Competent person – a person who has sufficient technical knowledge and/or experience to avoid danger, and who may handle specified safety documents when nominated by an appropriate officer of the utility.

Configuration – the connections between circuits and other plant items achieved by the open or closed states of circuit breakers and disconnectors; it can be applied to busbars (busbar configuration), substation or systems; additional descriptive terms will often by used, e.g.:

- radial configuration – a network in which no loops exist apart from those formed by circuits in parallel between the same nodes; this may be a feature of the construction or achieved by switching, by opening a circuit breaker or disconnector in a loop, for instance;
- meshed network configuration – a network in which loops exist.

Contingency – an event that has occurred or might occur. The term is often quantified e.g. double circuit contingency, often meaning faults on two circuits which exist simultaneously and are often implicitly taken as occurring virtually simultaneously. The implicit connotation of 'contingency' is that the event is (or would be) harmful to the viable operation of the system. A credible contingency is the coincident non-availability of plant on a local or global basis, as appropriate, which the system is designed to sustain frequently without disconnection of any consumers.

Control area – a part of an interconnected system whose operation is co-ordinated, often instructed, from one or a hierarchy of control centres operating in liaison with each other.

Control centre – a location at which facilities have been provided for the collection of data relating to the current operation of the system or a part of it (the control area), and/or for the issuing of instructions to operational personnel and automatic equipment in the control area.

Coupling factors or coefficients – variously called distribution factors, influence factors, coupling coefficients, etc.; these can be variously defined as:

- between circuits – the change in flow in one circuit when there is a unit change in flow in another circuit;
- between circuit and node/neutral – the change in flow in one circuit when there is a unit change of flow between another node and neutral;
- between circuit and node/node – the change in flow in one circuit when there is a unit change of flow between two nodes;
- between circuit and node – the change in flow in one circuit when there is a unit change of flow at one node, removed equally at all nodes.

These factors are often calculated using an active power model.

Coupling factors are particularly important when, for instance a sequence of $(n-1)$ circuit outage states have to be calculated or incremental transmission losses and flows estimated for a number of alternative patterns of generation. The factors are often calculated from the models, although one of the best known, the '*B*' coefficients used in transmission loss calculations, normally use a.c. models.

Credible . . . – a widely used, often imprecise, adjective to describe an event or state whose probability is considered high enough to justify action; thus a 'credible contingency' is an event whose occurrence is judged to be sufficiently likely to justify taking planning or operational measures against its occurrence or effects; a 'credible outage' is the non-availability of an item/s of plant considered sufficiently likely to justify action to ensure that viable operation of the system will continue if it should occur (see also 'Contingency').

Damping power – a measure of the ability of the system to suppress angular oscillations; related terms will be damper winding or amortiseur winding (short circuited winding on rotating machine rotors in which voltages are induced, hence currents flow and losses occur when the rotor oscillates with respect to the rotating magnetic field developed by balanced three phase currents flowing in the stator winding).

Defence plan – a term introduced by EdF to describe an integrated set of measures to restore a system to viable operation following a (generally) severe disturbance.

Derating – a reduction in the rating of generation or transmission plant caused by plant problems (e.g. failure of cooling fans) or environmental conditions (e.g.

extremely high ambient temperature leading to reduced thermal ratings of overhead conductors). The term is variously applied to thermal ratings, fault rupturing capacity, and occasionally voltage rating.

Diagram – in control room parlance, usually further qualified in terms of its physical construction (wall diagram, desk diagram); function (switching diagram, loading diagram, configuration diagram); whether animated or not; the function and animation will determine the system detail shown (e.g. substations as busbars or boxes), instrumentation (active and reactive power flows, voltages), circuit breakers plus isolators or line-end-open only. A 'disconnected diagram' is one in which the circuits between substations are not shown; for instance, a circuit from substation A to substation B could be shown as

- at substation A – an arrow from the appropriate plant item labelled '*S/Sn B*''
- at substation B – an arrow from the appropriate plant item labelled '*S/Sn A*'.

The 'open/closed' state of a circuit or 'on/off' state of an alarm will be indicated by an appropriately placed lamp or mechanical semaphore, including the so-called 'turnover aspirin' A series of lamps or semaphores can be used to indicate loading levels.

Direct current links – direct current connections between or within a.c. systems. Applications will include interconnecting asynchronous a.c. systems, long distance high power transmission, and connection within areas of high power density and hence providing one method of containing fault levels. D.C. links can be modelled approximately by placing active power generation and demand to represent the transfer over the link at the terminal nodes of the link since in general these can be specified independently of the operating conditions on the a.c. network).

Disaggregation – separation of functionally different parts of a vertically integrated utility (i.e. generation, transmission, distribution, supply) into the functional businesses (sometimes called unbundling).

Disconnector – alternatively 'disconnect switch', or in the past 'isolator'; a mechanical switching device used to change the connections in a system, including isolation of a plant item from all voltage sources. The switching capabilities of disconnectors will provide further classification, e.g. open/close only when the circuit is dead (the usual understanding if there is no further amplification), close onto fault, etc.

Dispatch (generation dispatch) – the determination and instruction of output to running plant; it may be qualified as active power dispatch, reactive power dispatch, economic dispatch; a 'secure' dispatch is one which satisfies the security criteria.

Earthing, earths – these terms describe the connection of items of plant which have been disconnected from the live system and are therefore dead, apart from electromagnetic and electrostatic effects, to earthed objects; there will be different types of earth, for instance portable earth (an earthing device which can be moved to different points on the system (e.g. in a substation)); primary earth (an earthing device placed at a position defined in a safety document); and drain earth (an earthing device supplied for protection against induced voltages).

El Nino – the extreme, periodic weather conditions which affect the Pacific and, to a lesser extent, south east Asia and the western seaboard of America.

Expert systems (ES) – An ES can be regarded as a means of recording and recalling at will human competence in a particular specialist, usually narrow field (often called domain). As such it can serve as an expert in that field. It will consist basically of six components (Figure A2.1) – knowledge base, inference engine, system database, algorithmic programs, diagnosis logic, man-machine interface.

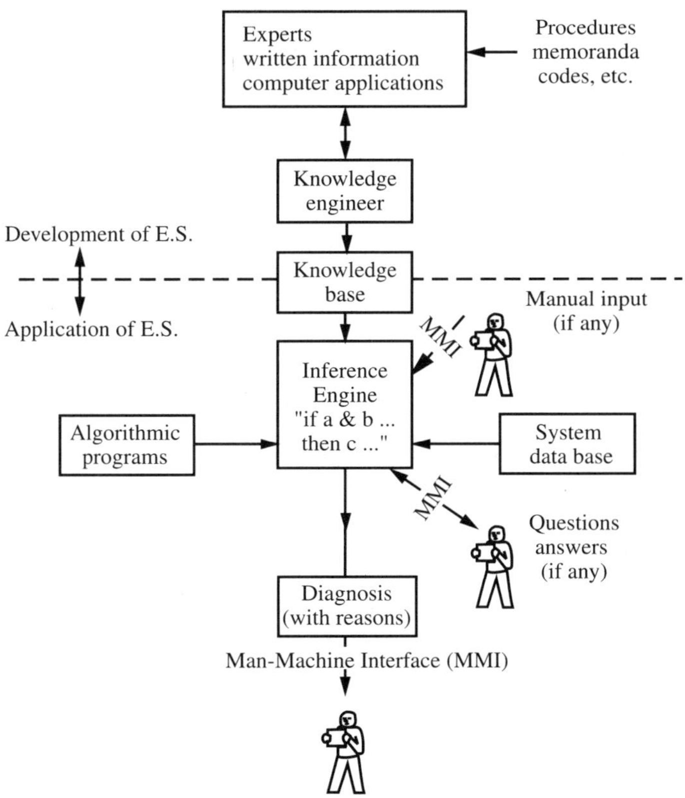

Figure A2.1 Components of an expert system

FACTS (Flexible A.C. Transmission System) – a transmission system employing various devices, usually with some element of electronic control, to increase the capability of the system.

Fault current limiter – a series element which limits the magnitude of current flowing on the occurrence of a fault; the term includes the recently developed FACTS devices whose impedance under normal system conditions will be small, but will increase automatically when the current flow increases on the occurrence of a fault.

Firm capacity – the minimum of the capacities (e.g. of a station or system) of all the credible outage states.

FERC – Federal Energy Regulating Commission (in the USA).

Franchise area – a part of an interconnected system in which a 'franchised monopoly' has an exclusive right to supply. Almost universal in the past, this system is now giving way to free trading in which customers can buy, from any supplier, facilitated by open access to networks.

Free flowing ties – interconnections between utilities in a power pool, the power flows in which are not directly controlled.

Frequency response (e.g. of demand) – the change of e.g. demand in response to a change in frequency.

Generation capacity – this term should be further qualified, e.g. nameplate rating, total (usually the generation available to supply both the system and the in-house service requirements of the station/unit), sent out (the generation available to the system after the house service needs have been met), firm (the sent out capacity of a station after the credible outage/s have been deducted).

Generation rejection – the disconnection or reduction in output of generation as a result of operation of a system protection scheme, or by operator action.

Generation reserve – this is the difference between the actual (or estimated) generation requirement and the actual (or estimated) available generation; the term 'generation margin' is sometimes used.

Generation response (to system frequency changes) – the change in the output of generation as a result of a change in frequency in the system to which the generation is connected. The response will depend upon the mode of boiler-turbine control in use. Under 'free governor action' deviations of system frequency from the frequency set point of the machine speeder motor will result in a change in the position of the turbine throttle valve and hence turbine and generation output. With 'sliding pressure' control, the speeder gear is at an extreme position and the turbine throttle valve fully open. Changes in output are effected by changing the boiler pressure.

Governor drop – the change in speed (or frequency) necessary to cause a change in output from zero to full load; related terms are deadband (the change in speed or frequency over which no change in output occurs) and system stiffness.

Heuristics (heuristic knowledge) – knowledge or understanding based on experience rather than being mathematically or physically provable. It will include rules of thumb, short cuts, informed guesses and the acceptability of approximations. In contrast, the term 'deep knowledge' is sometimes used to describe knowledge based on physical principles and physically or mathematically derived algorithms.

Integrated Gasification Combined (IGC) cycle generation – in this process, fuel such as coal or oil is heated with steam and oxygen. After removing the hydrogen and nitrogen components and particulates, the manufactured fuel gas is burnt in a combined cycle unit. The objective is to remove pollutants such as sulphur and nitrogen oxides and particulates from the flue gas at coal and oil fired stations. Nearly 5 GW of this type of plant was under construction in the mid-1990s.

Integrated resource planning – an extension of normal system planning to include the evaluation of costs and benefits incurred on both supply and consumer sides.

Intertripping – generally a mechanism to open a circuit breaker/s automatically as a consequence of the opening of another circuit breaker.

Interlock – typically a mechanical or electrical mechanism which prevents operation of another device (e.g. a disconnector) unless a specified configuration/s exists in associated parts of the system.

Interruptible demand – demand which can be disconnected by the system operator, usually in order to rectify an unacceptable operating state, sometimes to improve the operating economics of the system.

Investor-owned utility – a utility which is owned by its stakeholders.

Isokeraunic level – the number of days per year on which lightning occurs in an area.

LAMS, WAMS – acronyms used to describe monitoring systems extending over a local area (LAM) or wide area (WAM).

...limits – these are the upper and lower limits of a variable acceptable in planning studies or in operation; the limits may be hard (e.g. $x >$ limit) or soft (e.g. transgression of the limit incurs a penalty whose magnitude depends on the magnitude of the transgression; typically switchgear fault rating would be a hard limit and current ratings a soft limit, conditioned by the time for which the overload may exist).

Load following – in trading terms, the obligation of a utility which is wheeling power to provide from its own generation any deficit or surplus between the wheeled power and the instantaneous requirements of the customer or supplier of the wheeled power.

Load shedding (or demand shedding) – reduction in demand by disconnection; it may be initiated by an operator (system or substation) or automatically on detection of an abnormal operating condition, usually low frequency but occasionally low voltage. If a prolonged period of shedding is anticipated, this may be done on a rota – 'rotational shedding', usually implemented manually. Slight perturbations in the rota will ensure that times and days of week of disconnection will be varied.

Loop flows – flows that occur around closed loops in a power system and as such serve no useful purpose; have been called 'parasitic' flows in the past.

Loss of load probability – typically, the number of days on which the system demand will exceed the available system capacity, as a proportion of the total number of days (typically per year).

Marginal cost – the cost to a utility of providing the next unit of electricity; it should be further quantified, e.g. short run (the cost of fuel to provide the additional energy); long run (the cost of the system reinforcement and fuel to provide the additional energy).

Related terms are merit order table (a list of marginal costs of individual generators or stations in ascending order of magnitude); Table A cost (the cost of incremental generation including no load costs), Table B cost (the incremental cost of generation excluding no load costs) (Figure A2.2).

Marketer – an agent who sells energy on the behalf of a generator; he may also arrange the transmission and ancillary services required to deliver the energy to the consumer. The term has come into prominence with the advent of deregulation and open trading.

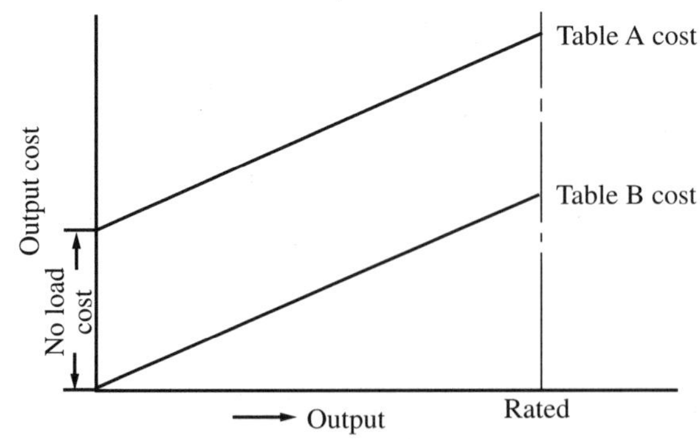

Figure A2.2 Table A and Table B costs

Mechanically Switched Capacitor (MSC) – a capacitor bank connected to the system through a mechanical circuit breaker or disconnector.

Monopoly (oligopoly) – one (a few) sellers exert market control.

$(n-1)$, $(n-2)$, etc. – abbreviations to describe the security criteria under which a system is planned or operated. $(n-1)$ means that a viable supply should be maintained with one of the total number n of circuits not available, $(n-2)$ two not available, etc. There can be confusion over n; is it the as constructed number, or the expected number after maintenance; does n refer to the number of lines or to the number of circuits (twice the number of lines if a double circuit construction is used)? The planning and operational criteria may be different.

Neural network (NN) – a NN consists of 'layers' of nodes called 'neurons' (Figure A2.3). The various physical quantities relevant to the process are connected (e.g. as scaled voltages) to the input layer. Weighted values of these are sent to some/all of the neurons of the first 'hidden' layer, where they are summated/processed and passed to the next hidden layer or to the output. It seems that in practice, neural networks rarely have more than one or two hidden layers. The advantages of a neural network will be its very fast response and, potentially, answers/relationships may be revealed that are not predicted. Its disadvantages are how to determine a suitable configuration (e.g. what inputs, interconnections, and forms of weighting functions), how to obtain optimum values for these (called 'training'). It seems that if known the performance equations of a system should materially help these tasks.

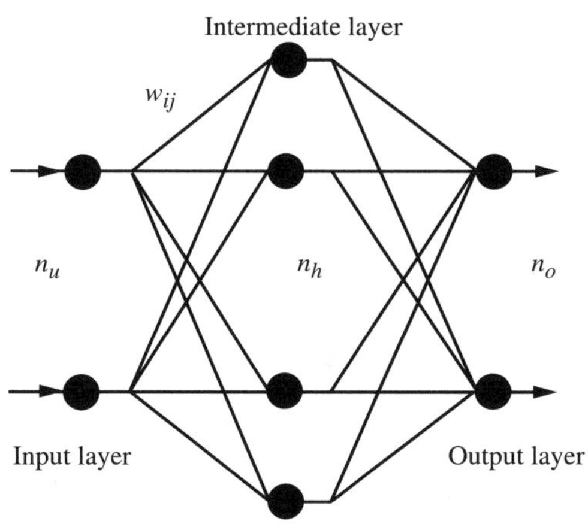

Figure A2.3 Structure of a feed-forward neural network with one hidden layer. W_{ij}: weighting factor for connection between nodes i and j; n_u: input neurons; n_o: output neurons; n_h: hidden neurons

Neutral earthing – several of the methods of earthing the neutral points of power systems are shown in Figure A2.4. The voltage levels at which these are typically used are indicated.

Nominal (voltages, ratings, etc.) – generally, the nameplate voltage, etc. of a plant item.

Non-firm service – a service (e.g. transmission) that is available to a customer when system conditions permit.

Non-utility generation – generation that is not owned exclusively by a utility, but which is connected to the utility system.

Non-simultaneous . . . (e.g. maximum demand) – the sum of the maximum values of the quantities included (e.g. the separate non-simultaneous maximum demands).

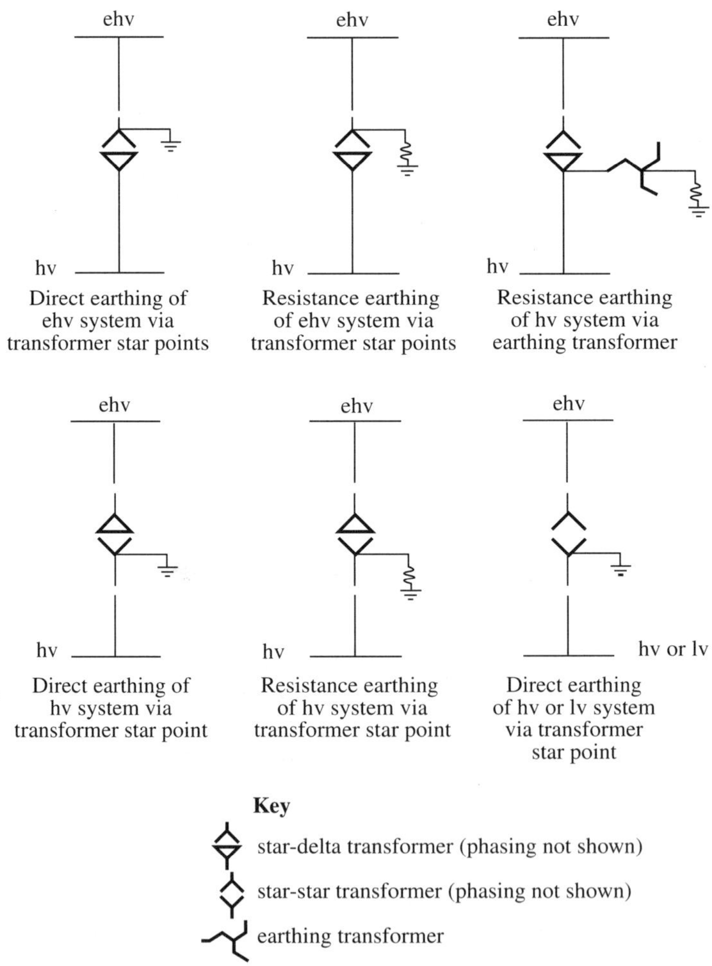

Figure A2.4 Some methods of neutral earthing

Normal – a widely used adjective (e.g. normal operation) to describe the state of a component or system when it is functioning as expected; the general connotation will be that the state is acceptable.

Normally open/normally closed – description of a network or substation switching state in which a circuit breaker or disconnector is, under normal conditions, open/closed.

Notice of Proposed Rulemaking – a term used by FERC stating its intention to introduce a rule/regulation on some aspect of power supply in the USA.

Obligation to supply (serve) – the supply regulations under which utilities in some countries are given licences to operate may include an obligation to offer supply (service) to appropriate consumers in the franchise areas.

Open access – as part of deregulation, some countries now require that authorities owning transmission should make this available to eligible Generators, Marketers, etc. needing such facilities. This term is used particularly in the USA.

Operational planning – the timing and content of activities covering such issues as: release of plant for maintenance, to accommodate new construction, or for repair; determining synchronizing and desynchronizing times of generators and loading profiles; determining optimum network configuration (e.g. for maximum transmission capacity, containment of fault levels, minimum transmission losses) or minimum generation costs; trading. The timescale for this work will be from hours to years ahead of the event. Some utilities will allocate it to the planning or system operation functions. Occasionally, the very short term work, up to a few days ahead, has been called operational programming.

Operational memoranda, operational procedures – in the interests of safety of personnel and plant and of security of supply, many utilities publish internally the rules and procedures relating to plant margins, operation of plant and safety (e.g. isolation and earthing) as a series of memoranda or procedures.

Operator training simulator – an integrated set of hardware and software on which 'hands on' training in the various aspects of control room work can be given. Simulators range from simple boxes containing a d.c. supply and variable resistors to represent generation infeeds, through workstations to teach switching duties up to a more or less completely equipped control room which can double as a standby control room and, with the addition of a system model, a comprehensive training facility.

Optimization – in mathematical terms, the act of determining a minimum or maximum value to a function (the objective function) subject to a set of constraints. For a single minimum/maximum to exist, the objective function must be convex (e.g. the inside of a saucer)/concave (e.g. top of a hill). This implies that the function is monotonic, i.e. it is steadily increasing/decreasing

Table A2.1 Broad Properties of Some Optimization Methods

Method (D) = direct (I) = indirect	Definition of system in optimization model	Restrictions on performance constraints	Restrictions on objective function	Restrictions on solution	Will solution be optimized	Comment
(1) Manual search (D)	Any available system performance equations can be used to test viability of proposed operating state	None	None	None	Not guaranteed	Labour intensive and hence probably limited field of search. May be possible to write computer program to generate system designs or operating states, thereby extending area and/or depth of search.
(2) Heuristic (D) or (I)	The underlying logic or physics of the process may be used in model development	None	None	None	Not guaranteed	Overlaps with computer aided implementation of (1)
(3) Gradient methods (D)	Equations or inequalities can be included as penalty terms in objective function	In some methods, must be differentiable over range considered	In some methods, must be differentiable over range considered	May only find local extreme, depending on the starting point	Not guaranteed	Likely to require some purpose built software
(4) Dynamic programming (D)	Any available system performance equations. Physical process must be Markovian	None	Any available method may be used to cost individual states	None	Within boundaries of search and search mesh used	Likely to require some purpose built software. Very powerful technique, but beware problems of high dimensionality.

(5) Linear programming (LP) (D)	By equations or inequalities	Must be linear (or linearized in the range of the study)	Must be linear (or linearized in the range of the study)	Only positive variables will be found and formulation must be tailored to this requirement	Within linearity and non-negativity constraints of the formulation	Very large problems can be solved. LP equations can be generated by computer. Sensitivities of solution to changes in constraints and objective function can be determined using shadow prices, parametric programming and cost ranging.
(6) Transportation problems						A special form of LP in which the coefficients of all variables in the constraints are 1 or − 1.
(7) Integer linear programming (D)	By equations or inequalities	Must be linear (or linearized in the range of the study)	Must be linear (or linearized in the range of the study)	Only positive integer solutions are sought	Within the linearity and integer requirements	
(8) Quadratic programming (D)	By equations or inequalities	Linear or linearized over the range of the study	May contain linear and quadratic terms	Only positive solutions for variables will be found and formulation must be tailored to meet this requirement	Within the constraints of the formulation	

(*continued*)

Table A2.1 *(continued)*

Method (D) = direct (I) = indirect	Definition of system in optimization model	Restrictions on performance constraints	Restrictions on objective function	Restrictions on solution	Will solution be optimized	Comment
(9) Differentiation (I)	Any available system performance equations included by successive elimination of variables or by Lagrangian multi pliers; inequalities by Kuhn-Tucker multipliers	Continuous over the range of variables studied	Convex/concave for minimization/ maximization and continuous over the range studied			
(10) Expert system (D)	By logical statements (e.g. heuristic) or performance equations	None	None	None	Not guaranteed	Potentially very flexible and often gives rapid solutions. Will usually require purpose built software
(11) Simulated annealing (D)	Any available system performance equations can be used to test viability of proposed operating states	None	None	None	Not guaranteed	Essentially an indirect and directed search about a previous solution. The techniques should avoid solutions becoming lodged at local minima.
(12) Genetic algorithm (D)	Any available system performance equations can be used to test viability of proposed operating states	None	None	None	Not guaranteed	An indirect and directed induction from previous solutions.

either side of a unique extreme value. The constraints will model the physical laws governing the system. Most optimization formulations require these to be linear, and if the actual system constraints are non-linear, approximations must be made, possibly linearizing the operation about a specified operating point. Alternatively, the constraints can be checked as a separate step between each optimum seeking change of solution variables, discarding those changes which would result in non-feasible operation.

The broad properties of some optimization methods are listed in Table A2.1. Each of these has its merits, for instance

- linear programming – computer codes are available for very large problems; formulation must be linear or be made so by approximation;
- search methods – powerful in practice, but will often need software as well as model development; optimum solutions are not guaranteed;
- heuristic methods – powerful and will tend to make use of the modeller's knowledge of the physics underlying the system behaviour; will need software development;
- dynamic programming – very powerful and tends to emulate human logical processes; the number of dimensions over which the search for the optima is made may have to be limited to avoid an exponential increase in computation.

A very important practical point is whether the solution must be integer (e.g. generators synchronized/not synchronized), continuous (e.g. output of a running generator) or mixed (e.g. generator synchronized/not synchronized and output if synchronized).

Outstation – a term mainly used in the communications field to describe a location on the power system with connections into the telemetry and communications systems.

Phase angle regulator (quadrature booster) – a phase angle regulator, sometimes called a quadrature booster, is a device which changes the phase angle of its output voltage with reference to its input voltage; when connected in a closed loop in a network, it will cause power to circulate around the loop. The mechanism is to inject a proportion of the sum of the phases *B* and *C* voltages into phase *A*, phases *C* and *A* into phase *B*, and phases *A* and *B* into phase *C* at the installation point of the booster.

Planned outage – a commonly used term to denote the disconnection of an item of plant from the system at a pre-arranged date and time, usually for an agreed duration.

Plant ordering – often called unit commitment, this is the determination and instruction of the synchronizing and desynchronizing times of the generating units.

Post-contingency operating procedures – operating procedures used by the system operator after a contingency has occurred, e.g. to contain overloads.

Power pool – all or part of a power system in which operations (and sometimes planning) of the individual utilities are co-ordinated for the mutual benefit of its members; the benefits may be in the form of reduced generation capacity, lower operating costs, better security, etc. Some pools will be 'tight pools', in which there will be a pool control centre which typically exercises considerable jurisdiction over the operating conditions of the individual members; others are 'loose pools', which operate under a common frequency bias tie line system with agreed primary and secondary regulation settings for all plant, agreed frequency and agreed external pool transfers.

Power quality – the standard of the supply in terms of the stability of the operating variables and their proximity to nominal values.

Power system monitoring – a widely used term covering many activities for instance manual inspection of telemetered data, automatic alarming of values outside limits, observation and recording of transients; actual, expected and post contingency values may be monitored.

Power system stabilizer – a device which reduces oscillations after a disturbance through the injection of signals of appropriate phase and magnitude into the generator control mechanisms.

Primary, secondary and tertiary regulation – primary regulation is the control of frequency provided typically by the combined actions of the turbine governors, and will be effective within a few seconds. Secondary regulation provides control of frequency and power transfers to external systems. It is frequently automatic, available in tens of seconds. Tertiary regulation is available from manual or automatic change of the set points of the secondary regulation controllers.

Protective gear, protective gear systems, protective relays, etc. – equipment to detect particular, usually abnormal, system conditions and initiate appropriate actions – for instance to give alarms, trip circuit breakers or start other sequences of protective gear operations. Some of the terms used and broad classifications for protective systems are:

- Voltage Transformers (VT), Current Transformers (CT) – instrument transformers connected to the system providing signals proportional to, typically, the phase to neutral voltages and the line currents at the point of connection;

- unit protection – this protection will only operate for faults within a specified section (zone) of the system, achieved by transmitting signals between relays connected to the CTs and VTs at the boundaries of the zone. Depending on the location of the fault, these signals will allow, cause or prevent tripping of the circuit breakers at the boundaries of the zones;
- intertrip signal – a signal which causes a breaker at one point on the system to operate on its receipt from another point;
- impedance protection – a protection system which measures the loop impedance (e.g. phase to neutral) and operates to trip a line circuit breaker when the impedance falls below a set value. Admittance protection operates on similar principles. The term 'distance protection' is sometimes used. This can be set to operate for faults within a specified distance of the protected circuit breaker, but not beyond that distance. Additional features will be that it is directional (operates for faults into or beyond the protected feeder, and not behind the relay location), it can be time delayed and it can be multizone. Thus distance protection can be applied to several lines in series as illustrated in Figure A2.5(a). To minimize overall clearance times 'acceleration signals' can be sent from an operated first zone relay at one line end to a, so far, unoperated second or third zone relay at the other end, as illustrated in Figure A2.5(b). This then gives practically first zone clearance time for the whole line.
- overcurrent protection – an 'overcurrent relay' operates when a current in excess of its setting value flows through it. For protection against phase faults the relay will be connected to the CTs in the system phase conductors (e.g. in a circuit breaker). These relays may incorporate a time delay feature so that their operating time will depend upon the magnitude of the fault current (e.g. Figure A2.6 for an IDMTL (Inverse Definite Minimum Time Delay) relay). Taking the simplest case of a radial feeder in several sections fed from a power source (Figure A2.7), the requirement will be that if a fault occurs on one section, the infeeding circuit breaker to that section should operate first to clear the fault. This is achieved by 'time grading' the settings (e.g. the relay on the breaker most remote from the supply point would operate in say 0.4 seconds, the next one in 0.9 seconds, the next 1.4 seconds, and the final one (at the supply point) in 1.9 seconds). The operating times will vary with the magnitude of the fault current, but the principle of discrimination between the operating times will be maintained.
- Some of the precautions to be noted when designing/setting relay installations are:
  - mutual coupling between parallel circuits can affect the reach of distance protection (setting is the calculation and setting on site of relax operating parameters such as current, time, impedance reach, etc)

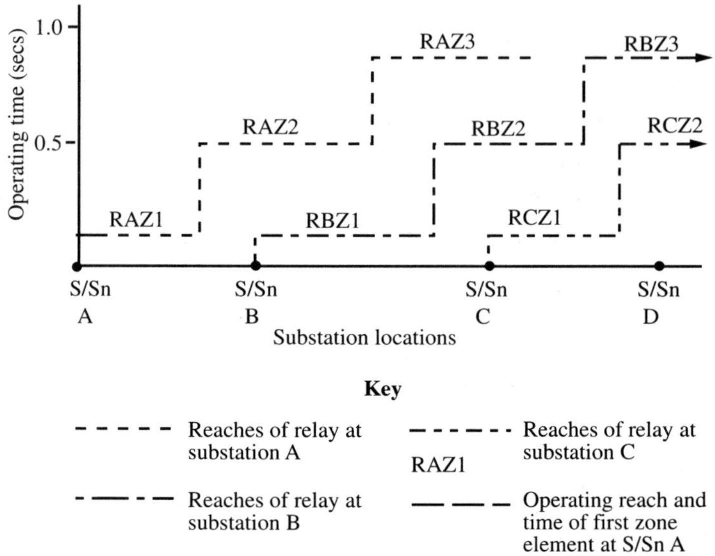

(a)

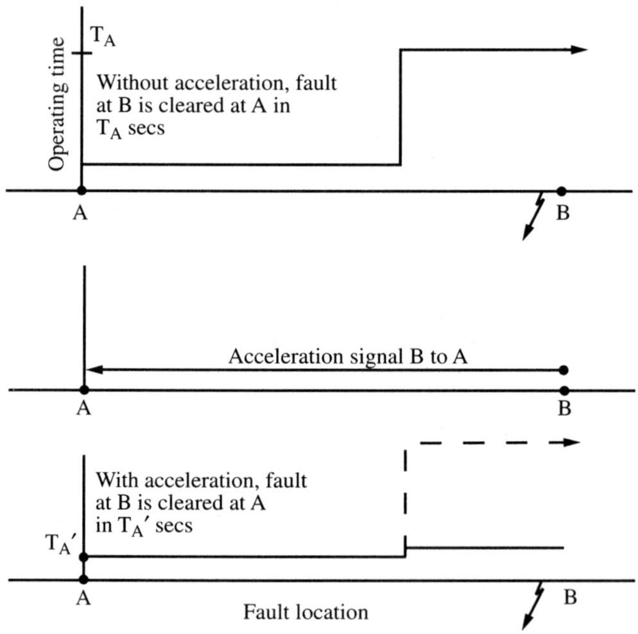

(b)

Figure A2.5 (a) Application of distance protection; (b) distance protection with acceleration

– infeeds into the protected circuit not measured by the CTs of the protected circuit, as in Figure A2.8(a) will shorten the reach of distance protection; earth fault current flowing into the star point of a star-delta transformer will affect the reach of earth fault distance protection (Figure A2.8(b));

– the magnitude of minimum fault current in relation to maximum normal currents; it may be difficult to protect the system against minimum-current faults and, at the same time, allow maximum normal values of current flows.

- Minimum fault current – this is the minimum current that can be expected to flow on the occurrence of a specified fault, sometimes at a specific location and sometimes, e.g. when considering the suitability of a particular type of protection, anywhere on a system.
- High-set overcurrent protection – this is a simple but very useful form of protection in appropriate circumstances. It depends upon an abrupt change in the magnitude of fault current between adjacent points on the system. Thus, in Figure A2.9, currents for faults between the substation circuit breaker *A* and the transformer will be considerably higher than those on the lower voltage

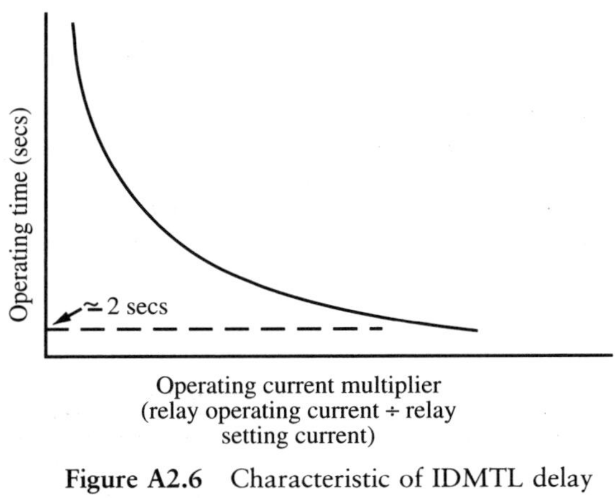

Figure A2.6 Characteristic of IDMTL delay

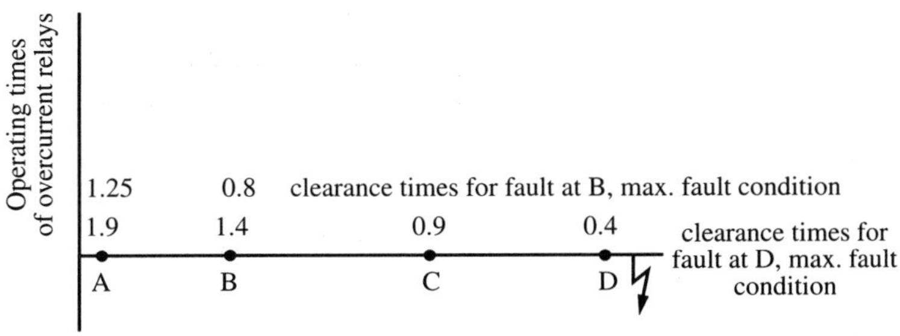

Figure A2.7 Discrimination between overcurrent relays on a radial feeder

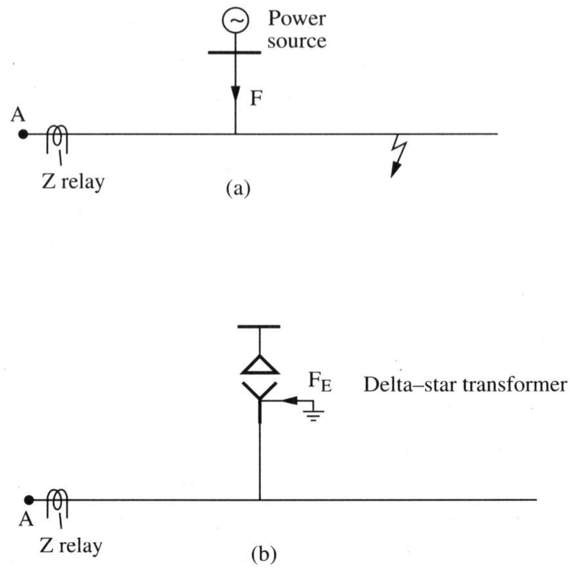

Figure A2.8 Problems caused by mid-circuit fault infeeds. (a) Phase fault: infeed at F not measured by distance relays at A; (b) Earth fault: infeed F_E not measured by distance relays at A

side of the transformer. Setting a directional overcurrent relay at *A* between these values means that it will only operate for faults between *A* and *B*.

- Directional overcurrent – overcurrent protection which only operates for MVA flows in a specified direction (normally away from the busbar).
- Earth fault protection – protection which operates on the detection of current flow in neutral connections (see also 'Neutral earthing').
- Busbar protection – usually, a form of unit protection to provide discriminative protection on sections of busbars.
- Main protection – the protection system installed as the principal means to detect and disconnect faulted plant. It will usually be a unit form of protection. On important circuit, two main protection systems may be provided, both of

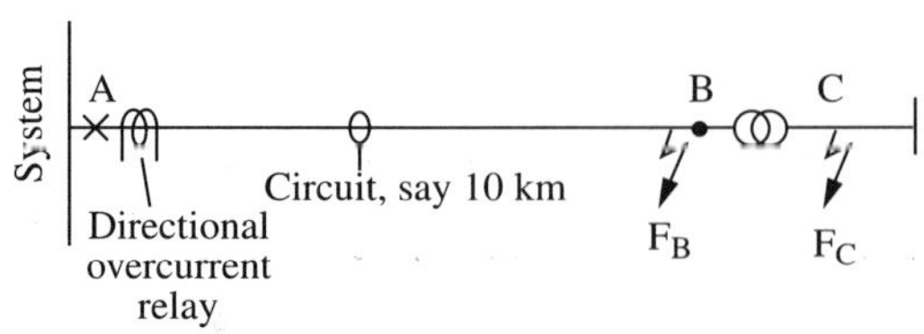

Figure A2.9 Principle of high-set o/c relay. Because of the impedance of the transformer, the fault current at B may be several times that at C

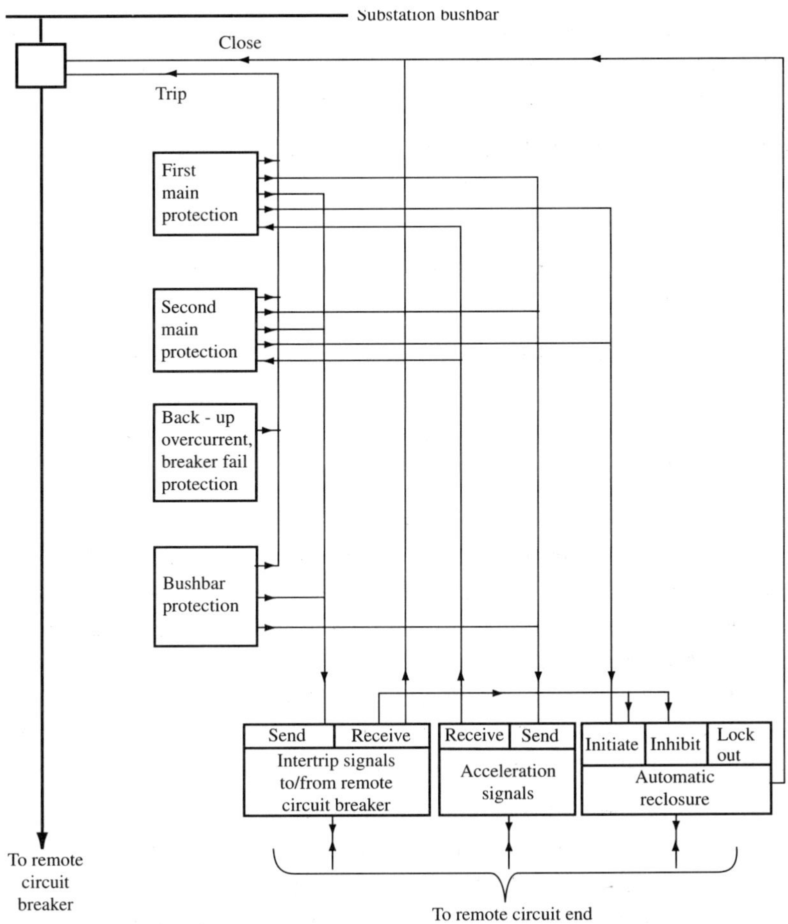

(a) Signal paths

Relay control system	Operational Functions: Trip local circuit breaker	Trip remote circuit breaker	Close local circuit breaker	Close remote Circuit breaker
First main prtection	✔	✔		
Second main protection	✔	✔		
Busbar protection	✔	✔		
Back-up protection	✔			
Auto-reclose			✔	
Control operations	✔	✔	✔	✔

(b) Protection functions

Figure A2.10 Protection and control on a main circuit

which must operate to trip the associated circuit. The elements of a protection system which might be installed on a very important circuit are shown in Figure A2.10, based on information in *Modern Power Practice*, Vol K.

- Back-up protection – a protection system installed in addition to the main protection system/s to ensure operation (and fault clearance) in the event that the main system/s fail.

PURPA – the Public Utility Regulatory Policies Act (1978); the Act regulating public utilities in the USA.

Real-time pricing – pricing electricity based on its delivered cost at the time of use.

Regulating capacity – the system capacity required/available to follow the very short term, stochastic, changes in demand.

Reliability [3] – a definition of reliability when applied to a power system is its ability to meet the demands of consumers within acceptable values of frequency and voltage. Numerous definitions have been proposed for individual supply points, for instance probability of failure, expected frequency of failure, expected frequency of demand curtailment, expected energy not supplied, expected duration of demand curtailment, maximum (or average) demand curtailed, maximum (or average) energy curtailed, maximum (or average) duration of demand curtailment.

Indices can also be defined for the whole system, for instance the total demand curtailed/system peak demand (MW/MW peak), the total energy not supplied/system peak demand (MWh/MW peak), maximum system demand (or energy) not supplied under any contingency condition (MW/MWh), average number of BSP disturbances per year, average values of some of these indices per supply point, etc.

Surveys of actual reliability of supply have been prepared [4]. In this the measures used are the number of disturbances experienced by utilities related to their severities, frequency and duration of interruption, restoration time per interruption, supply points affected.

Surveys have also been made of the performance of BES control aids [4].

Remote Terminal Unit (RTU) – the interface between the telemetry and the protection, control and metering equipment at an outstation.

Reserve – a widely used term denoting the margin between required and available quantities of a resource. It will require qualification in regard to parameter (e.g. generation reserve) and possibly response time (e.g. fast reserve).

Reserved capacity – capacity (e.g. of generation or transmission) reserved for a specific customer.

Restoration – the process of restoring normal supply to an area which has suffered disconnections or disturbed operating conditions.

Richter scale – the logarithmic scale (1–10) used to quantify the severity of an earthquake. Severe earthquakes will be in the range 7.5 +, and very severe 8 +.

Safety procedures, rules and codes of practice – rules to ensure the safety of personnel required to carry out work on potentially live equipment.

Scheduled . . . – a quantity or action that has been included in a programme of actions (e.g. scheduled outage, scheduled transfer).

Security (e.g. of supply), secure (e.g. a supply) – another term widely but often imprecisely used to describe the ability of a system or supply to withstand unexpected events.

Security standards, security criteria – a statement of the plant outages following which it should still be possible to provide a supply within the maximum specified voltage and frequency tolerances, and plant loading conditions.

SC, DC1, DC2, etc. – abbreviations to describe the construction of an overhead line, occasionally of a cable route, as follows:

- SC: a line with three conductors (for a three phase supply);
- DC2: a line with two sets of three conductors, providing two three phase circuits;
- DC1: a line constructed to carry two sets of three conductors, but with only one set strung, hence providing a single three phase circuit.

Cable runs may also be described as single circuit, etc.

Shadow price – a term used in economics and mathematical programming.

Short circuit current, fault current – the current which flows when a conductor at one voltage level touches or arcs over to a conductor or conductive material at another voltage, usually zero or near zero with respect to earth; related terms are earth-fault current (conductor touches conductive earthed object), three-phase fault current, etc. Faults in this context are usually defined in terms of the phases involved (e.g. three-phase fault, phase to phase fault, phase to earth fault, etc.).

Simultaneous tap change – an operational procedure in which the operating voltage of a transmission network is changed to meet weather, possibly loading, conditions by pre-arranged and virtually simultaneous tap changes on all the generator transformers.

Special protection scheme – a purpose built protection system designed to adjust generation and/or transmission operating parameters so as to achieve or maintain a viable operating state.

Stability/instability – several forms of stability (or its reverse instability) are distinguished:

- steady state stability – the ability of all generators to remain in synchronism following a very small increase in power transfer about the operating point;
- transient stability – the ability of the system to regain synchronism following a large disturbance (e.g. a sudden increase in transfer impedance or power flow across the system);
- dynamic stability – the ability of the system to remain stable and for oscillations to die out following a small signal disturbance about the operating point;
- voltage stability – the ability of a system to maintain an acceptable voltage profile following a small increase in demand and/or credible configuration change.

Standby supply, standby service – a supply or service available when required through a normally open connection (circuit breaker or disconnector); occasionally, a standby supply in rural areas might be provided by a portable generator.

Standing instruction – an instruction concerning operation of the power system which must always be followed.

Static Var Compensator (SVC) – a shunt connected static generator and/or absorber of reactive power; some of the principal types are thyristor controlled or switched reactor, thyristor switched capacitor; more generally, a static var system is a number of static capacitors and reactors connected to the power system via steady state devices, and providing a rapidly controllable source of reactive power.

Stranded investments/costs – investment costs of plant which system developments have rendered useless; a 'sunk cost' is one already committed.

Stuck breaker – a breaker which does not operate when required to do so, for instance because of loss of air or power supplies.

Sub-synchronous resonance – an oscillatory condition which occurs at frequencies below nominal, caused by the interchange of energy between series capacitors and the inductance of the transmission system.

Synchronizing power – a measure of the ability of a synchronous machine to retain synchronism after being subjected to a disturbance.

System state – as used in this book, the operational viability of the system in terms of its operating parameters and ability to withstand contingencies. Five states are defined: normal, normal (alert), alert, emergency and restoration:

(a) Normal – plant loadings within the continuous capabilities, voltages and frequency within operational limits, conditions following a credible contingency are acceptable.

(b) Normal (alert) – following a credible contingency, action can be taken within the timescales allowed by the plant capabilities to restore the system to an acceptable state. Very rapid action is not necessary.

(c) Alert – rapid or immediate action required. If a credible contingency then occurs, the system will enter the emergency state; alternatively, action must be take rapidly to prevent unacceptable overloads, voltage or frequency conditions, or protective gear operations.

(d) Emergency – unacceptable plant loading, voltage or frequency conditions exist or demand has been lost or the system is split. Immediate action is necessary to restore the system to an acceptable state.

(e) Restoration – the system is in the process of being restored from some abnormal state (d, c or b above) to normal.

Although not normally included, the correction of time errors could also be classified as a restorative action.

The term 'adequacy' has been used in the past to describe the ability of a system to supply the power and energy demands placed on it at all times.

Tariff – a statement of prices and conditions to provide all or components of an electrical supply.

Transfer capability – the amount of power which can be transmitted from one area to another within the system; in approximate analysis, such as might be used subconsciously by operators judging the viability of a system from a mimic diagram display, this may be taken as relatively constant, but it will in fact vary with loading and switching conditions on the system.

Transmission costing – one of the consequences of unbundling has been that the services provided by a vertically integrated utility must be charged separately. One of these services will be transmission, and the task of determining an appropriate charge is called 'transmission costing'.

Two shifting – depending on the form of the input/output characteristics of a generating unit, it may be more economic to operate it either at full load (or most economic load) or zero load, adjusting the total generation input to the system by connecting/disconnecting generating units on the system. This has been called ‘two shifting’ a generator.

Unbundling – separating the services provided by a utility into its main components – generation, transmission, distribution and supply; these may or may not be parts of the same utility.

Under-frequency relay – a relay which detects the existence of a low system frequency; such relays may be used variously to initiate an alarm, disconnect demand or act as a starting relay for other protection. ‘Rate of change of frequency’ and ‘frequency trend’ relays have also been installed to detect frequency changes indicating the onset of dangerous conditions, and either trip demand or plant, or act as starting relays for other protection.

Voltage collapse – a condition in which control actions such as tap changing on transformers are inadequate to stop further voltage decline; it can occur at the power receiving end of a single line or in power receiving areas of the system.

Voltage flicker – repetitive, irregular fluctuations in a light source caused by variations in the voltage supply to the source.

Wayleave – in the UK, the authority granted by an owner of land to build and use an overhead line. It will include restrictions on the use of land each side of the line (e.g. tree growing) so as to ensure safe operation of the line; wayleaves will also be required for cables. The equivalent term in North America is ‘right-of-way’.

Wheeling – the transmission of electricity by a transmission owning utility on behalf of another utility; this utility will normally pay ‘wheeling’ charges to the owning utility. A related term in North America is a ‘wires charge’, charges levied by owners on users of the transmission and distribution facilities.

Wholesale power market – a group of utilities and marketers which buy and sell electricity and the associated ancillary services between themselves and to outside customers.

REFERENCES

A2.1. IEEE Power Engineering Society: ‘*Glossary of Terms and Definitions Concerning Electric Power Transmission Systems Access and Wheeling*’, IEEE, **96 TP** 110–0.

A2.2. Edison Electric Institute Glossary of Electric Utility Terms (brochure).

A2.3. Allan, R. N., Billington R., 1992. ‘Power system reliability and its assessment Part 1: Background and generating capacity’, *Power Engineering Journal*, **6** (4), 191–196. ‘Part 2: Composite generation and transmission system’, *ibid.* **6** (6), 291–297. ‘Part 3: Distribution systems and economic considerations’, *ibid.* 7, 1993.

A2.4. Winter W. H., 'Cigre brochure on bulk electricity option operational performance: measurement options and survey results', Cigre Working Group 39.05 *Cigue Electra* **131** 185–191.

A2.5. Schaffer G., 1996. 'User experience with EMS functions', *Cigre Electra* **164** February.

A2.6. Lon, P. V., Bore, D., Kirschen, D., 1997. 'Innovations in the control centre due to open trading', Paper 39-02-02 Cigre Symposium, Tours, June.

Appendix 3
Some Useful Mathematical and Modelling Techniques in Power Systems Studies

The determination of a 'best' solution can often be framed as a mathematical optimization problem, essentially minimizing the use of resources (the objective function) whilst satisfying constraints imposed by nature or man on the behaviour of the system. The resources will depend on the problem – capital or operating cost, or their equivalent such as man years in many, but say elapsed time in others, as would be the case in determining actions to minimize the duration of a failure of supply.

Optimization seems a fundamental process in nature, in the sense that equilibrium states are usually, if not always, states of minimum stored energy or energy dissipation. This property can be used to determine optimum solutions. Less fundamental are the direct and indirect approaches. In the direct approach, successive feasible solutions are computed until some parameter of the objective function indicates that an optimum has been reached; Linear Programming (LP), Quadratic Programming (QP), and hill climbing techniques are examples. In the indirect approach, a set of equations characterizing the optimum, for example that the first derivatives of the objective function are zero, are solved. Dynamic Programming (DP) is a very important technique, and is essentially a form of ordered search. Generically, these techniques are termed mathematical programming, and comments on these follow.

Some reference texts are listed at the end of this appendix. The author has also taken material from his book *Power Systems Engineering and Mathematics* [1]. This reviews the basic process of planning and design of engineering systems, and describes applications to power system studies.

A3.1 LINEAR PROGRAMMING

With varying degrees of approximation, the cost or profit of a physical system can often be modelled as a linear function of the activities (the objective function) and the physical interaction between the activities as a set of linear equations or inequalities (constraints). Mathematically, inequalities can be converted to equations by the addition of variables which, provided they are included at zero costs in the objective function, will allow the take up of slack between the physical variables and the constraint limits without affecting the optimum solution for these activities. Thus

$$a_{11}x_1 + a_{21}x_2 \leq b_1 \quad \mathrm{Max}\, C_1x_1 + C_2x_2$$

is equivalent to

$$a_{11}x_1 + a_{21}x_2 + x_3 = b_1 \quad \mathrm{Max}\, C_1x_1 + C_2x_2 + Ox_3$$

x_3 is known as a 'slack' variable.

Linear programming provides theory and algorithms, whereby a linear objective function can be minimized or maximized subject to a set of linear constraints when the number of physical and slack variables exceeds the number of equations, and the variables are to be non-negative.

Hence, the general form of a linear program in n variables, r constraints is

$$\text{Minimize/maximize} \sum_{i=1}^{n} C_i x_i \tag{A3.1}$$

$$\text{subject to} \qquad b_j^m \leq \sum_{i=1}^{n} a_{ij}x_i \leq b_j^M \quad (j = 1, \ldots, r) \tag{A3.2}$$

with the understanding that $x_i \geqslant 0$.

The solution of a linear program involves an iterative series of matrix manipulations. In the original Simplex method, selected dependent variables (called basic variables and equal in number to the number of constraints) are expressed in terms of the remaining independent (called non-basic) variables, which are zero in value.

In practice, an 'artificial variable' may be added to each equality. A suitable starting-point is obtained by inspection by treating these and the slack variables as basic variables in the first matrix. Optimization is achieved by interchanging basic and non-basic variables between iterations. The form of the matrix (often called a 'Simplex tableau') at each step indicates which variables are to be interchanged, and when an optimum has been reached, the numerical values of the basic variables then being the solution sought. Flexible computer programs for the solution of very large linear programs with 1000 + constraints, and an unlimited number of variables are available.

Considerably more than a single optimum solution may be required – for instance, over what ranges of system parameters does the solution hold, what is the physical interaction that produces the optimum? Fortunately, considerable help can be obtained from linear programming; some additional information inherent in the basic solution or available from further computation is as follows:

- *Shadow prices*. These are the additional costs incurred as each element of the requirements vector (the b_j^m, b_j^M in equation (A3.2)) is changed in turn by one unit.
- *Cost ranging*. This indicates the range over which the cost of each basic variable in the optimum solution can be changed without its deletion from the solution.
- *Parametric programming*. This enables the solution changes to be followed as the coefficients in the objective function, the requirements vector or the constraint matrix are changed.

Turning to problem formulation, it is well worth studying both the general literature and the user manuals before attempting a problem of any size. For instance, stepped-cost functions can perhaps be dealt with more conveniently than by assigning bounded variables to each cost step. Variables may be used to define strategies rather than component activities. Frequently, such devices amount to trading number of constraints against number of variables. External analysis can sometimes be used to reduce the system detail within the linear programming (e.g. constraints based on coupling factors rather than the basic network laws).

An authority on mathematical programming, Dr. S. Vajda, once suggested that the very existence of a large LP problem implies an ordered structure of constraints. It may be possible, therefore, to generate the constraints by computer program, as has been done by the author for configuration design problems. In the event one needs to establish whether the data is best ordered by constraints (i.e. by defining the variables in each constraint), or by variables (i.e. by defining the constraints in which each variable appears).

Finally, it may be that the problem is better suited to a dual formulation. Corresponding to every LP problem there is a dual problem in which the coefficients of the original objective function become the requirements vector in the dual, and the original requirements vector forms the coefficients in the dual objective function. Problems in which there are more inequality constraints than physical variables, for example configuration design or switching problems, can with advantage be formulated in the dual form.

A3.2 SOME SPECIAL FORMS AND EXTENSIONS OF LINEAR PROGRAMMING

A3.2.1 Transportation

Suppose the resource allocation problem is to transport coal from three pits a, b and c to three power stations A, B and C at minimum cost, the coal available at pits P_a, P_b, P_c being equal in total, to that required at stations S_A, S_B, S_C (Figure A3.1). If C_{aA} is the cost per unit transported from a to A and x_{aA} the quantity moved, etc., the linear program is:

$$
\begin{bmatrix}
x_{aA} + x_{aB} + x_{aC} & & \\
& x_{bA} + x_{bB} + x_{bC} & \\
& & x_{cA} + x_{cB} + x_{cC} \\
-x_{aA} \qquad\qquad & -x_{bA} \qquad\qquad & -x_{cA} \qquad\qquad \\
\quad -x_{aB} \qquad & \quad -x_{bB} \qquad & \quad -x_{bC} \qquad \\
\qquad\qquad -x_{aC} & \qquad\qquad -x_{bC} & \qquad\qquad -x_{cC}
\end{bmatrix}
\begin{matrix}
= P_a \\ = P_b \\ = P_c \\ = -S_A \\ = -S_b \\ = -S_c
\end{matrix}
$$

Minimize $C_{aA}x_{aA} + \cdots + C_{cC}x_{cC}$.

These constraint equations have a special form. The coefficients of all variables are 1 or -1. Each variable occurs not more than once in each equation. It occurs twice in total, once with coefficient $+1$, and once with coefficient -1.

For obvious reasons, this special form of linear program is called a *transportation problem*, and is important because the numerical solution is much simpler than for the general LP. Computer programs are available to solve such problems in many thousands of variables.

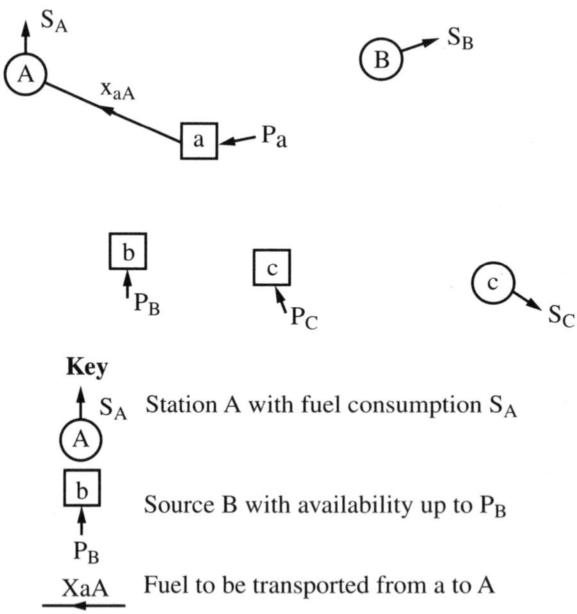

Figure A3.1 The transportation problem

The formulation can be extended to deal with a surplus or deficit of resources by adding dummy sinks or sources; with transhipment by treating each transhipment point as both source and sink; and with restriction on flows.

Solutions to transportation problems are integer if the resources and requirements are integer.

A.3.2.2 Integer Linear Programming

Integer linear programming is an extension of LP in which the variables are constrained to have integer values. In mixed integer LP, only some of the variables are constrained to have integer values. Some of the applications are:

- allocation problems in which subdivision of resource units is not meaningful (e.g. half an aeroplane in a transport problem);
- allocation problems in which the cost of each activity includes fixed costs F_i in addition to a running cost C_i proportional to the level of activity*. Introducing integer variables V_i (to equal 1 or 0, indicating use/non-use of the activity) the formulation will be:

$$\text{Minimize} \quad \sum_{i=1}^{n} C_i x_i + V_i F_i \tag{A3.3}$$

$$\text{subject to} \quad b_j^m \leq \sum_{i=1}^{n} a_{ij} x_i \leq b_j^M (j = 1, \ldots, r) \tag{A3.4}$$

$$M_i V_i \geq x_i \; (i = 1, \ldots, n) \tag{A3.5}$$

$$V_i \text{ to be integer.}$$

M_i is chosen so that $M_i >$ maximum allowable value of x_i.

- problems in which variables may hold only one of a number of discrete values. If x_i is constrained to be $m_{t1}, m_{i2}, \ldots m_{ip}$, additional equations will be necessary:

$$x_i = V_{i1} m_{i1} + \cdots + V_{i1} m_{ip}, \tag{A3.6}$$

$$V_{i1} + \cdots + V_{ip} = 1, \tag{A3.7}$$

Algorithms for discrete LP generally start with a continuous (i.e. normal LP) solution to the same problem. In one class of methods, equations are added as the tableaux progress, starting with that for the continuous solution, which force

*In ordinary LP, the coefficients in the objective function will be the marginal cost of increasing the activities by one unit. Fixed costs can be included by dividing this over the expected level of activity prior to the solution, with iteration if a bad guess is made.

successive variables to take integer values. 'Branch and bound' methods have been widely used. In these the immediate-integer bounds are placed progressively on non-integer variables required to be integer, and successive linear programs are computed which explore a range of solutions until an optimum is reached with the requisite variables integer.

A3.2.3 Quadratic Programming

Quadratic programming is the term applied to the optimization of an objective function containing quadratic terms subject to linear constraints. If the quadratic terms can be expressed as the sum of squares of linear expressions all with positive/negative coefficients, the function is convex/concave and a global minimum/ maximum can be found[‡].

A procedure similar to the Simplex method can be used for numerical solutions. In the linear case, the strategy from iteration to iteration is determined by observing the coefficients in the objective function, these being, of course, the partial derivatives of this function. Similar use can be made of the corresponding partial derivatives, no longer constant, in the quadratic programming case.

Computer programs are available for the solution of quadratic programs involving some hundreds of constraints plus variables.

A3.3 NON-LINEAR PROGRAMMING

In the general resource allocation problem, both objective function and constraints may be non-linear. Several approaches have been used.

A3.3.1 The Indirect Approach Using Lagrangian and Kuhn–Tucker Multipliers

The minimum of a convex function $f(x)$ of n variables $x_1, \ldots, x_n$ is obtained when

$$\frac{\partial f(x)}{\partial x_i} = 0 \quad (i = 1, \ldots, n) \tag{A3.8}$$

If the x_i are constrained by r equations $g_j(x) = 0$, these can sometimes be used to eliminate r variables in the objective function, followed by its differentiation with

[‡]Alternatively, a function is convex/concave if it is never/always under-estimated by linear interpolation.

respect to the remaining $(n-r)$ variables. Alternatively, the Lagrangian multiplier method can be used. This states that a minimum to $f(x)$ is found when

$$\frac{\partial f(x)}{\partial x_i} + \sum_{j=1}^{r} \lambda_j \frac{\partial g_j(x)}{\partial x_i} = 0 \quad (i = 1, \ldots, n) \tag{A3.9}$$

and*

$$g_j(x) = 0 \quad (j = 1, \ldots, r) \tag{A3.10}$$

Kuhn and Tucker established the conditions under which a minimum exists when inequality constraints are present. If $f(x)$ is a convex function of n variables which are subject to constraints

$$g_j(x) = 0 \quad (j = 1, \ldots, r) \tag{A3.11}$$

$$h_k(x) \leq 0 \quad (k = 1, \ldots, p) \tag{A3.12}$$

A minimum to $f(x)$ is found when

$$\frac{\partial f(x)}{\partial x_i} + \sum_{j=1}^{r} \lambda_j \frac{\partial g_j(x)}{\partial x_i} + \sum_{k=1}^{p} \mu_k \frac{\partial h_k(x)}{\partial x_i} = 0 \quad (i = 1, \ldots, n) \tag{A3.13}$$

$$\mu_k h_k(x) = 0 \quad \text{and} \quad \mu_k \geq 0 \quad (k = 1, \ldots, p) \tag{A3.14/3.15}$$

$$g_j(x) = 0 \quad (j = 1, \ldots, r) \tag{A3.16}$$

$$h_k(x) \leq 0 \quad (k = 1, \ldots, p) \tag{A3.17}$$

The μ_k are called Kuhn–Tucker variables.

The optimization problem resolves then mainly to the solution of the simultaneous, probably non-linear, equations resulting from (A3.8); from (A3.9) and (A.3.10); or with added non-equality conditions, from (A.3.11)–(A.3.15). Gauss, Gauss–Seidel and Newton–Raphson techniques can be used. In the first two, in each iteration each equation in turn is used to establish an improved value of its dominant variable

The latter is also an iterative procedure based on Taylor's expansion $(f(a+h) = f(a) + hf'(a) + h^2 f''(a)/2! + \ldots \cong f(a) + hf'(a))$. If the solution of $f(a) = 0$ is required and a^0 is an assumed solution, a better approximation is $a^1 = a^0 + (\Delta a)^0$, where

$$f(a^1) \approx 0 = f(a^0) + (\Delta a)^0 f'(a^0)$$

*Note that $[\partial g_j(x)/\partial x_i]$ is the transposed Jacobian of $g(x)$.

or

$$(\Delta a)^0 = -f(a^0)/f'(a^0)$$

At the kth iteration $(\Delta a)^k = -f(a^k)/f'(a^k)$.

Generalizing, suppose the equations to be solved are

$$f_i(y_1, \ldots, y_n) = 0 \quad (i = 1, \ldots, n) \tag{A3.18}$$

An initial solution $Y^0 = [y_1^0, \ldots, y_n^0]$ is assumed, and the following equations formed:

$$f_i(Y^0) + \sum_{j=1}^{n} \frac{\partial f_i(Y^0)}{\partial y_j} (\Delta y_j)^0 = 0 \quad (i = 1, \ldots, n) \tag{A3.19}$$

The linear equations (A3.19) are solved for the $(\Delta y_j)^0$, when

$$Y^1 = [y_1^0 + (\Delta y_1)^0, \ldots, y_n^0 + (\Delta y_n)^0] \tag{A3.20}$$

Equations (A3.19) are reformed at the solution point Y^1, and solved for increments $(\Delta y_j)^1$. The process is repeated until the increments $(\Delta y_j)^k$ are sufficiently small.

General purpose computer programs for the solution of non-linear simultaneous equations are available. The user specifies the equations and a suitable starting point, the program generating and solving equation (A3.19), etc.

A3.3.2 The Direct Approach Using Gradient Methods

The basic idea of these methods is, starting with some arbitrary values of the variables in the objective function, to determine changes in these variables which will yield an improved value of the function, and to repeat this process from successive improved values until an optimum is reached[§] Thus, at each iteration two steps are involved – determination of the direction and magnitude of change. Constraints generally can be dealt with by inclusion of 'penalty factors' in the objective function, so as to increase its cost rapidly with transgression of the constraints; or by limiting the direction of moves to be in line with a constraint which has become operative.

Considering the unconstrained optimization of $f(x) = f(x_1, \ldots, x_n)$, the simplest procedure is to change each variable in turn, reducing the value of $f(x)$ as far as possible before passing on to the next. Intuitively, this method may

[§] In the general case, this will be a local extremum, the one found depending on the starting values assumed.

produce an oscillating solution if there is interaction between the variables, and one might consider changing each variable in turn by a small fixed amount, retaining only those changes which improve the objective function. Alternatively, the effect on the objective function of changing each variable in turn by a small amount, with the others held constant, could be determined, and only that variable change producing the biggest improvement retained for the next iteration. Clearly, there is no difficulty in constraining individual variables between limits.

In the method of steepest descent, the gradient of the objective function is used to determine proportionate changes in the variables at each iteration. If at the ath iteration the variables have values $X^a = (x_1^a, \ldots, x_n^a)$, the components of the gradient will be

$$G_i^a = -\left(\frac{\partial f(x)}{\partial x_i} \Bigg/ \left\{ \sum_{i=1}^{n} \left(\frac{\partial f(x)}{\partial x_i} \right)^2 \right\}^{\frac{1}{2}} \right)_{x=X_1} \quad (i = 1, 2, \ldots, n) \tag{A3.21}$$

For minimization of $f(x)$, the new values of the variables will be

$$x_i^{a+1} = x_i^a + t^a G_i^a \quad (i = 1, \ldots, n) \tag{A3.22}$$

with t^a an arbitrary factor, or such that $f(X^{a+1})$ is a minimum, established, say, by incremental changes in t^a.

First-order gradient methods of this type may only converge slowly, since the gradient of a function near its extreme values may be very small. Second-order gradient methods, making use of the second derivatives of the function, are much more efficient.

Turning to constrained optimization of $f(x)$, penalty factor methods are simple in concept. Methods of 'feasible directions' in which only changes in variables are allowed which satisfy constraints, have been described.

A3.4 DYNAMIC PROGRAMMING

In the techniques described so far, a solution to the whole problem exists at each stage of the calculation, although that solution may not be feasible, and certainly will not be optimal until the final iteration is made. In dynamic programming, feasible and optimum solutions to parts of the problem are established, and progress is made by taking more and more of the problem into account until the whole is covered. Conceptually, this multi-stage decision process is obviously applicable to the allocation of resources over a span of time, when, say, decisions have to be made on what plant to provide in successive years, but it can equally be used to determine allocations for a fixed time, say capital investment in alternative production facilities.

Terms used commonly in dynamic programming are:

- Stage – the time intervals, or component parts, into which the total system has been decomposed for study are called 'stages'.
- State and state variables – the set of variables defining a stage appropriately for the study are called 'state variables', and together constitute the system 'state'. There may be a number of possible states in each stage.
- Policy – a sequence of decisions leading to the adoption of specific states at each stage is a 'policy'. A policy which optimises the objective function for the total system is an 'optimal policy'.

Bellman's Principle of Optimality and Markovian Systems

Dynamic programming depends on Bellman's Principle of Optimality – 'an optimal policy has the property that whatever the initial state and initial decision are, the remaining decisions must constitute an optimal policy with regard to the state resulting from the first decision'. [2, 3] For Markovian systems, this means that for any state i in the kth stage it is only necessary to consider the state in the $(k-1)$th stage, which leads to state i of stage k in an optimal manner. Hence, if there are m possible states in the $(k+1)$th stage and n in the kth stage, direct enumeration would require mn possibilities to be carried forward to the $(k+1)$th stage, whereas dynamic programming requires only n, each being derived from its optimal state in the $(k-1)$th stage. Expressed mathematically:

If $C_k(\Omega i)$ is the cost of state i in stage k, $C^k_{k-1}(\Omega j^i)$ the cost of transformation from state j in stage $(k-1)$ to state i in stage k, and $T_{k-1}(\Omega_j)$ the accumulated cost of arriving at the end of stage $(k-1)$ in state j, the recurrence equation defining the sequence of optimum transformations and states is

$$T_k(\Omega_i) = \underset{j}{\text{Min}}\,(T_{k-1}(\Omega_j) + C_k(\Omega_j) + C^k_{k-1}(\Omega^i_j)) \qquad \text{(A3.23)}$$

In an alternative formulation which introduces a method of including constraints, suppose a total resource X is available which can be used in activities or stages $1, 2, \ldots, N$, the profit if x_i is used for activity j being $P_j(x_i)$. The formulation is then

$$\begin{aligned} &\text{Maximize} \sum_{j=1}^{N} P_j(x_i) \\ &\text{subject to} \sum_{i=1}^{N} x_i = X \end{aligned} \qquad \text{(A3.24)}$$

The procedure is first to calculate the profits from allocating various levels of resource between 0 and X to activity 1. Various proportions of the remaining resource $(X - x_1)$, for each allocation to activity 1, are then allocated to activity 2, and hence profits from the first two activities found for an optimal allocation of any level of resource to activities 1 and 2. Activities $3, 4, \ldots, N$ are included in turn. If $T_j(x_k)$ is the profit from allocating x_k units of resource optimally to activities $1, 2, \ldots, j$, the recurrence equation will be

$$T_j(x_k)_{x \geq x_k > 0} = \underset{k_k > x_i > 0}{\text{Max}} \left(P_j(x_i) + T_{j-1}(x_k - x_i)\right) \tag{A3.25}$$

for

$$j = 2, 3, \ldots, N \text{ successively}$$

with

$$T_1(x_k) = P_1(x_k)$$

Dynamic programming can yield an immense saving over straightforward evaluation of alternatives; if m states are possible in each of n stages, dynamic programming requires the evaluation of $m^2(n-1)$ possibilities, against m^n for enumeration. Nevertheless, what Bellman has called the 'curse of dimensionality' may limit its applications. If each state is defined by s state variables, each at one of v possible values, the number of states to consider at each stage will be $m = v^s$. This will be a measure of the storage capacity and computation needed per stage, and is critically dependent on the number of state variables or dimension of the problem. Since constraints other than those on single variables are equivalent to additional dimensions, dynamic programming is also only applicable to problems with few multi-variable constraints.

A3.5 OPERATING COSTS

The prediction of operating costs is an essential component of planning and operational planning work. It requires simulation of the operation of the system over the requisite period of time. 'Loading simulation' or 'production costing' programmes are used. In one method which is capable of high accuracy, the study period is decomposed into a number of representative periods, which may not be contiguous in time, over which the demand is sensibly constant and to estimate for each period the minimum cost pattern of generation from the total available. Costs over the total time will then be estimated period by period.

In another method, more approximate but requiring less computation and appropriate to extended study durations, the demand cycle is represented as a demand duration curve in which the number of hours for which the load exceeds any given level is plotted (Figure A3.2), or in histogram form in Figure A3.2(b). These will be based on representative daily load curves over the study duration.

The estimation of the generation pattern, and hence fuel consumption and cost, is essentially an exercise in plant scheduling and dispatching, and similar techniques will be used with probably some simplifications in the interests of computing time. The concepts of incremental cost of generation and the listing of such costs by order of magnitude into 'merit orders' are very useful here. The incremental costs of a unit is the slope of its output curve. Those typical of different types of plant have been indicated in Figure 5.1, and will be quoted as cost or heat used/unit of energy (e.g. £/MWhr) at a given output. Generally, two merit orders are distinguished – one in which the incremental costs or heat rates are inclusive of fixed heat requirements, and which should be used in scheduling calculations (i.e. choice of plant to run); and one in accordance with the definition above which excludes fixed heat requirements, and which should be used for dispatching calculations (i.e. the allocation of output to running plant, or, more precisely, plant with no off-load cost).

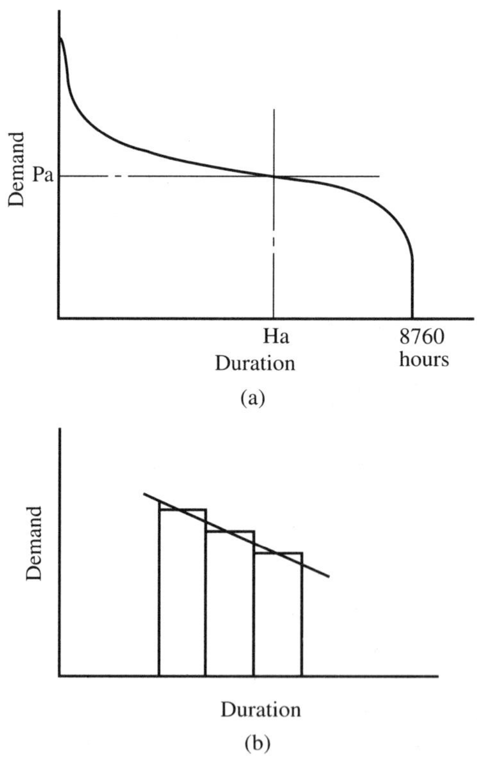

Figure A3.2 (a) Annual demand duration curve, (b) part of annual demand duration curve in histogram form

The core of a loading simulation program may then contain the following steps for each main time period – week or weekend, summer or winter, etc.

(a) Compute the incremental costs $\rho_i = dF_i(g_i^M)/dg_i + F_{fi}/g_i^M$ of each unit assumed available at its maximum output g_i^M. Availability may be specified discretely for each unit, or as average figures for types of plant, in which case the set capabilites will be written down by the availability factor. $F_i(g_i)$ is the running cost of the unit at output g_i and F_{fi} its no load cost.

(b) List the ρ_i in ascending order of magnitude with corresponding running generation total.

(c) Select the generation total in this list just greater than the required generation commitment (equal to expected demand plus spare plus external transfer). Plant above this point will be taken as on load.

(d) If the sets have characteristics as Figure 5.1(c), compute for this committed plant the incremental costs $\rho_i = dF(g_t)/dg_t$, at the maximum and minimum generation figures and list in ascending order.

(e) Determine for reasonable values ρ_R of ρ (i.e. probably those nearing the higher cost end of this list) the active power output of each unit in accordance with the following criteria (the superscripts M and m indicate incremental costs at maximum and minimum outputs, respectively):

(i) $g_i = g_i^M$ if $\rho_R \geq \rho_i^M$,

(ii) $g_i = g_i^m$ if $\rho_R \leq \rho_i^m$,

(iii) $g_i = \dfrac{(\rho_R - \rho_i^m)}{\rho_i^M - \rho_i^m}(\rho_i^M - \rho_i^m) + g_i^m$.

These equations assume that the cost – output functions are quadratic.

(f) Summate g_i, for all committed plant and repeat from (e) until $\sum_i g_i$, equals the expected demand plus transfer.

(g) Summate the operating costs $\sum_i(F_{fi} + F_i(g_i))$.

If the sets have characteristics as in Figure 5.1(a) or (b), the procedure can be simplified. For steps (a)–(c), the incremental costs at maximum output (Figure A5.1(a)) or at the economic and maximum outputs (Figure 5.2(b)) are listed in ascending order with corresponding outputs. For steps (d)–(f) minimum generation is taken on the selected sets and the difference between expected demand plus transfer, and the sum of the minimum generations is taken up by summation down a merit order of outputs less minimum generations.

Such procedures can be modified to produce ever closer simulations of actual operation. For instance, if the demand is represented by average daily curves, unit start up and shut down sequences will be automatically obtained, thus allowing start-up and banking costs to be included explicitly. Transmission losses can be included, then requiring some form of network or penalty factor calculation. Following completion of the steps outlined above, the spinning spare capacity can be examined and, if it is necessary to increase this, the loading of the highest cost sets increased with corresponding decrease in that of the slightly lower cost ones. Transmission limitations can be included as group constraints. As these factors are introduced, the computations will tend towards the types used in day-to-day operation. In the ultimate, the main difference will be in the number of system states studie – say between 100 and 500 in simulation of a year's operation as against 25 000 upwards in actual operation.

A3.6 POWER SYSTEM ANALYSIS

In spite of the growth in the capability and availability of computers, there is still scope for tailoring models to the tasks to be done. In practical terms, group transfer analysis, d.c. load flows and a.c. load flows will each have a place as will the equal area criterion and step-by-step analysis in stability studies. Some of the principal exact and approximate models used in power system analysis are outlined below.

A3.6.1 Power Flows and Voltages

The determination of circuit power flows for given nodal conditions is the commonest analytical requirement. The problem is usually solved in terms of voltages between each node and a reference node, which in Figure A3.3 has been

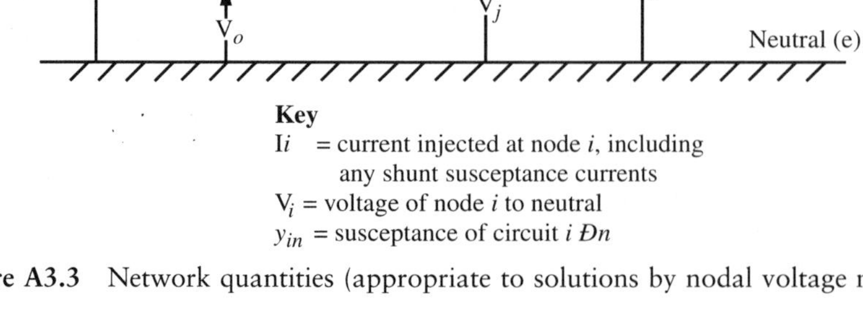

Key
I_i = current injected at node i, including any shunt susceptance currents
V_i = voltage of node i to neutral
y_{in} = susceptance of circuit i Ðn

Figure A3.3 Network quantities (appropriate to solutions by nodal voltage methods)

taken as neutral. The current injected at node i, I_i will be equal to the sum of currents in circuits, including any shunt susceptance $\bar{y}_{ie}$ representing generally equivalent π-capacitances, connected to that node. Hence

$$\begin{aligned} I_i &= I_{i0} + \cdots + I_{ie} + \cdots + I_{iN} \\ &= y_{i0}(V_i - V_o) + y_{i1}(V_i - V_1) + \cdots + y_{ie}V_i + \cdots + y_{iN}(V_i - V_N) \\ &= -y_{i0}V_o - y_{i1}V_1 \cdots + (y_{i0} + y_{i1} + \cdots + y_{ie} + \cdots + y_{iN})V_i \\ &\cdots - y_{iN}V_N = \sum_{j=0}^{N} Y_{ij}V_j \end{aligned} \quad \text{(A3.26)}$$

where

$$Y_{ii} = (y_{i0} + y_{i1} + \cdots + y_{ie} + \cdots + y_{iN}) \quad \text{and} \quad Y_{ij} = -y_{ij} \quad \text{(A3.27)}$$

There will be $(N + 1)$ complex equations (A3.26). The apparent power is only known or implied at N nodes, since the network losses are unknown until the solution is obtained. At the remaining slack node (taken as node 0), only the voltage is specified. Hence, the equation for this node is superfluous, and in the other N equations, the terms $Y_{i0}V_o$ are constant. The non-redundant set of equations is therefore

$$I_i - Y_{i0}V_0 = \sum_{j=1}^{N} Y_{ij}V_j \quad \text{for} \quad i = 1, 2, \ldots, N \quad \text{(A3.28)}$$

with the branch flows found from

$$I_{ij} = y_{ij}(V_i - V_j) \quad \text{(A3.29)}$$

or, including branch capacitances,

$$y_{ij}(V_i - V_j) + V_i b_{ij} \quad \text{(A3.30)}$$

In matrix notation

$$[I] - [Y_o][V_o] = [Y][V] \quad \text{(A3.31)}$$

$[Y]$ is an $(N \times N)$ matrix for nodes 1 to N. The diagonal element Y_{ii} is the sum of admittances connected to node i and the off-diagonal element Y_{ji} the negated admittances between nodes i and j.[¶]

In practice the voltage at consumers' terminals must be within a small tolerance of declared value and this is achieved by tap changing on transformers between the transmission, sub-transmission and distribution network. The effect

[¶] $[Y]$ can also be formed from the network connection matrix $[C]$ and the matrix of branch admittance $[y]$

$$[Y] = [C]_t [y][C]$$

$[C]$ has m rows (branches) and N columns (nodes excluding reference) with he nodal connection of each branch defined by $(+1)$ and (-1) in the appropriate row. The numbering must be consistent, e.g. $+1$ at the higher numbered node. $[y]$ is an $(m \times m)$ matrix, with diagonal terms equal to the branch admittances and off-diagonal terms the mutual admittances, usually zero.

is to maintain the apparent power S_i constant, irrespective of variation in V_i. Hence, $I_i = S_i/V_i^*$ and (A3.28) becomes

$$\frac{S_i}{V_i^*} - Y_{i0}V_0 = \sum_{j=1}^{N} Y_{ij}V_j \quad \text{for } i = 1, 2, \ldots \tag{A3.32}$$

or in matrix notation

$$\left[\frac{S}{V}\right] - [Y_o V_o] = [Y][V] \tag{A3.33}$$

A widely used approximation, the so-called 'd.c. solution', is considered very briefly below.

A3.7 THE D.C. APPROXIMATION

The active power flow between two nodes at voltage $V_i/\delta i$, $V_j/\delta j$ connected by an impedance z_{ij}, for which $x_{ij} \gg r_{ij}$ is approximately

$$\begin{aligned} P_{ij} &= \frac{V_i V_j}{x_{ij}} \sin(\delta_i - \delta_j) \\ &\approx b_{ij}(\delta_i - \delta_j) \end{aligned} \tag{A3.34}$$

where V_i, $V_j \approx 1$ and $(\delta_i - \delta_j)$ is small.

This approximation is used in the 'd.c.' load flow. Circuits are represented by their reactances, and nodal transfers by the active power components. The result is an estimate of active power flows.

In Figure A3.4 for node i,

$$\begin{aligned} P_i &= b_{i1}(\delta_i - \delta_1) + \ldots + b_{i0}\delta_i + \ldots + b_{iN}(\delta_i - \delta_N) \\ &= -b_{i1}\delta_1 \ldots + (b_{i0} + b_{i1} + \ldots + b_{iN})\delta_i \ldots - b_{iN}\delta_N \end{aligned}$$

The set of N equations, for an $(N + 1)$ node network (node 0 being the reference node) is similar in form to (A3.26). It can be written in matrix form as

$$[P] = [B][\delta] \quad \text{or} \quad [\delta] = [B]^{-1}[P] = [X][P] \tag{A3.35}$$

which can be solved for δ by iterative or matrix techniques. The equations are linear, however, and no iterations are required in the latter case. $[B]$ is an $(N \times N)$ matrix with diagonal terms B_{ii} equal to the sum of the series susceptances of the branches connected to node i, and off-diagonal terms B_{ij} equal to the negated series susceptance of branch ij.

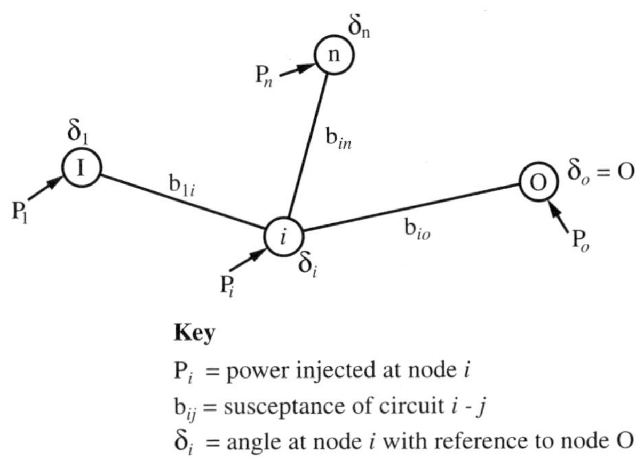

Figure A3.4 Network quantities (the 'd.c.' approximation)

Surprisingly, in view of its extensive use, the accuracy of the d.c. load flow does not seem to have been studied extensively. Studies made when the first on-line security assessment facility was being developed indicated that at the higher, in relation to circuit rating, power flows, the active power flows computed from a d.c. power flow and those from a full a.c. solution agreed within 1 or 2 percent.

REFERENCES

1. Knight, U. G., 1972. *Power Systems Engineering and Mathematics*. Pergamon, Oxford.
2. Bellman, R. E., and Dreyfus, S. E., 1962. *Applied Dynamic Programming*. Princeton University Press.
3. Bellman, R. E., 1961. *Adaptive Control Processes: A guided tour*. Princeton University Press.

FURTHER READING

Craven, B. D., 1978. *Mathematical Programming and Control Theory*, Longman.
Dantzig, G. B., 1963. *Linear Programming and Extensions*. Rand Corporation of America.
Fletcher, R., 1979. *Possible Methods of Optimisation*. Wiley.
Luo, Z-P., Pang, J-S. and Ralph, D., 1956. *Mathematical Programming with Equilibrium Constraints*. Cambridge University Press.
Schrijver, A., 1986. *Theory of Linear and Integer Programming*. Wiley.
Vajda, S., 1941. *Mathematical Programming*. Addison Wesley.

Index